Thomas Craig

A Treatise on Projections

Thomas Craig

A Treatise on Projections

ISBN/EAN: 9783743354203

Manufactured in Europe, USA, Canada, Australia, Japa

Cover: Foto ©ninafisch / pixelio.de

Manufactured and distributed by brebook publishing software (www.brebook.com)

Thomas Craig

A Treatise on Projections

UNITED STATES
COAST AND GEODETIC SURVEY

CARLILE P. PATTERSON
SUPERINTENDENT

A TREATISE

ON

PROJECTIONS

BY

THOMAS CRAIG

WASHINGTON
GOVERNMENT PRINTING OFFICE
1882

PART I.

MATHEMATICAL THEORY OF PROJECTIONS.

CONTENTS.

PART I.

§ I.

PERSPECTIVE PROJECTION.

§ II.

ORTHOMORPHIC PROJECTION.

§ III.

ORTHOMORPHIC PROJECTION (Continued).

§ IV.

PROJECTIONS BY DEVELOPMENT.

§ V.

ZENITHAL PROJECTIONS.

§ VI.

EQUIVALENT PROJECTION.

§ VII.

ON THE GENERAL THEORY OF ORTHOMORPHIC PROJECTION.

§ VIII.

GENERAL THEORY OF EQUIVALENT PROJECTIONS.

§ IX.

GENERAL THEORY OF PROJECTIONS BY DEVELOPMENT.

PART II.

CONSTRUCTION OF PROJECTIONS.

PREFACE.

In the following paper an attempt has been made to give a sufficiently comprehensive account of the theory of projections to answer the requirements of the ordinary student of that subject. The literature of projections is very large, and its history presents the names of many of the most eminent mathematicians that have lived between the time of Ptolemy and the present day. In the great mass of papers, memoirs, &c., which have been written upon projections there is much that is of the highest value and much that, though interesting, is trifling and unimportant. Thus many projections have been devised for map construction which are merely elegant geometrical trifles. Although in what follows the author has taken up every method of projection with which he is acquainted, he has not thought it necessary in the cases referred to to do more than mention them and give references to the papers or books in which they may be found fully treated.

As the different conditions which projections for particular purposes have to satisfy are so wholly unlike, it is necessary, of course, to have a different method of treatment for the various cases. Thus no general theory underlying the whole subject of projections can be given. Perhaps the only division of the subject—omitting the simple case of perspective projection—that has ever been fully treated is that of projection by similarity of infinitely small areas. This is a most important case, the general theory of which, for the representation of any surface upon any other, has been given by Gauss. The mathematical difficulties in the way of such a treatment of equivalent projections and projections by development seem to be insurmountable, but certainly offer a most attractive field for mathematical research. The author has attempted to add a little to what is already known on these subjects, but feels that what he has done is of little consequence unless, indeed, it should tempt some abler mathematician to take up the subject and develop it as it deserves. A few of the solutions of simple problems in the paper, it is believed by the author, are new and simpler than any he was able to find in the writings of others. The solution of the problem of the projection of an ellipsoid of three unequal axes upon a sphere by Gauss's method is also believed to be new. With these few exceptions there is no claim to originality in what follows; the attempt having simply been made to present in as simple and natural a form as possible what others have done. The two treatises on projections from which much aid has been obtained are those by Littrow and Germain. Littrow's Chorographie, which appeared in Vienna in 1833, was at that time a most valuable work, but is at the present day too limited in its scope to be of very much use to the student. Unquestionably the most important treatise on the subject at this time is Germain's "*Traité des projections*," which contains an account of almost every projection that has ever been invented. The author is under much obligation to this work, both for references to original sources and for solutions of particular problems. In cases where processes or diagrams are taken from this work that are by the author supposed to have been original with M. Germain, special mention of them is made in the text; when, however, Germain has drawn from earlier sources no mention is made of his book, but as far as possible references to the original papers are given. The opening brief chapter on conic sections has been taken in great part from Salmon's Conic Sections. The object of that chapter is only to give in a simple manner some of the more important and elementary properties of the curves of the second order, so that convenient reference could be made in the subsequent part of the paper to the various formulas connected with these curves, and also simple means given for constructing them. At the request of Superintendent Carlile P. Patterson the paper has been divided into two parts. The first part contains the mathematical theory of projections, while the second part contains merely such a sufficient account of the various projections as will enable the draughtsman to construct them. The principal papers from which excerpts have been made are the following:

Lagrange: "Sur la construction des cartes géographiques." Nouveaux Mémoires de l'Académie de Berlin, 1779.

Gauss: "Allgemeine Auflösung der Aufgabe, die Theile einer gegebenen Fläche auf einer andern gegebenen Fläche so abzubilden, dass die Abbildung dem Abgebildeten in den kleinsten Theilen ähnlich wird." Gesammelte Werke. Göttingen edition.

D'Avezac: "Coup d'œil historique sur la projection des cartes géographiques." Société de géographie de Paris, 1863.

Tchebychef: "Sur la construction des cartes géographiques." Academy of Sciences of St. Petersburg, 1853.

Collignon: Journal de l'École polytechnique, 41e cahier.

Mollweide: Zach's Monatliche Correspondenz, 1805.

James and Clarke: "On projections for maps applying to a very large extent of the earth's surface." Philosophical Magazine, 1865.

Airy: "Explanation of a projection by Balance of Errors applying to a very large extent of the earth's surface," &c. Philosophical Magazine, 1861.

Tissot: "Trouver le meilleur mode de projection pour chaque contrée particulière." Comptes-Rendus, 1860.

An immense list of papers bearing on the subject of projections might be given, but it hardly seems necessary. The above list includes all from which anything of importance has been taken. When minor papers are quoted reference is always made to them in the same place. Special reference may be made to a paper by Mr. C. A. Schott, Assistant United States Coast and Geodetic Survey. This paper is a resumé in very compact form of most that is of importance in the subject of projections together with a comparison of the principal methods of projection in use at the present day. This paper forms Appendix No. 15 in the annual report of the Coast and Geodetic Survey for 1880. In conclusion, the author may say that although he has endeavored to give full credit to previous writers on the subject, still it is possible that some reference has been omitted. This should, however, be taken as an unintentional oversight, or due to the fact that the author has not been able to trace back to its original source the solution of process in question, and not in any case to a desire to withhold from any other author his full measure of credit.

THOMAS CRAIG.

COAST AND GEODETIC SURVEY OFFICE,
August 19, 1880.

INTRODUCTION.

The history of any science is a history of very gradual evolution; so slow at times is the course of this evolution that often the thread of tradition seems to be broken, and we are left to grope in the darkness of historical uncertainty for the path by which we are to be conducted to the full daylight, which always, sooner or later, meets the patient inquirer after truth.

The origin of a science is usually to be sought for not in any systematic treatise, but in the investigation and solution of some particular problem. This is especially the case in the ordinary history of the great improvements in any department of mathematical science. Some problem, mathematical or physical, is proposed, which is found to be insoluble by known methods. This condition of insolubility may arise from one of two causes: Either there exists no machinery powerful enough to effect the required reduction, or the workmen are not sufficiently expert to employ their tools in the performance of an entirely new piece of work. The problem proposed is, however, finally solved, and in its solution some new principle, or new application of old principles, is necessarily introduced. If a principle is brought to light it is soon found that in its application it is not of necessity limited to the particular question which occasioned its discovery, and it is then stated in an abstract form and applied to problems of gradually increasing generality.

Other principles, similar in their nature, are added, and the original principle itself receives such modifications and extensions as are from time to time deemed necessary. The same is true of the new application of old principles; the application is at first thought to be merely confined to a particular problem, but it is soon recognized that this problem is but one, and generally a very simple one, out of a large class, to which the same processes of investigation and solution are applicable. The result in both of these cases is the same. A time comes when these several problems, solutions, and principles are grouped together and found to produce an entirely new and consistent method; a nomenclature and uniform system of notation is adopted, and the principles of the new method become entitled to rank as a distinct science.

In examining the laws which regulate the progress of the human mind in the discovery of truth, the most important points of evidence and data are derived from sciences which have been at their first promulgation the most incomplete, and have owed their subsequent advancement to the successive labors of several, rather than to the unaided efforts of a single mind. It very seldom happens that an individual discoverer gives us the results of his labors in the same form in which they originally presented themselves to his own mind. Still more rarely are the successive steps by which the original investigator conducts the mind of his reader to the perception of new truth identical with those which he himself took in first arriving at it. The early history of any science which has been slow in developing shows us how tedious and inelegant first methods generally are, and also shows how natural it is for the investigator to replace his rough-hewn highway when steadily, and without stopping to admire the beauty of the surrounding landscape, he has driven over all obstacles, by a beautiful avenue which by easy and natural stages conducts to the desired goal and affords glimpses here and there along its course of immense possibilities in, as yet, unexplored regions. Hence it is easy to see the historical importance of those sciences whose principles have been given to the public, not in a complete and systematic form, but gradually, and by methods more or less tedious and imperfect. No science can furnish a better example of this than the science of geography, and in particular that department of it which is the subject of the following paper—the Theory of Projections.

The name *projection* itself has a history, and it would be curious to trace the development which has occurred in its applications. Borrowed by geographers from geometers, it has come gradually to signify *any* method of representation of the surface of the earth upon a plane. In all rigor the use of the term projection ought to be confined to representations obtained directly according to the laws of perspective; but it has been extended to take into account representations by development and by other and purely conventional methods. Its mathematical significance is even more extended, as there it is not confined in its application to the representation of the sphere, or spheroid, upon a plane, but of any curve or surface upon any other.

The sphere being non-developable, the exact representation of its surface, or even a portion of its surface, upon a plane, is impossible. Certain conditions can, however, be fulfilled in any projection which will render it sufficiently exact for any particular purpose. The areas may be preserved, *i. e.*, all areas on the sphere may be reduced in the same proportion, in which case we have an equivalent projection; or the angles may be preserved, in which case we have an orthomorphic projection. The exigencies of any particular use for which the projection is designed give rise to an immense number of other conditions corresponding to which projections have from time to time been invented. It frequently happens that a projection having been constructed to satisfy one impressed condition, is also found to satisfy a number of others; for instance, the stereographic projection at the same time represents the parallels of latitude in their true form, and preserves the angles between the meridians themselves and between these and the parallels; thus this projection, which was originally constructed as a perspective projection, also fulfills the condition of being orthomorphic. The history of projection has been, in consequence of the impossibility of producing a perfect solution of the problem, peculiarly a history of the solution of more or less independent problems.

A method of projection which will answer for a country whose extent in latitude is small will not at all answer for another country of great length in a north and south direction; a projection which serves admirably for the representation of the polar regions is not at all applicable for countries near the equator; a projection which is the most convenient for the purposes of the navigator is of little or no value to the geodesist; and so throughout the entire range of the subject particular conditions have constantly to be satisfied, and special rather than general problems to be solved.

It is not, however, the intention in this introductory sketch to give a historical account of the subject, as it would be neither appropriate nor necessary. The complete historical account by M. D'Avezac, in the "Bulletin de la Société de Géographie," of Paris, for 1863, leaves absolutely nothing to be said on the subject.

The reader will, however, find in the references to § VII (see Coast and Geodetic Survey Report for 1880, Appendix No. 15) a valuable bibliography of the subject. This section* was written by Mr. Charles A. Schott, assistant, United States Coast and Geodetic Survey, with the express object of giving an account of the method of polyconic projection employed in constructing the charts and maps of the Survey. It subserves, however, a double purpose, as it contains a succinct and valuable résumé of much that precedes, with full references to the original sources from which information had to be compiled, and also gives a scientific account of the polyconic projection and a comparison of this, illustrated by examples, with a number of other projections most frequently met with. It is accompanied by six plates and a chart.

We have already spoken of the orthomorphic and equivalent projections, and we may mention in connection with these other general methods and the order in which they are treated in the following pages. This order is not altogether the most scientific, but seems to be better designed for the gradual introduction of the reader to the difficulties of the subject than any other. Dividing the general topic into the following heads:

I. Orthomorphic Projection,
II. Equivalent Projection,
III. Zenithal Projection,
IV. Projection by Development,

the arrangement has been as follows:

§ I. After a brief introductory account of the principal properties of the conic sections the subject of perspective projection is taken up. Strictly speaking this should fall under the head of zenithal projection, that is, projections which can be regarded as the geometrical representations

* The paper is not reproduced here, but the reader is referred to Coast and Geodetic Survey Report for 1880, Appendix No. 15.

of the sphere upon the plane of the horizon of any place. The theory of perspective projections is, however, *per se* quite self-contained, and is withal the most natural and simple method of representation, so it was thought desirable to open with that method rather than a more general and philosophical, but, at the same time, more difficult method. Sections II and III treat of the different methods of orthomorphic projection; § IV treats of projections by development; § V gives an account of zenithal projections, § VI of equivalent projections, and § VII has already been referred to (and for which see the report of 1880, Appendix No. 15) as containing Mr. Schott's account of the polyconic projection and its comparison with several other methods. These sections can for the most part be read by any one possessing a fair acquaintance with the methods of ordinary analytic geometry and the elements of the differential and integral calculus. The next three sections are extremely general in their nature, and will require a rather more extensive mathematical knowledge. They were designed to connect the particular problem of the plane representation of a sphere with the much more comprehensive methods of representation of one surface upon another which have engaged the attention of the most brilliant mathematicians. Keeping in mind, however, that the book is designed for the use of students, the author has only stepped across the threshold which leads to the purely transcendental portion of the subject, and has only given just enough to awaken in the mind of the reader, who has a real interest in the general theory, a desire to go himself to the original memoirs for fuller information. A brief section is given on the spheroidal form of the earth. On this subject very little was either required or desirable; it was not necessary to say much, because the student interested in this subject would naturally seek for information in a treatise on geodesy rather than in one on projections; it was not desirable to discuss it very fully, because present existing theories, both as to the figure and size of the earth, seem to be in a transition state.

The United States Coast and Geodetic Survey will undoubtedly soon be able to produce a much better value of the ellipticity than has yet been given. In view of that fact, and also of the fact that the greatest possible change that may take place in the present assigned value of the ellipticity will produce differences in the tables which would be almost inappreciable, it has not been deemed necessary to make any new tables even in the place of old ones, which have been computed on the supposition of an ellipticity as small as $\frac{1}{302}$.

It is readily seen that the general theory of projections touches upon a great number of other subjects in such a way as to make it a little difficult to decide what is and what is not necessary to incorporate in a treatise having this for its title. Even confining oneself to papers and books entitled "projections, &c.," it is not easy to sift out only that which is of primary importance to the beginner from the immense mass of work—good and bad—that has been done upon this subject.

Few departments of mathematics contain more eminent names among those of their founders.

From the time of Ptolemy until the present day the most profound mathematicians have devoted time and attention to this subject. The large majority of investigators have, however, had in view the attaining of some particular end, and have devised ingenious, but in most cases, rather forced methods of arriving at the desired result. Others, such as Lambert, Lagrange, Euler, Gauss, and Littrow, have treated the question from a more general theoretical point of view; but even in these cases, as only particular divisions of the subject were taken up, the results, as constituting general theories of projection, were very incomplete. The name of Lambert occurs most frequently in the history of this branch of geography, and it is an unquestionable fact that he has done more for the advancement of the subject in the way of inventing ingenious and useful methods than all of those who either preceded or have followed him. The greatest credit is, however, due to those princes of the realm of mathematics who, like Euler, Lagrange, and Gauss have done so much for the advancement of the theory. Lagrange proposes and resolves the problem of "the representation of a sphere upon a plane in such a way that the smallest parts of the projection shall be similar to the corresponding elements of the sphere, and in which the meridians and parallels shall be represented upon the map by circles." Gauss solves a far more general problem in a manner so perfect that it leaves it impossible to add a word to his general theory of orthomorphic projection. Lambert, Bonne, Mercator, Mollweide, Collignon, Airy, and James are but a few of those who have produced a marked progress in the theory of projections.

In concluding this brief introductory note, the author can do no better than again to refer the reader, desirous of fuller historical information, to D'Avezac's very valuable memoir, and to mention the following most important treatises and memoirs which, having appeared either within the present or towards the close of the last century, are comparatively easy of access:

TREATISES ON PROJECTIONS.

Chorographie oder Anleitung, aller Arten von Land- See- und Himmels-Karten. Littrow. Vienna, 1833.

Traité des projections des cartes géographiques. Germain. Paris, 1865.

Lehrbuch der Karten-Projection. Gretschel. Weimar, 1873.

MEMOIRS.

Gauss. Algemeine Auflösung der Aufgabe, die Theile einer Fläche so abzubilden, &c. Schumacher's Astronomischen Abhandlungen. Altona, 1825. Also in the Gottingen edition of Gauss's works.

Lagrange. Sur la construction des cartes géographiques. Mém., de Berlin, 1779.

Henry. Mémoire sur la projection des cartes. Paris, 1810.

Puissant. Supplément au second livre du Traité de topographie. Paris, 1810.

Euler de repræsentatione superficiei sphæricæ super plano. Acta Acad. Petrop., 1777, pars i.

Lambert. Beyträge zum Gebrauche der Mathematik. Berlin, 1772.

Murdoch. Phil. Trans. Vol. I.

Schmidt. Lehrbuch der mathematischen und physischen Geographie. Gottingen, 1829.

Zach's Monat. Corresp. Vols. 11, 12, 13, 14, 18, 25, 28 contain many papers by Mollweide, Albers, Textor, and many others.

Crelle's Journal für die reine und angewandte Mathematik, the Mathematische Annalen, the Annali di Matematica, the Comptes-Rendus of the French Academy, and the Bulletin of the Academy at St. Petersburg, all contain very valuable papers. The same may be said of the Journal de l'École Polytechnique and the Journal de l'École Normale Supérieur of Paris. All of these sources have been consulted in the preparation of the following treatise, and it does not seem to the author as if anything worthy of preservation has been overlooked or left out from any cause whatever.

A TREATISE ON PROJECTIONS.

MATHEMATICAL THEORY OF PROJECTIONS.

§ 1.

PERSPECTIVE PROJECTION.

A surface in perspective projection has the same appearance as that which it would present to the eye of an observer situated at any determinate point of space. The right line drawn from the eye of the observer to the center, in the case of a central surface, is normal to the plane of projection which may intersect this line at any point of its length. The projection of any point of the surface under consideration is then the point of intersection with this plane of the line joining the given point to the eye. We will now confine ourselves to the surface of the sphere. Imagine any line drawn on the surface and every point of the line joined to the eye by a straight line; the aggregate of lines will form the surface of a cone, and the intersection of this cone with the plane will be the projection of the curve drawn upon the sphere. If the cone so formed is of the nth order its intersection with the sphere will be of the $2n$th order and with the plane will be of the nth order, so that in general the degree of the curve is lowered by projection. If the cone is a circular cone, its intersection with the sphere will in general be a sphero-conic and will be projected in a conic section, as the intersection of a cone of the second degree with a plane is a curve of the second degree or a conic section.

We will in general only be concerned with the projections of circles of the sphere, and they will be projected in conic sections; so before proceeding further with the subject it will be convenient to give a brief statement of the more important properties of these curves as deduced by a study of their equations.

The general equation of the second degree in two variables is the equation of a conic section. This equation in its most general form may be written

$$Ax^2+2\,Hxy+By^2+2\,Gx+2\,Fy+C=0$$

containing five independent constants, viz, the ratios $\frac{A}{C}, \frac{2\,H}{C}$, &c. Five relations between the coefficients are sufficient to determine a curve of the second degree; for, though the general equation contains six constants, the nature of the curve depends not on the absolute magnitude of these but on their mutual ratios, since if we multiply or divide the equation by any constant it remains unaltered. We may, therefore, divide the equation by the quantity C, making the absolute term equal to 1, and there will remain but five constants to be determined.

Transformation to new co-ordinate axes frequently has the effect of simplifying very much the equations with which we are dealing; and it will be useful here to find what the general equation becomes on being transformed to a new set of axes, assuming, for a first transformation, that the new axes are parallel to the old.

For this purpose make $x=x+x'$, $y=y+y'$; x', y' being the co-ordinates of the new origin. We will find that the coefficients of x^2, xy, and y^2 remain as before, A, 2 H, and B; that

the new G is $G'=Ax'+Hy'+G$

the new F is $F'=Hx'+By'+F$

the new constant term is $C'=Ax'^2+2\,Hx'y'+By'^2+2\,Gx'+2\,Fy'+C$

Suppose that we again transform the original equation, this time, however, to polar co-ordinates, by making $x=\rho\cos\theta$, $y=\rho\sin\theta$; the equation so transformed is

$$(A\cos^2\theta+2H\cos\theta\sin\theta+B\sin^2\theta)\rho^2+2(G\cos\theta+F\sin\theta)\rho+C=0$$

Write this for a moment as

$$\gamma\left(\frac{1}{\rho}\right)^2+2\beta\left(\frac{1}{\rho}\right)+\alpha=0$$

α being of course $=A\cos^2\theta+2H\sin\theta\cos\theta+B\sin^2\theta$, &c. The roots of this equation are

$$\frac{1}{\rho}=\frac{-\beta\pm\sqrt{\beta^2-\alpha\gamma}}{\gamma}$$

that is, the straight line ρ drawn from the origin meets the conic in two different points.

Suppose here that $\alpha=0$; then for one root we have $\frac{1}{\rho}=0$ or $\rho=\infty$; that is, if the coefficient of $\rho^2=0$ the line drawn from the origin meets the curve in two points, one of which lies at infinity The coefficient of ρ^2 is, however,

$$A\cos^2\theta+2H\sin\theta\cos\theta+B\sin^2\theta$$

This equated to zero gives

$$A+2H\tan\theta+B\tan^2\theta=0$$

a quadratic in $\tan\theta$, and consequently we have that there can be drawn through the origin two real, coincident, or imaginary lines, which will meet the curve at an infinite distance; each of which lines also meets the curve at one finite point determined by

$$2\rho(G\cos\theta+F\sin\theta)+C=0$$

We will now seek the test which will tell us what class of locus is represented by a given equation of the second degree; or we wish to ascertain the *form* of the curve, whether it is limited in any way or extends to infinity in any direction. Of course if the curve be limited in every direction, no radius vector can be drawn which will meet it at infinity. For an infinite value of the radius vector we must have

$$A+2H\tan\theta+B\tan^2\theta=0$$

If $H^2-AB<0$, the roots of this equation will be imaginary, or no real value of θ can be found which will make

$$A\cos^2\theta+2H\sin\theta\cos\theta+B\sin^2\theta=0$$

The curve in this case is limited in every direction, and is an Ellipse.

If $H^2-AB>0$, there are two real roots to the equation

$$A+2H\tan\theta+B\tan^2\theta=0$$

consequently two real values of θ, corresponding to which two lines can be drawn from the origin meeting the curve at infinity. This curve is the Hyperbola.

If $H^2-AB=0$, the roots of the quadratic are equal and consequently the two straight lines which can be drawn to meet the curve at infinity will coincide. The curve in this case is the Parabola.

If in a quadratic $\alpha x^2+2\beta x+\gamma=0$ the coefficient β vanishes, the roots are equal with contrary signs. Thus then if in the transformed equation

$$G\cos\theta+F\sin\theta=0$$

the two values found for ρ will be equal and opposite in sign. The points answering to the equal and opposite values of ρ are equidistant from the origin and on opposite sides of it, and so we have that the chord represented by $Gx+Fy=0$ is bisected at the origin. If we had $G=0$ and $F=0$, then whatever be the value of θ we should always have

$$G\cos\theta+F\sin\theta=0$$

or all chords drawn through the origin would be bisected.

Now, by transformation to suitable axes, we can in general cause the coefficients of x and y to vanish. Thus equating to zero the coefficients of x and y obtained by a transformation to a new origin, we find that the co-ordinates of this new origin must fulfill the conditions

$$Ax'+Hy'+G=0 \qquad Hx'+By'+F=0$$

in order that all chords drawn through that point may be then bisected. The point thus determined is called the *center* of the curve. As these equations for determining the center are linear there can exist only one center to any conic section. The co-ordinates of the center are found to be

$$x'=\frac{BG-HF}{H^2-AB} \qquad y'=\frac{AF-HG}{H^2-AB}$$

For $H^2-AB=0$ the center lies at infinity, which is the case of the parabola, and this curve is in consequence called a *non-central* curve. Obviously the centers of the ellipse and hyperbola lie at a finite distance.

We have seen that a chord through the origin is bisected if $G \cos \theta + F \sin \theta = 0$. Now transforming the origin to any point, it appears that a parallel chord will be bisected at the new origin if

$$G' \cos \theta + F' \sin \theta = 0$$

or if

$$(Ax'+Hy'+G) \cos \theta + (Hx'+By'+F) \sin \theta = 0$$

This, therefore, is a relation which must be satisfied by the co-ordinates of the new origin if it be the middle point of a chord making with the axis of x the angle θ. Hence the locus of the middle points of parallel chords is

$$(Ax+Hy+G) \cos \theta + (Hx+By+F) \sin \theta = 0$$

This line bisecting a system of parallel chords is called a diameter, and we see that it passes through the intersection of

$$Ax+Hy+G=0 \qquad Hx+By+F=0$$

Therefore every diameter passes through the center of the curve.

If two diameters of a conic be such that one of them bisects all chords parallel to the other, then, conversely, the second will bisect all chords parallel to the first.

The equation of a diameter which bisects chords making an angle θ with the axis of x is

$$Ax+Hy+G+(Hx+By+F) \tan \theta = 0$$

Calling θ' the angle, which this line makes with x, we have

$$\tan \theta' = -\frac{A+H \tan \theta}{H+B \tan \theta}$$

whence

$$B \tan \theta \tan \theta' + H (\tan \theta + \tan \theta') + A = 0$$

And the symmetry of the equation shows that the chords making an angle θ' are also bisected by the diameter making an angle θ with x.

Diameters so related that each bisects all chords parallel to the other are called *conjugate diameters.*

The general equation of the second degree in two variables is now, when transformed to the center,

$$Ax^2+2Hxy+By^2+C'=0$$

where C' is readily found to be equal to

$$\frac{ABC+2FGH-AF^2-BG^2-CH^2}{AB-H^2}$$

If the numerator of this fraction equals 0, the equation would become

$$Ax^2+2Hxy+By^2=0$$

the equation of a pair of right lines; the condition, then, that the general equation should represent a pair of right lines is

$$ABC+2FGH-AF^2-BG^2-CH^2=0$$

or, in determinant form,

$$\begin{vmatrix} A, & H, & G, \\ H, & B, & F, \\ G, & F, & C, \end{vmatrix}=0$$

The angles that two conjugate diameters make with the axis of x are connected by the relation—

$$B\tan\theta\tan\theta'+H(\tan\theta+\tan\theta')+A=0$$

If the diameters are at right angles,

$$\tan\theta'=-\frac{1}{\tan\theta}$$

Hence—

$$H\tan^2\theta+(A-B)\tan\theta-H=0$$

a quadratic equation for the determination of θ. Transforming back to rectangular co-ordinates, this is—

$$Hx^2-(A-B)xy-Hy^2=0$$

the well-known equation of two real lines at right angles to each other. These rectangular diameters are called the axes of the curve.

We have seen that when

$$A\cos^2\theta+2H\sin\theta\cos\theta+B\sin^2\theta=0$$

the radius vector meets the curve at infinity, and also in one other point determined by

$$\rho=-\frac{C}{G\cos\theta+F\sin\theta}$$

But if the origin be the center, we have $G=0$ and $F=0$; hence this distance will also become infinite. Hence two lines can be drawn from the center and meeting the curve in two *coincident* points at infinity; these lines are called the *asymptotes* of the curve and are real in the case of the hyperbola and imaginary in the case of the ellipse. The equation of the axes was found to be

$$Hx^2-(A-B)xy-Hy^2=0$$

This is the equation of a pair of lines bisecting the angle between the lines

$$Ax^2+2Hxy+By^2=0$$

Therefore, the axes of the curve bisect the angle between the asymptotes.

The preceding results might all have been obtained by a simple transformation of co-ordinates. Suppose that, our original axes being rectangular, we turn the system round through an angle ω, *i. e.*, make

$$x=x\cos\omega-y\sin\omega \qquad y=x\sin\omega+y\cos\omega$$

the new coefficient of x^2 will now be

$$A'=A\cos^2\omega+2H\cos\omega\sin\omega+B\sin^2\omega$$

also

$$H'=B\sin\omega\cos\omega+H(\cos^2\omega-\sin^2\omega)-A\sin\omega\cos\omega$$

$$B'=A\sin^2\omega-2H\sin\omega\cos\omega+B\cos^2\omega$$

By putting $H'=0$ we get the same equation as before for determining $\tan \omega$, and in fact this gives us

$$\tan 2\theta = \frac{2H}{A-B}$$

for the tangent of the angle made with the given axes by either axis of the curve. Add together $A'+B'$ and we have—

$$A'+B'=A+B$$

Again, write

$$2A'=A+B+2H\sin 2\omega+(A-B)\cos 2\omega$$
$$2B'=A+B-2H\sin 2\omega-(A-B)\cos 2\omega$$

hence

$$4A'B'=(A+B)^2-[2H\sin 2\omega+(A-B)\cos 2\omega]^2$$

but

$$4H'^2=[2H\cos 2\theta-(A-B)\sin 2\theta]^2$$

therefore

$$4(A'B'-H'^2)=4(AB-H^2)$$

or

$$A'B'-H'^2=AB-H^2$$

When, therefore, we transpose an equation of the second degree from one set of rectangular axes to another, the quantities $A+B$ and $AB-H^2$ remain unaltered.

When, therefore, we want the equation transformed to the *axes*, we have the new $H=0$, and

$$A+B=A'+B' \qquad AB-H^2=A'B'$$

From these we can form a quadratic equation to find A' and B'.

We have now for the equation referred to the center and axes

$$A'x^2+B'y^2+C'=0.$$

Let the intercepts made by the ellipse (or hyperbola) on the axes be a on x and b on y. Then

$$A'=\frac{c'}{a^2} \qquad B'=\frac{c'}{b^2}$$

and the equation becomes

$$\frac{x^2}{a^2}+\frac{y^2}{b^2}=1$$

and for the hyperbola

$$\frac{x^2}{a^2}-\frac{y^2}{b^2}=1$$

as the equation of the hyperbola only differs from the ellipse (in this case of transformation to the axes) in the sign of the coefficient of y^2.

For the polar equation of the ellipse write

$$x=\rho\cos\lambda \qquad y=\rho\sin\lambda$$

and the equation simply becomes

$$\rho^2=\frac{a^2b^2}{a^2-(a^2-b^2)\cos^2\lambda}$$

or, by making

$$\frac{a^2-b^2}{a^2}=e^2,\ (e<1)$$

$$\rho^2=\frac{b^2}{1-e^2\cos^2\lambda}$$

e being the *eccentricity* of the ellipse.

In like manner we find for the polar equation of the hyperbola, the center being the pole,

$$\rho^2=\frac{a^2b^2}{(a^2+b^2)\cos^2\lambda-a^2}$$

Making

$$\frac{a^2+b^2}{a^2}=e^2,\ (e>1)$$

this is

$$\rho^2=\frac{b^2}{e^2\cos^2\lambda-1}$$

In the case of the ellipse the points on the major axis at the distance $\sqrt{a^2-b^2}=ae$ from the center are called the *foci*. In the case of the hyperbola these are at the distance $\sqrt{a^2+b^2}$ from the center.

In the polar equation of the ellipse we see that the least value the denominator $b^2+(a^2-b^2)\sin^2\lambda$ can have is when $\lambda=0$; therefore the greatest value of ρ is when $\lambda=0$ and is equal to a. Similarly, we find that the least value of ρ is for $\lambda=\frac{\pi}{2}$ and this value is equal to b. These two lines are the axis *major* and the axis *minor* of the curve. It is also clear that the smaller λ is, the larger ρ will be; hence, the nearer any diameter is to the axis major the greater it will be. If $\lambda=\lambda$, or $\lambda=-\lambda$, we will find the same value of ρ; hence two diameters which make equal angles with the axis will be equal. The figure of the ellipse is clearly that given in the figure

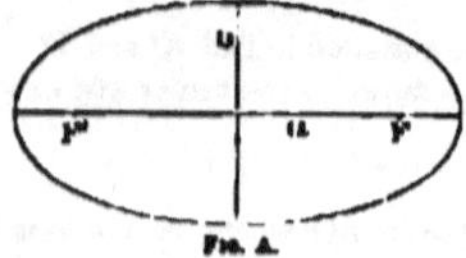

Fig. A.

F and F′ denoting the positions of the foci.

If we solve the equation of the ellipse for y we get

$$y=\frac{b}{a}\sqrt{a^2-x^2}$$

Now, if we describe a concentric circle with radius a, its equation will be

$$y=\sqrt{a^2-x^2}$$

Hence, we have the following construction:

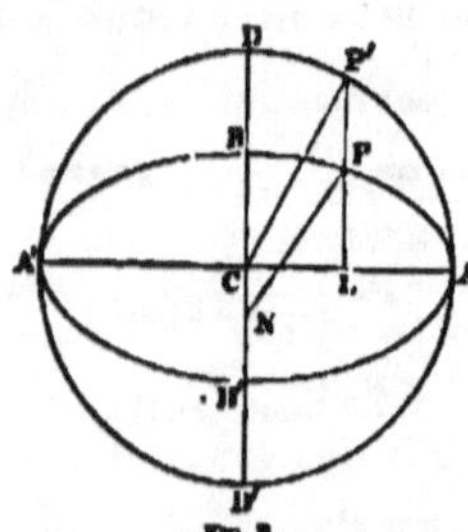

Fig. B.

Describe a circle with radius a and take on each ordinate LP′ a length LP, such that $\frac{LP}{LP'}=\frac{b}{a}$; then will P be a point of the ellipse. A similar construction holds for the minor axis, only in that case the ratio of the ordinate of the ellipse to that of the circle equals $\frac{a}{b}$. The construction is arrived at simply as follows:

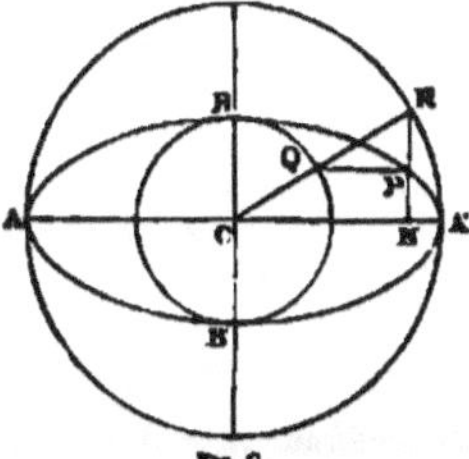

FIG. C.

Describe concentric circles with radii a and b respectively. Draw any radius CR of the large circle and from the point Q, where it cuts the small circle, draw QP parallel to the axis of x; the point of intersection of this line with the line RN, drawn perpendicular to the axis of x, gives us P, a point of the ellipse. Similarly, any number of points can be obtained and the ellipse drawn through them. Or, again, suppose that we have any line AB constant in length, which moves so that the point B shall always lie on the axis of x and A on the axis of y.

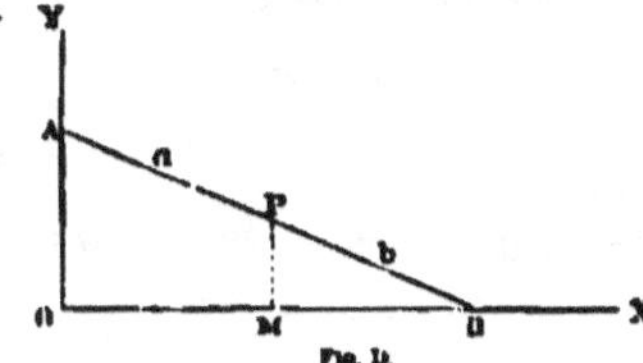

FIG. D.

Now, assume any point P, either between A and B or on the prolongation of the line AB, such that AP$=a$ and BP$=b$; then the locus of P is an ellipse; for calling x and y the co-ordinates of P, we have

$$\frac{x}{a}=\cos \text{ABO} \qquad \frac{y}{b}=\sin \text{ABO}$$

therefore

$$\frac{x^2}{a^2}+\frac{y^2}{b^2}=1$$

Tangent.

The equation of the chord joining any two points $x'y'$ and $x''y''$ on the curve is

$$\frac{(x-x')(x-x'')}{a^2}+\frac{(y-y')(y-y'')}{b^2}=\frac{x^2}{a^2}+\frac{y^2}{b^2}-1$$

or

$$\frac{(x'+x'')x}{a^2}+\frac{(y'+y'')y}{b^2}=\frac{x'x''}{a^2}+\frac{y'y''}{b^2}-1$$

which, when $x'y'$ and $x''y''$ approach indefinitely near to each other, becomes

$$\frac{xx'}{a^2}+\frac{yy'}{b^2}=1$$

the equation of the tangent to the ellipse. For the hyperbola the corresponding equation is

$$\frac{xx'}{a^2}-\frac{yy'}{b^2}=1$$

The intercept of the tangent on the axis of x is $=\frac{a^2}{x'}$.

The *subtangent* is the distance from the foot of the ordinate at the point of tangency to the point of intersection of the tangent and the axis of x. Therefore, subtangent is $=\frac{a^2-x'^2}{x'}$.

The quantity $\frac{a^2}{x'}$, being independent of b, gives a simple means of drawing a tangent to the ellipse. If we describe a circle of radius a, and at that point of the circle which has x' for abscissa draw a tangent, it will intersect the axis of x at the same point as the tangent to the ellipse, so that joining the point thus graphically found with point $x'y'$ of the ellipse, we will have the tangent to this curve.

Normal.

Forming the equation of the perpendicular to the line $\frac{xx'}{a^2}+\frac{yy'}{b^2}=1$ at the point $x'y'$, we have

$$\frac{x'}{a^2}(y-y')=\frac{y'}{b^2}(x-x')$$

or

$$\frac{a^2x}{x'}-\frac{b^2y}{y'}=a^2-b^2=c^2$$

the equation of the normal to the ellipse.

The intercept of the normal on the axis of x equals

$$\frac{c^2}{a^2}x'=e^2x'$$

We can thus draw a normal to an ellipse, for given the intercept of the normal on the axis of x we can find x', the abscissa of the point through which the normal is drawn.

The *subnormal* is the portion intercepted on the axis between the normal and the ordinate, and equals

$$x'-\frac{c^2}{a^2}x'=\frac{b^2}{a^2}x'.$$

Foci.

The square of the distance from any point $x'y'$ of the ellipse to the focus is equal to

$$(x'-c)^2+y'^2=x'^2+y'^2-2\,cx'+c^2$$

since by definition the co-ordinates of the focus are $x=c$, $y=0$. But

$$x'^2+y'^2=x'^2+\frac{b^2}{a^2}(a^2-x'^2)=b^2+e^2x'^2$$

and

$$b^2+c^2=a^2$$

Hence the distance which we may call FP equals

$$\sqrt{a^2-2\,cx'+e^2x'^2}$$

Hence

$$FP=a-ex'$$

We reject the negative value $ex'-a$, as we are only concerned with the absolute magnitude of FP and not its direction. Similarly, for the distance from the other focus $(-c, 0)$ we find

$$F'P=a+ex'$$

Hence

$$FP+F'P=2a$$

or for a fundamental property of the ellipse we have *the sum of the distances from any point of the ellipse to the focus is constant and equal to the major axis of the ellipse.*

It is not difficult to show that for the hyperbola

$$FP - F'P = 2a$$

or *in the hyperbola the difference of the distances from the foci to any point of the curve is equal to the major axis.*

By help of these theorems the ellipse or hyperbola can be described mechanically.

If the extremities of a thread be fastened at two fixed points F and F', it is plain that a pencil moved about so as to keep the thread always stretched will describe an ellipse whose major axis is the length of the thread. In order to describe a hyperbola, let a ruler be fastened at one extremity

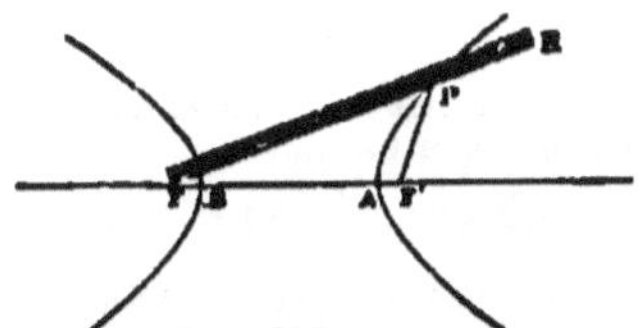

FIG. 5.

F, and capable of moving round it, then if a thread, fastened to a fixed point F' and also to a fixed point on the ruler R, be kept stretched by a ring at P, as the ruler is moved round the point P will describe a hyperbola; for, since the sum of F'P and PR is constant, the difference of FP and F'P will be constant.

Directrix.

The directrix is a line perpendicular to the axis major at a distance from the center $= \pm \frac{a^2}{c}$. The distance of the directrix from any point of the curve is

$$\frac{a^2}{c} - x' = \frac{a}{c}(a - ex') = \frac{1}{e}(a - ex')$$

Hence we have another fundamental property of these curves which would enable us to construct the curve, viz, that the distance of any point on the curve from the focus is to its distance from the directrix as e is to 1.

The length of the focal radius vector we have found to equal $a - ex'$; but x' (being measured from the center) equals $\rho \cos \lambda + c$. Hence

$$\rho = a - e \rho \cos \lambda - ec$$

or, solving for ρ and replacing c by its value ae

$$\rho = \frac{a(1-e^2)}{1 + e \cos \lambda} = \frac{b^2}{a} \frac{1}{1 + e \cos \lambda}$$

The double ordinate at the focus is called the *parameter* or *Latus-Rectum*; its half is found, by making $\lambda = \frac{\pi}{2}$, to be $= \frac{b^2}{a} = a(1-e^2)$. Denoting the parameter by p the equation may be written

$$\rho = \frac{p}{2} \frac{1}{1 + e \cos \lambda}$$

The properties that we have discussed so far have been common to both the ellipse and hyperbola, but the hyperbola by virtue of its real asymptotes possesses properties which the ellipse does not possess. We saw that in the general case the equation of the asymptotes was obtained by placing the highest powers of the variables $= 0$, the center being the origin. Thus we have for the equation of the asymptotes to the hyperbola

$$\frac{x^2}{a^2} - \frac{y^2}{b^2} = 0$$

or

$$\frac{x}{a'}-\frac{y}{b'}=0 \qquad \frac{x}{a'}+\frac{y}{b'}=0$$

a' and b' being any pair of conjugate diameters. Hence the asymptotes are parallel to the diagonals of any parallelogram whose sides are any pair of conjugate diameters.

Parabola.

We have already seen that when the equation of the second degree represents a parabola, we must have

$$H^2-AB=0$$

This is clearly only the condition that the first three terms of the general equation should constitute a perfect square, or that the equation might be written

$$(ax+\beta y)^2+2\,Gx+2\,Fy+C=0$$

Transformation of axes will greatly simplify this equation. Suppose that we take for new axes the line $ax+\beta y$ and the perpendicular on it $\beta x-ay$. Now in the equation of the curve we know that $ax+\beta y$ and $2\,Gx+2\,Fy+c$ are respectively proportional to the lengths of perpendiculars let fall from the point xy to the lines

$$ax+\beta y=0 \qquad 2\,Gx+2\,Fy+C=0$$

Hence the equation of the curve asserts that the square of the perpendicular from any point of the curve on the first of these lines is proportional to the perpendicular from the same point on the second line. Now, since the new co-ordinates x' and y' are to denote the lengths of perpendiculars from any point on the new axes, we have

$$x'=\frac{\beta x-ay}{\sqrt{a^2+\beta^2}} \qquad y'=\frac{ax+\beta y}{\sqrt{a^2+\beta^2}}$$

Make $a^2+\beta^2=\gamma^2$; and we have

$$\gamma x'=\beta x-ay \qquad \gamma x=ay'+\beta x'$$
$$\gamma y'=ax+\beta y \qquad \gamma y=\beta y'-ax'$$

Making these substitutions in the equation of the curve, it becomes

$$\gamma^3y'^2+2\,(G\beta-Fa)\,x'+2\,(Ga+F\beta)\,y'+\gamma C=0$$

Or by simply turning the axes through a certain angle we have reduced the equation to the form

$$B'y^2+2\,G'x+2F'y+C'=0$$

Again, change to parallel axes through a new origin $x'y'$; the equation now becomes

$$B'y^2+2\,G'x+2\,(B'y'+F')y+B'y'^2+2\,G'x'+2\,F'y'+C'=0$$

As the coefficient of x has remained unchanged, we evidently cannot make it vanish by this kind of transformation. But we can determine $x'y'$ so that the coefficient of y and the absolute term shall vanish. Take for the co-ordinates of the new origin

$$x'=\frac{F'^2-B'C'}{2\,G'B'} \qquad y'=-\frac{F'}{B'}$$

then the equation reduces to

$$y^2=-\frac{2\,G'}{B'}x$$

or simply

$$y^2=px$$

when we have

$$p=\frac{2\,(F\alpha-G\beta)}{(\alpha^2+\beta^2)^{\frac{1}{2}}}$$

The quantity p is, in the assumed case of rectangular co-ordinates, called the ***principal parameter*** of the curve. Since every value of x gives two equal and opposite values for y, the curve must be symmetrical with respect to the axis of x. None of the curve can lie on the negative side of the origin, since a negative value of x will give imaginary values of y. The figure of the curve is that here represented.

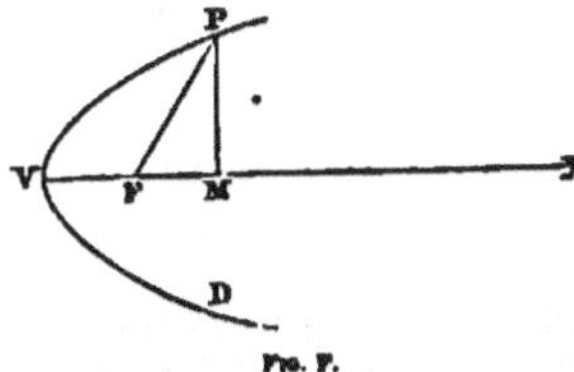

FIG. F.

The equation of the chord joining any two points on the curve is

$$(y-y')\,(y-y'')=y^2-px$$

or

$$(y'+y'')\,y=px+y'y''$$

Make $y'=y''$ and write $y'^2=px'$ and the equation of the tangent to the parabola is

$$2\,y'y=p\,(x+x')$$

For the intercept on the axis of x we have $x=-x'$; that is, the distance from the foot of the ordinate of contact to the point of intersection of the tangent with the axis of the curve is bisected at the vertex, or, simply, the ***subtangent*** is bisected at the vertex.

Normal.

The equation of the normal is

$$p\,(y-y')+2\,y'(x-x')=0$$

Its intercept on the axis is

$$x=x'+\tfrac{1}{2}p$$

The ***subnormal*** being defined as before by the relation

$$\text{Subnor.}=x-x'$$

we have for the parabola that the subnormal is constant and equals $\frac{1}{2}p$ or $\frac{1}{2}$ parameter.

Focus.

The focus of the parabola is a point situated on the axis of the curve and at a distance from the vertex equal to one-fourth of the principal parameter. Calling m this distance, we have for the square of the distance of any point of the curve from the focus

$$(x'-m^2)+y'^2=x'^2-2\,mx'+m^2+4\,mx'=(x'+m)^2$$

Hence the distance of any point from the focus equals $x'+m$.

The directrix of the parabola is a straight line perpendicular to the axis and at a distance of vertex outside of the curve equal to m; hence the distance of any point on the curve from the directrix must equal $x'+m$. We have, then, as a fundamental property of the parabola, that ***the distance of any point of the curve from the focus is equal to its distance from the directrix.***

The equation of the ellipse is satisfied by making $x=a\sin\chi$, $y=b\cos\chi$, when χ is the complement of the angle P''cL (Fig. B), or is the complement of the eccentric anomaly. We have from these values of x and y

$$dx=a\cos\chi d\chi \qquad dy=-b\sin\chi d\chi$$

and for the element of arc

$$ds=\sqrt{dx^2+dy^2}=\sqrt{a^2\cos^2\chi+b^2\sin^2\chi}\,d\chi=\sqrt{a^2-(a^2-b^2)\sin^2\chi}\,d\chi$$

Taking the eccentricity of the ellipse $\frac{\sqrt{a^2-b^2}}{a}$ as the modulus of an elliptic integral, we have at once for the length of the entire ellipse,

$$s=4a\int_0^{\frac{\pi}{2}}\sqrt{1-k^2\sin^2\chi}\,d\chi=4a\int_0^{\frac{\pi}{2}}\Delta(k\chi)d\chi$$

or

$$s=4aE_1k$$

E_1k denoting the complete elliptic integral of the second kind. If the eccentricity is small or the ellipse is nearly a circle, the function E_1k has for value (Cayley's Elliptic Functions, page 46)

$$E_1k=\frac{\pi}{2}\left(1-\frac{1}{2^2}k^2-\frac{1^2.3}{2^2.4^2}k^4-\frac{1^2.3^2.5}{2^2.4^2.6^2}k^6-\ldots.-\frac{1^2.3^2.\ldots.(2i-3)^2(2i-1)}{2^2.4^2.\ldots.(2i)^2}k^{2i}-\right)$$

The area of the ellipse is well known to be πab. It can be obtained readily by integrating $\int\int dxdy$, the limits of x and y being taken from the equation

$$\frac{x^2}{a^2}+\frac{y^2}{b^2}=1$$

For the hyperbola

$$\frac{x^2}{a^2}-\frac{y^2}{b^2}=1$$

write $x=a\sec\upsilon$, $y=b\tan\upsilon$, where υ is the eccentric anomaly

$$dx=a\sec\upsilon\tan\upsilon d\upsilon \qquad dy=a\sec^2\upsilon d\upsilon$$

and thence

$$ds=\frac{d\upsilon}{\cos\upsilon}\sqrt{b^2+a^2\sin^2\upsilon}$$

Here take

$$k=\frac{a}{\sqrt{a^2+b^2}}$$

the reciprocal of the eccentricity; then k' the complementary modulus equals $\frac{b}{\sqrt{a^2+b^2}}$

Assume an angle μ such that $\tan\upsilon=k'\mu$. Then

$$x=\frac{a}{\cos\mu}\Delta\mu \qquad dx=\frac{ak'^2\sin\mu d\mu}{\cos^2\mu\Delta\mu}$$

$$y=bk'\tan\mu \qquad dy=\frac{bk'd\mu}{\cos^2\mu}$$

and thence

$$ds=\frac{bk'd\mu}{\cos^2\mu\Delta\mu}$$

By differentiation of $\Delta\mu\tan\mu$ we find

$$d.\Delta\mu\tan\mu=\left(\frac{k'^2}{\cos^2\mu\Delta\mu}-\frac{k'^2}{\Delta\mu}+\Delta\mu\right)d\mu$$

and conversely integrating from zero we find

$$k'^2 \int \frac{d\mu}{\cos^2 \mu \Delta \mu} = \Delta \mu \tan \mu + k'^2 F\mu + E\mu$$

Substituting this in the expression for s, and remembering that $bk = ak'$ we have

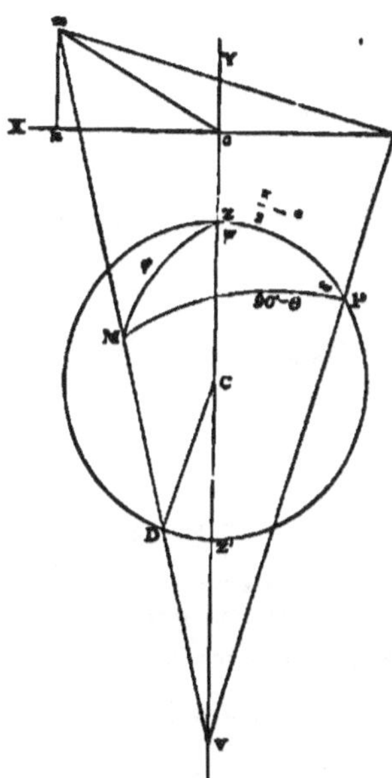

Fig. 1.

$$s = \frac{a}{k}(\tan \mu \Delta \mu + k'^2 F\mu - E\mu$$

where s denotes the length of an arc of the hyperbola measured from the vertex.

The incongruity of explaining the elementary principles of Conic Sections and assuming a knowledge of the more difficult one of Elliptic Functions will perhaps strike all readers; but the object of the explanation in the former case was not so much to teach conics to one who had not studied the subject as it was to give a brief *résumé* of the more important elementary principles, which would afford means of practically *drawing* these curves. Hereafter elementary explanations will not be given except in particular cases where it may be desirable to bring out some important fact in the process.

We will take up now the subject of *perspective projection*—taking the plane of projection at first as outside of the sphere. Let C (Fig. 1) denote the center of the sphere, V the point of sight, Op the trace of the plane of projection upon the plane of the paper, P the pole of the equator, and M any other point on the surface of the sphere, having θ for latitude and ω for longitude, PZZ' being the first meridian. Then we have

$$\omega = ZPM, \text{ to these add } PZ = \frac{\pi}{2} - \alpha$$

$$90^\circ - \theta = PM \qquad MZ = \varphi$$

$$\gamma = CZ \qquad PZM = \psi$$

and also assume $VC = c$, $VO = c'$. The projection of Z is o and of P is p, these being the points in which the projecting lines pierce the plane of projection. Assume m as the projection of M, then the position of this point must be determined with reference to some system of co-ordinates. The most convenient system to adopt will be the rectangular system OX, OY, then $On = x$, $nm = y$, and we have to determine x and y as functions of the given constants c, c', r and the angular magnitudes θ and ω.

Equating the sum of the three angles about the point C to the sum of the three angles of the triangle CDM we have, since this triangle is isosceles,

$$MDC = \frac{\varphi + VCD}{2}$$

also

$$MVZ = \frac{ZM - DC}{2} = \frac{\varphi - VCD}{2}$$

combining these two results, we have

$$MDC = \varphi - V; \ ADC = \pi - (\varphi - V)$$

Again, in the triangle VDC

$$\frac{\sin D}{\sin V} = \frac{c}{r}$$

or

$$\sin(\varphi - V) = \frac{c}{r} \sin V;$$

Solving this for tan V, we find

$$\tan V = \frac{r \sin \varphi}{c + r \cos \varphi}$$

and consequently

$$OM = OV \tan V = \frac{c' r \sin \varphi}{c + r \cos \varphi}$$

Observing now that $MOP' = \psi$, we at once obtain for x and y the values

$$x = \frac{-c' r \sin \varphi \cos \psi}{c + r \cos \varphi} \qquad y = \frac{c' r \sin \varphi \sin \psi}{c + r \cos \varphi}$$

We have now to determine φ and ψ in terms of θ and ω. In the spherical triangle PZM we have

$$\sin \varphi \sin \psi = \cos \theta \sin \omega$$

$$\sin \varphi \cos \psi = \frac{\sin \theta - \sin \alpha \cos \varphi}{\cos \alpha}$$

$$\cos \varphi = \sin \alpha \sin \theta + \cos \alpha \cos \theta \cos \omega$$

Combination of the last two gives

$$\sin \varphi \cos \psi = \cos \alpha \sin \theta - \sin \alpha \cos \theta \cos \omega$$

Substituting these values of $\cos \varphi$, $\sin \varphi \cos \psi$ and $\sin \varphi \sin \psi$ in the values of x and y, these become

$$(1) \qquad x = \frac{c' r (\sin \alpha \cos \theta \cos \omega - \cos \alpha \sin \theta)}{c + r (\cos \alpha \cos \theta \cos \omega + \sin \alpha \sin \theta)} \qquad y = \frac{c' r \cos \theta \sin \omega}{c + r (\cos \alpha \cos \theta \cos \omega + \sin \alpha \sin \theta)}$$

Upon these two equations depends the entire construction of the perspective projection of a sphere upon a plane.

For $c' = c + r$ the plane of projection becomes the tangent plane at the point z; for $c' = c$ the plane passes through the center of the sphere, and the great circle so cut out will be the limiting line of the projection. For the last case the equations become

$$(2) \qquad x = \frac{cr (\sin \alpha \cos \theta \cos \omega - \cos \alpha \sin \theta)}{c + r (\cos \alpha \cos \theta \cos \omega + \sin \alpha \sin \theta)} \qquad y = \frac{cr \cos \theta \sin \omega}{c + r (\cos \alpha \cos \theta \cos \omega + \sin \alpha \sin \theta)}$$

If the eye be conceived as situated upon the prolongation of the axis of the earth the plane of projection will coincide with that of the equator and we have an *equatorial projection*. The values of x and y for this case are found by making $\alpha = \frac{\pi}{2}$ in the last formula, thus:

$$x = \frac{cr \cos \theta \cos \omega}{c + r \sin \theta} \qquad y = \frac{cr \cos \theta \sin \omega}{c + r \sin \theta}$$

If the eye is placed in the plane of the equator the plane of projection will pass through a meridian, and the projection is said to be a *meridian projection*. For this case we have $\alpha = 0$.

$$x = \frac{-cr \sin \theta}{c + r \cos \theta \cos \omega} \qquad y = \frac{cr \cos \theta \sin \omega}{c + r \cos \theta \cos \omega}$$

By making $\theta = 0$ in equations 2, and giving ω a series of values, we will determine as many points of the projection of the equator as will be necessary to draw that line. And, in like manner, by giving θ any constant value, and giving ω any series of values, we can determine the projections of the intersections of all the meridians with the assumed parallel of θ. This process is, however, a very lengthy and inelegant one. As we are concerned only with the projections of circles of the sphere we know that, the projecting curve being of the second degree, these projections will be curves of the second degree. It will consequently be desirable to find the equations of these curves, and from the equations construct the projections.

To obtain the equation of the projections of the meridians, it is only necessary to eliminate the latitude v between equations 2; dividing the second of these equations by the first, we have

$$\frac{y}{x}=\frac{\cos\theta\sin\omega}{\sin\alpha\cos\theta\cos\omega-\cos\alpha\sin\theta}=\frac{\sin\omega}{\sin\alpha\cos\omega-\cos\alpha\tan\theta}\quad.$$

from which

$$\tan\theta=\frac{y\sin\alpha\cos\omega-x\sin\omega}{y\cos\alpha}$$

or

$$\sin^2\theta=\frac{y\sin^2\alpha\cos^2\omega-2xy\sin\alpha\sin\omega\cos\omega+x^2\sin^2\omega}{y^2\cos^2\alpha+y^2\sin^2\alpha\cos^2\omega-2xy\sin\alpha\sin\omega\cos\omega+x^2\sin^2\omega}$$

$$\cos^2\theta=\frac{y^2\cos^2\alpha}{y^2\cos^2\alpha+y^2\sin^2\alpha\cos^2\omega-2xy\sin\alpha\sin\omega\cos\omega+x^2\sin^2\omega}$$

Substituting these in the values of y gives us the general equation of the projection of meridians.

(3) $$x^2(c^2-r^2\sin^2\alpha)\sin^2\omega+xy(r^2-c^2)\sin\alpha\sin 2\omega+y^2(c^2-c^2\sin^2\alpha\sin^2\omega-r^2\cos^2\omega)$$
$$-x(r^2c\sin 2\alpha\sin^2\omega+yr^2c\cos\alpha\sin 2\omega-r^2c^2\cos^2\alpha\sin^2\omega=0$$

This is the equation of an ellipse whose semi-axes a and b are given by

(4) $$a=\frac{rc}{\sqrt{c^2-r^2(1-\cos^2\alpha\sin^2\omega)}}\qquad b=\frac{rc^2\cos\alpha\sin\omega}{c^2-r^2(1-\cos^2\alpha\sin^2\omega)}$$

and whose center is at the point

(5) $$\xi=\frac{cr^2\sin^2\alpha\sin^2\omega}{2[c^2-r^2(1-\cos^2\alpha\sin^2\omega)]}\qquad \eta=\frac{-cr^2\cos\alpha\sin^2\omega}{2[c^2-r^2(1-\cos^2\alpha\sin^2\omega)]}$$

For the direction of the axes we have, ω being the angle that the major axes makes with the axis of x,

$$\tan 2\omega=\frac{\sin\alpha\sin^2\omega}{\cos^2\omega-\sin^2\alpha\sin^2\omega}$$

and from this, by means of the formulas,

$$\tan 2\omega=\frac{2\tan\omega}{1-\tan^2\omega}\qquad \sin 2\omega=2\sin\omega\cos\omega$$

Then easily follows

(6)* $$\tan^2\omega=\frac{-1}{\sin\alpha}$$

The quantities ξ and η have different values for every meridian, *i. e.*, for every value of ω; if, then, we eliminate ω between the two equations giving ξ and η, we will obtain the locus of the centers of all the ellipses

$$\frac{\xi}{\eta}=-\sin\alpha\tan\omega$$

or

$$\tan\omega=-\frac{\xi}{\eta}\sin\alpha$$

from which

$$\sin^2\omega=\frac{\xi^2}{\xi^2+\eta^2\sin^2\alpha}$$

Substituting this in the equation giving ξ, it becomes

$$(7)\qquad \eta^2(c^2-r^2)\sin^2\alpha+\xi^2(c^2-r^2\sin^2\alpha)-\xi cr^2\sin\alpha\cos\alpha=0$$

the equation of an ellipse having ξ and η for its current co-ordinates. The ellipse also passes through the origin of co-ordinates as it lacks an absolute term. The center of the ellipse lies on the axis of x and is given by

$$(8)\qquad \xi=\frac{r^2c\sin\alpha\cos\alpha}{2(c^2-r^2\sin^2\alpha)}\qquad \eta'=0$$

and its semi-axes a' and b' are given by

$$(9)\qquad a'=\frac{r^2c\sin\alpha\cos\alpha}{2(c^2-r^2\sin^2\alpha)}\qquad b'=\frac{r^2c\cos\alpha}{2\sqrt{(c^2-r^2)(c^2-r^2\sin^2\alpha)}}$$

It is obvious that the major axis coincides with the axis of x, and consequently $\omega=0$.

PROJECTIONS OF THE PARALLELS.

To obtain these projections it is only necessary to eliminate ω between equations 2. We have

$$\cos\omega=\frac{cx+rc\cos\alpha\sin\theta}{cr\sin\alpha\cos\theta-rx\cos\alpha\cos\theta}$$

Dividing the second of equations 2 by the first gives

$$\frac{y}{x}=\frac{\sin\omega}{\sin\alpha\cos\omega-\cos\alpha\tan\theta}$$

From this we can readily obtain

$$y\sin\alpha\cos\theta\cos\omega-y\cos\alpha\sin\theta=x\cos\theta\sin\omega$$

Square this in order to get rid of $\sin\omega$ and we have

$$(y^2\sin^2\alpha\cos^2\theta+x^2\cos^2\theta)\cos^2\omega-2y^2\sin\alpha\cos\alpha\sin\theta\cos\theta\cos\omega-x^2\cos^2\theta+y^2\cos^2\alpha\sin^2\theta=0$$

Substituting in this the value of $\cos\omega$ given above, and performing several easy but tedious reductions, we come finally to the equation of the projections of the parallels in the form

$$(10)\qquad x^2[c^2+2rc\sin\alpha\sin\theta-r^2\cos(\alpha-\theta)\cos(\alpha+\theta)]+y^2[c\sin\alpha+r\sin\theta]^2+2rcx(c\cos\alpha\sin\theta+r\sin\alpha\cos\alpha)-c^2r^2\sin(\alpha-\theta)\sin(\alpha+\theta)=0$$

The curve is an

$$\left.\begin{array}{l}\text{Ellipse}\\ \text{Hyperbola}\\ \text{Parabola}\end{array}\right\}\text{according as } c^2+2rc\sin\alpha\sin\theta-r^2\cos(\alpha-\theta)\cos(\alpha+\theta)\left\{\begin{array}{l}>0\\ <0\\ =0\end{array}\right.$$

the quantity H^2-AB being here replaced merely by B since H=0 and A is a perfect square and positive.

By the usual process of transformation to the center and axes we find

$$(11)\qquad b'=\frac{rc\cos\theta(c\sin\alpha+a\sin\theta)}{c^2+2rc\sin\alpha\sin\theta+r^2\sin^2\theta-r^2\cos^2\alpha}$$

$$ba'=\frac{rc\cos\theta}{\sqrt{c^2+2rc\sin\alpha\sin\theta+r^2\sin^2\theta-r^2\cos^2\alpha}}$$

for the axes, and

$$\xi'=\frac{cr\cos\alpha(c\sin\theta+r\sin\alpha)}{c^2+2cr\sin\alpha\sin\theta+c^2\sin^2\theta-r^2\cos^2\alpha}$$

$$(12)\qquad \eta'=0\qquad \omega'=0$$

for the center and direction of the major axis. It follows, then, that the centers of the projections of the parallels all lie upon the axis of x.

For the projection of the equator $\theta=0$

(13) $$x^2\frac{(c^2-r^2\cos^2\alpha)}{c^2\sin^2\alpha}+y^2+\frac{2r^2\cot\alpha}{c}x-r^2=0$$

obviously an ellipse whose axes are

(14) $$a_1'=\frac{rc^2\sin\alpha}{c^2-r^2\cos^2\alpha} \qquad b_1'=\frac{cr}{\sqrt{c^2-r^2\cos^2\alpha}}$$

and whose center is at

(15) $$\xi_1'=-\frac{r^2c\sin\alpha\cos\alpha}{c^2-r^2\cos^2\alpha} \qquad \eta_1'=0$$

The distance p of the projection of the pole from the center of the entire projection is found by making $\theta=\pm\frac{\pi}{2}$ in the expression for ξ', and we have thus

(16) $$p=-\frac{cr\cos\alpha}{c+r\sin\alpha} \qquad p'=\frac{cr\cos\alpha}{c-r\sin\alpha}$$

two points on the axis of x.

EQUATORIAL PROJECTION.

For $\alpha=\frac{\pi}{2}$ the general equations become

(17) $$x=\frac{cr\cos\theta\cos\omega}{c+r\sin\theta} \qquad y=\frac{cr\cos\theta\sin\omega}{c+r\sin\theta}$$

We have then for the general equation of the projections of the parallels

(18) $$x^2+y^2=\frac{c^2r^2\cos^2\theta}{(c+r\sin\theta)^2}$$

ω being eliminated by the simple process of squaring and adding. This is the equation of a circle whose center is at the origin of co-ordinates. For the elimination of θ it is only necessary to divide y by x, thus

$$\frac{y}{x}=\tan\omega$$

We see from this that the meridians are projected in straight lines, and that the angle included between the projections of any two meridians is equal to the angle between the meridians themselves. Fig. 2 gives an idea of this projection.

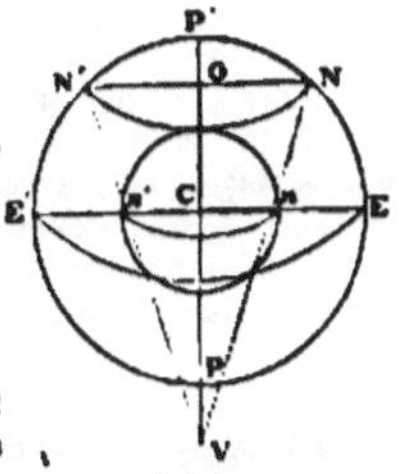

Fig. 2.

In the case where the point of sight is without the sphere, *i. e.*, where $c>r$ the projection will extend from the equator to the parallel which passes through the point where the tangent from the point of sight meets the meridian PEP'E'; this latitude is given by

$$\tan\theta_1=\frac{r}{\sqrt{c^2-r^2}}$$

and the radius of its projection is equal to

$$\frac{cr\cos\theta_1}{c-r\sin\theta_1}$$

Divide now the circle of projection into degrees, and count upon it from the same point E the latitude and longitude, and upon the line of the poles PP' lay off $cV=c$; join V with the extremities of any parallel which it is desired to construct and the intersections of these projecting lines with the diameter EE', viz, n and n', are points in the circumference of the circle into which the given parallel is projected; we have then merely with c as a center to describe circles passing through these points, and they will be the projections of the parallels. The meridians are of course constructed by merely drawing the diameters of the circle of projection.

MERIDIAN PROJECTION.

We have already found for this case

(19)
$$x=-\frac{cr \sin \theta}{c+r \cos \theta \cos \omega} \qquad y=\frac{cr \cos \theta \sin \omega}{c+r \cos \theta \cos \omega}$$

from which

$$x=-\frac{y \tan \theta}{\sin \omega}$$

It is to be observed that for the negative co-ordinates the values of x only change sign, while for negative longitudes the x remains unchanged and y changes its sign. For the projections of the meridians eliminate θ_1.

$$\tan \theta=-\frac{x \sin \omega}{y}$$

from which follows

$$\sin^2 \theta=\frac{x^2 \sin^2 \omega}{y^2+x^2 \sin^2 \omega} \qquad \cos^2 \theta=\frac{y^2}{y^2+x^2 \sin^2 \omega}$$

which being substituted in either of the expressions for x or y would give us the equations of the projections of the meridians. Similarly the projections of the parallels may be found by eliminating ω between the expressions for x and y. A further consideration of this projection in the general case would be productive of but little that could interest, so we shall leave the subject here, taking it up, however, in the various special cases of perspective projection that we shall study.

ORTHOGRAPHIC PROJECTION.

In the case of orthographic projection the eye is supposed to be placed at an infinite distance from the center of the sphere, *i. e.*, $c=\infty$. The projecting cone becomes then in this case a cylinder, the right section of which is a great circle of the sphere. Here there can no parabolas or hyperbolas occur as the projection of any circle of the sphere, but all circles will be projected in circles, ellipses, or straight lines according to the inclination of their planes to the axis of the cylinder. This projection is not used for geographical purposes, though it has been for celestial charts, and is commonly employed for architectural and mechanical drawings.

For $c=\infty$ equation 2 gives

(20)
$$x=r(\sin \alpha \cos \theta \cos \omega-\cos \alpha \sin \theta) \qquad y=r \cos \theta \sin \omega$$

The general equation of meridians now becomes

(21)
$$x^2 \sin^2 \omega-xy \sin \alpha \sin 2\omega+y^2(1-\sin^2 \alpha \sin^2 \omega)-r^2 \cos^2 \alpha \sin^2 \omega=0$$

The equation of an ellipse whose semi-axes (equations 3) are

(22)
$$a=r \qquad b=r \cos \alpha \sin \omega$$

for the center $\xi=\eta=0$, and for the direction of the major axis

$$\tan^2 \omega=\frac{-1}{\sin \alpha}$$

The equation of parallels is now (*vide* equation 9).

(23)
$$x^2+y^2 \sin^2 \alpha+2rx \cos \alpha \sin \theta-r^2 \sin(\alpha-\theta) \sin(\alpha+\theta)=0$$

an ellipse whose semi-axes are

$$a'=r \cos \theta \qquad b'=r \cos \theta \sin \alpha$$

The center is at

$$\xi=r \cos \alpha \sin \theta \qquad \eta=0$$

and the direction of the major axis $\omega=0$.

The equation of the projection of the equator is

(24) $$x^2+y^2\sin^2\alpha-r^2\sin^2\alpha=0$$

for which

$$a_1=r \qquad b_1=r\sin\alpha \qquad \xi=\eta=0$$

And finally for the pole

(25) $$p=-r\cos\alpha.$$

These expressions are sufficient to determine the orthographic projection for any position of the eye.

ORTHOGRAPHIC EQUATORIAL PROJECTION.

The condition that the eye should be on the axis of the earth, and the plane of the equator that of projection, is arrived at, as in the general case, by making $\alpha=\frac{\pi}{2}$; we find then

(26) $$x=r\cos\theta\cos\omega \qquad y=r\cos\theta\sin\omega$$

and, eliminating θ, the meridians are given by

$$\frac{y}{x}=\tan\omega$$

and the parallels by

(27) $$x^2+y^2=r^2\cos^2\theta$$

Thus the meridians are projected in straight lines passing through the center of projection and the parallels are projected into their true sizes as concentric circles.

If the celestial sphere be thus projected it will be desirable to find the ecliptic. This is simply a great circle whose plane has an inclination of 23° 28′ to that of the equator. Their line of section has the longitude 0° or 180°. It is obvious that the required projection is an ellipse whose major axis $=2r$ and is coincident with the projection of the first meridian, and whose minor axis $=2r\cos 23^\circ 28'$ and is coincident with the projection of the meridian of 90°.

MERIDIAN PROJECTION.

In this case the eye is in the plane of the equator, usually also in that of the first meridian; here then $\alpha=0$ and

(28) $$x=r\sin\theta \qquad y=r\cos\theta\sin\omega$$

The equation of the projection of meridians is

(29) $$\frac{x^2}{a^2}+\frac{y^2}{r^2\sin^2\omega}=1$$

the equation of an ellipse for which

$$a=r \qquad b=r\sin t \qquad \xi=0 \qquad \eta=0$$

The equation

(30) $$x=r\sin\theta$$

being independent of ω, is the equation of the parallels; *i. e.*, the parallels are projected into right lines parallel to the axis of y, or, the same thing, parallel to the equator.

For celestial charts the plane of projection is usually that containing the axis of the equator and of the ecliptic, or simply the solstitial colure. The projections of the equator and ecliptic and all parallels to either will in this case be right lines. The center of the projection will represent the equinoctial points, and the solstices are projected in the extremities of the ecliptic. Declination circles of right ascension $\frac{\pi}{2}-\alpha$ and meridians of celestial longitude ω are projected in ellipses whose major axis equals $2r$ and whose minor axes respectively equal $r\cos\alpha$ and $r\sin\omega$.

The orthographic projection has the disadvantage of giving the natural sizes only at the center of the chart. Towards the outside of the projection the portions of the earth's projection are much too small, and at the limit are infinitely small. Moreover, only one hemisphere can be represented upon a single chart.

STEREOGRAPHIC PROJECTION.

In this case the eye is on the surface of the sphere; *i. e.*, in equations 2 we have $c=r$, and in consequence

(31) $$x=\frac{r(\sin\alpha\cos\theta\cos\omega-\cos\alpha\sin\theta)}{1+\cos\alpha\cos\theta\cos\omega+\sin\alpha\sin\theta}\qquad y=\frac{r\cos\theta\sin\omega}{1+\cos\alpha\cos\theta\cos\omega+\sin\alpha\sin\theta}$$

Our general equation of meridians becomes

(32) $$x^2+y^2-2xr\tan\alpha-2yr\frac{\cot\omega}{\cos\alpha}-r^2=0$$

Here H=O and A=B, the well known conditions that the general equation should represent a circle. The center of the circle is at the point

$$\xi=r\tan\alpha\qquad\qquad \eta=-r\frac{\cot\omega}{\cos\alpha}$$

and the radius is

$$R=\frac{r}{\cos\alpha\sin\omega}$$

For the projection of the pole we have

$$p=-\frac{r\cos\alpha}{1+\sin\alpha}=-r\cot\frac{\alpha}{2}$$

a point through which the projections of all the meridians must pass. The equation of the locus of centers of meridians is in this case

(33) $$\xi=r\tan\alpha$$

a straight line parallel to the axis of x at a distance from it $=r\tan\alpha$.

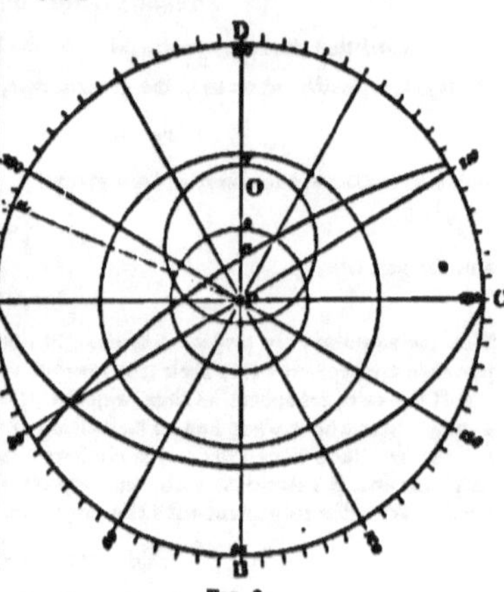

FIG. 2.

The equation of parallels becomes

(34) $$x^2+y^2+2rx\frac{\cos\alpha}{\sin\alpha+\sin\theta}+\frac{r^2(\sin\theta-\sin\alpha)}{\sin\theta+\sin\alpha}=0$$

a circle whose center is at

$$\xi=-\frac{r\cos\alpha}{\sin\alpha+\sin\theta},\qquad \eta=0$$

whose radius is

$$R'=\frac{r\cos\theta}{\sin\alpha+\sin\theta}$$

The equation of the equator is $\theta=0$.

(35) $$x^2+y^2+2rx\cot\alpha-r^2=0$$

the position and magnitude of this projection being given by

$$R'=r\operatorname{cosec}\alpha\qquad \xi=-r\cot\alpha\qquad \eta=0$$

We thus see that both meridians and parallels are projected in circles for this kind of projection, and since, by varying the angle α, we can cause the plane of projection to assume any position relatively to the equator and parallels, it follows *that all circles of the sphere are projected in circles.*

STEREOGRAPHIC EQUATORIAL PROJECTION.

The plane of the equator is here taken for the plane of projection and so $\alpha=\frac{\pi}{2}$; this gives

(36) $$x=\frac{r\cos\theta\cos\omega}{1+\sin\theta}\qquad\qquad y=\frac{r\cos\theta\sin\omega}{1+\sin\theta}$$

Calling ζ the complement of the latitude θ, we have from these equations

$$(37) \qquad \frac{y}{x}=\tan \omega \qquad\qquad x^2+y^2=r^2 \tan^2 \frac{\zeta}{2}$$

The meridians are thus projected in straight lines passing through the origin, or simply in the diameter of the equator. The parallels are projected in circles of radius $=r \tan \frac{\zeta}{2}$. For the equator itself $\zeta=\frac{\pi}{2}$ and the radius of this projection $=r$, the radius of the sphere. Fig. 3 represents this projection, the eye being placed at the south pole; P is the north pole, and ABCD is the equator. To draw this a circle of radius, r is described about any point P and its circumference divided into equal portions of 5° or 10°, or whatever may be most desirable. The diameters AP 180, 30 P 210, 90° P 270 are the meridians of 0°, 30°, and 90°. The parallels are all drawn about P as a center with radii $=r \tan \frac{\zeta}{2}$. Table I, which is constructed by means of this formula, gives the value of ρ (the radius) for every 5° of latitude θ $(=90°-\zeta)$; in the table r is assumed $=1$.

If a perpendicular be erected at the extremity A of the diameter AC the tangents and secants of all the angles necessary to construct the chart may be laid off on it. If, for example, the angle $APa=23°$, then $Aa'=r \tan 23$, $Pa'=r \sec 23°$. If r be taken as unity, the construction will of course be quite simple.

TO DRAW THE ECLIPTIC WITH ITS PARALLELS AND CIRCLES OF LONGITUDE.

In order to do this it is necessary to remember that the stereographic projection of every circle is a circle. Draw now, as in Fig. 4, two diameters of a circle perpendicular to one another as AB and CD and the chords DE and DF cutting AB in e and f, then fe is the stereographic projection of the arc or chord FE. Now, the angle FED is measured by one-half of the arc FD and

$$\text{angle } efD=\frac{BD+AF}{2}=\frac{AD+AF}{2}=\tfrac{1}{2}\,FD$$

Therefore, $FED=efD$ and $EFD=feD$.

If, therefore, FE be the diameter of a circle on the surface of the sphere, the surface of a cone DEF will cut a plane through AB perpendicular to DC, in the figure fe. Since, however, the plane of this section, on account of the equality of the angles FED and efD, makes a subcontrary section, the curve of intersection fe is a circle. The distance in the plane of projection from the center of this circle to the origin O is

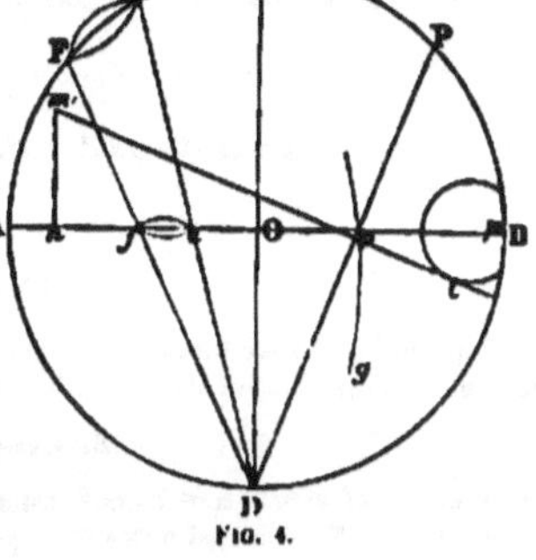

FIG. 4.

$$\delta=\frac{Of+Oe}{2}$$

and the radius

$$R=\frac{Of-Oe}{2}$$

Let λ denote the distance of the pole of FE from the point of sight C, and ρ the distance of the pole from the circumference of FE. Then we have

$$CF=\lambda+\rho \qquad\qquad CE=\lambda-\rho$$

and from the triangles DOf and DOe

$$Of=r \tan \tfrac{1}{2}(\lambda+\rho) \qquad\qquad Oe=r \tan \tfrac{1}{2}(\lambda-\rho)$$

Substituting these in the above value of δ, this becomes

$$\delta=\frac{r}{2}\left\{\tan\tfrac{1}{2}(\lambda+\mu)+\tan\tfrac{1}{2}(\lambda-\mu)\right\}$$

or

$$\delta=\frac{r\sin\lambda}{\cos\lambda+\cos\mu}$$

and, in like manner,

$$R=\frac{r\sin\mu}{\cos\lambda+\cos\mu}$$

Example.—Fig. 3.

If it is desired to draw the parallel to the ecliptic, which is 30° distant from its pole, it is only necessary to lay off from P, on the diameter PD, the distance PO

$$=\delta=\frac{r\sin 23^\circ 28'}{\cos 23^\circ 28'+\cos 30^\circ}$$

from o thus obtained as a center and with a radius oq

$$=R=\frac{r\sin 30^\circ}{\cos 23^\circ 28'+\cos 30^\circ}$$

describe a circle; this is the required projection. For the ecliptic itself $\mu=90^\circ$, and the distance to its center is PO

$$=r\tan 23^\circ 28'$$

and its radius OQ

$$=r\sec 23^\circ 28'$$

AQC is this projection. For the pole of the ecliptic $\mu=0$; whence

$$Pp=\delta=\frac{r\sin 23^\circ 28'}{1+\cos 23^\circ 28'}=r\tan 11^\circ 44'$$

Further, the distances of the point P from the two points in which the diameter BD is cut by the parallel to the

$$\delta+R=r\tan\tfrac{1}{2}(\lambda+\mu) \qquad \delta-R=r\tan\tfrac{1}{2}(\lambda-\mu)$$

GENERALIZED DISCUSSION.

We shall now take up the problem of perspective projection from a more general point of view. Until now the position of the variable point M has been determined by means of the quantities

$$ZPM=\omega \qquad PM=90^\circ-\phi$$

The position of M will now be determined with reference to *any* fixed point, say L, on the surface of the sphere. To this end write $ML=\chi$, $MLP=\Psi$. To determine with respect to P the quantities

$$ZPL=H \text{ and } LP=90^\circ-\theta$$

Let, for example, P denote the pole of the equator and L that of the ecliptic; then $90^\circ-\theta=$ obliquity of the ecliptic. For a star M the longitude $=90^\circ-\Psi$, the latitude $=90-\chi$, the declination $=\phi$, and the right ascension $=MPL-90^\circ$ or $=\beta-90^\circ$, denoting MPL by β.

TO DETERMINE THE VALUES OF ω AND ϕ IN TERMS OF χ AND Ψ, THE LATITUDE AND LONGITUDE OF M.

In the spherical triangle PLM two sides $PL=90^\circ-\theta$ and $LM=\chi$, and the included angle $MLP=\Psi$ are known, and from them we have for the determination of $MPL=\beta$ and $PM=90^\circ-\phi$ the formulas

(38)
$$\begin{aligned}\cos\phi\sin\beta&=\sin\chi\sin\Psi\\ \cos\phi\cos\beta&=\cos\theta\cos\chi-\sin\theta\sin\chi\cos\Psi\\ \sin\phi&=\sin\theta\cos\chi+\cos\theta\sin\chi\cos\Psi\end{aligned}$$

β and θ being found from these, ω is given by the relation

$$\omega=\Pi+\beta$$

The values of these quantities ω and θ must then be substituted in equations 2 to find the values of x and y, the co-ordinates of the projection of M in terms of the new variables.

APPLICATION TO THE STEREOGRAPHIC EQUATORIAL PROJECTION OF PARALLELS TO THE ECLIPTIC.

For this case $\alpha=90°$ and the distance of the eye from the plane of projection$=r$. The values of x and y already found are

$$(39)\qquad x=\frac{r\cos\theta\cos\omega}{1+\sin\theta}\qquad\qquad y=\frac{r\cos\sin\omega}{1+\sin\theta}$$

Now since, according to assumption, the pole L lies in the circle whose plane passes through BD perpendicular to the plane of the paper, we have

$$\Pi=90°\qquad\qquad 90°-\theta=\varepsilon$$

ε denoting the obliquity of the ecliptic. The distance of the given parallel from the pole is χ, and for the points in which it is intersected by the circle perpendicular to BD, $\varpi=0°$ and $180°$. The above equations become by the substitution of these values

$$(40)\qquad \begin{aligned}\cos\theta\sin\beta&=0\\ \cos\theta\cos\beta&=\sin\varepsilon\cos\chi-\cos\varepsilon\sin\chi\\ \sin\theta&=\cos\varepsilon\cos\chi+\sin\varepsilon\sin\chi\end{aligned}$$

Whence it follows that

$$(41)\qquad \beta=0\qquad\qquad \sin\theta=\cos(\varepsilon-\chi)\qquad\qquad \theta=90°-(\varepsilon-\chi)$$

Similarly for $\varpi=180°$

$$(42)\qquad \beta=0\qquad\qquad \theta=90-(\varepsilon+\chi)$$

In general for the two points

$$(43)\qquad \theta=90°-(\varepsilon\pm\chi)\qquad\qquad \omega=\Pi+\beta=90°$$

Substituting these values of θ and ω in the expressions for x and y we have for these co-ordinates

$$(44)\qquad x=0\qquad\qquad y=\frac{-r\sin(\varepsilon\pm\chi)}{1+\cos(\varepsilon\pm\chi)}=-r\tan\tfrac{1}{2}(\varepsilon\pm\chi)$$

For the pole of the ecliptic $\chi=0$ and

$$(45)\qquad x=0\qquad\qquad y=r\tan\frac{\varepsilon}{2}$$

Since in stereographic projection all circles of the sphere are projected as circles, it will only be necessary to find the projections of any three points of the sphere which lie in a circle to be able to determine the center and radius of the projection of the circle. Calling R the radius of the projection, and ξ and η the co-ordinates of its center, also $x_1\,y_1$ x_2y_2, x_3y_3, three points of the circumference, we have then

$$(46)\qquad \begin{aligned}(x_1-\xi)^2+(y_1-\eta)^2&=R^2\\ (x_2-\xi)^2+(y_2-\eta)^2&=R^2\\ (x_3-\xi)^2+(y_3-\eta)^2&=R^2\end{aligned}$$

from which

$$\xi=\frac{E\,(y_3-y_2)-G\,(y_2-y_1)}{(x_2-x_1)\,(y_3-y_1)-(x_3-x_1)\,(y_2-y_1)}$$

(47)
$$\eta = \frac{G\,(x_2-x_1)-E\,(x_3-x_1)}{(x_2-x_1)\,(y_3-y_1)-(x_3-x_1)\,(y_2-y_1)}$$

$$R = \sqrt{(x_1-\xi)^2+(y_1-\eta)^2}$$

where

$$2\,E = (x^2_2+y^2_2)-(x^2_1+y^2_1) \qquad 2\,G = (x^2_3+y^2_3)-(x^2_1+y^2_1).$$

Of course this lengthy analytical process need not be employed, for, having found the three points 1, 2, 3, it is only necessary to draw lines 1–2, 2–3, 3–1, and bisect any two of them by perpendiculars which will meet at the center of the circle.

Two points are sufficient to determine the projection if they lie at the extremities of a diameter of the sphere. For example: if two points are known on the sphere whose angular distances from the fixed point L are ψ and $180°-\psi$ we have for the determination of β, θ, ω

(48)
$$\begin{aligned} \cos\theta \sin\beta &= \sin\chi \sin\psi \\ \cos\theta \cos\beta &= \cos\Theta \cos\chi \pm \sin\Theta \sin\chi \cos\psi \\ \sin\theta &= \sin\Theta \cos\chi \pm \cos\Theta \sin\chi \cos\psi \end{aligned}$$

The upper and lower signs give us two values of β and θ, also $\omega = H+\beta$. Now, from the known expressions for x and y, viz:

(49)
$$x = \frac{r\cos\theta\cos\omega}{1+\sin\theta} \qquad y = \frac{r\cos\theta\sin\omega}{1+\sin\theta}$$

we will be able to determine two points $x_1 y_1$, $x_2 y_2$. The co-ordinates, then, of the center of the projected circle are

$$\xi = \tfrac{1}{2}\,(x_2-x_1) \qquad \eta = \tfrac{1}{2}\,(y_2-y_1)$$

and for the radius

$$R = \tfrac{1}{2}\sqrt{(x_2-x_1)^2+(y_2-y_1)^2}$$

STEREOGRAPHIC MERIDIAN PROJECTION.

In this projection, which is the one commonly employed when a complete hemisphere is to be projected, the eye is placed at any point of the equator, and the plane of the meridian 90° distant from the eye is taken as the plane of projection.

For terrestrial charts the plane of the meridian at Greenwich is usually taken as the plane of projection, and the eye will then be at the point whose longitude is 90° or 270°. But here, as in the former cases, the meridian passing through the eye is to be taken as the first meridian in the reckoning of longitude.

This projection will give the means of representing the two terrestrial hemispheres upon two separate charts. If it is desired to obtain maps of the polar regions the stereographic equatorial projection should be employed. For this case we have $\chi=0$, and equations 31 become

(51)
$$x = \frac{-r\sin\theta}{1+\cos\theta\cos\omega} \qquad y = \frac{r\cos\theta\sin\omega}{1+\cos\theta\cos\omega}$$

The equation of meridians thus becomes (equation) 32)

(52)
$$x^2+y^2+2\,yr\cot\omega-r^2=0$$

a circle whose center is at

$$\xi=0 \quad \eta=-r\cot\omega$$

and whose radius $R=r\,\text{cosec}\,\omega$. For the bounding meridian $\omega=90°$, and $R=r$, which determine the bounding circle of the chart. For the meridian passing through the eye we have $\omega=0$, therefore, $R=\eta=\infty$, and $\xi=0$; this meridian is thus projected in a straight line, which is of course obvious without any proof. For the distance from the center of the map to the intersection of the meridian under consideration with the equator it is easy to see that we have $\delta=r\tan\frac{\omega}{2}$

For the parallels equation 34 becomes

$$(53) \qquad x^2+y^2+2\,rx\operatorname{cosec}\theta+r^2=0$$

a circle whose center is given by

$$\xi'=-r\operatorname{cosec}\theta \qquad \eta'=0$$

and whose radius is $R'=r\cot\theta$.

The equator is projected in a straight line, as is obvious from the conditions

$$\theta=0 \qquad R'=\xi'=\infty \qquad \eta'=0$$

For the distance from the center of the map to the point of intersection of the parallel under consideration with the first meridian, we have $\delta'=r\tan\frac{\theta}{2}$. To construct this projection, draw a circle with radius BA=BC$=r$ to any convenient scale. The equator and meridian passing through the eye are projected in a pair of rectangular diameters. Take AC and DE for these lines, D and E are, of course, the poles of the equator. Lay off on AC, in opposite directions from B, the distances

$$\eta=\pm r\cot\omega$$

giving ω any convenient series of values. The points thus obtained are the centers of projections of meridians. The values $-r\cot\omega$ giving the centers of meridians that lie on the + side of DE and the values $+r\cot\omega$ giving the centers of the meridians that lie on the − side of DE. With these points as centers, draw circles of radii $=r\operatorname{cosec}\omega$ and the meridian projections will be constructed.

Similarly for the parallels we lay off distances above and below B on DE$=\pm r\operatorname{cosec}\theta$ and with these points as centers draw circles of radii $=r\cot\theta$; these will be the projections of the parallels. Here, however, the $\pm r\operatorname{cosec}\theta$ give the centers of the circles lying on the $\pm$ side of AC respectively, which is the opposite of what held in drawing the meridians.

TO PROJECT THE ECLIPTIC AND ITS PARALLELS.

For this case we have

$$e=0 \qquad H=0 \qquad \theta=90^\circ-\epsilon$$

Letting χ denote the distance of the given parallel from the pole of the ecliptic and $\Upsilon=0$ and 180°, equations 38 give

$$(54) \qquad \beta=0 \qquad \omega=0 \qquad \theta=90^\circ-(\epsilon\pm\chi)$$

Substituting these values in the above values of x and y we obtain

$$(55) \qquad x=-r\tan\frac{90^\circ-(\epsilon\pm\chi)}{2} \qquad y=0$$

Take, therefore, from the point D, on the diameter DE, the distances

$$(56) \qquad x_1=r\tan\tfrac{1}{2}[90^\circ-(\epsilon+\chi)] \qquad x_2=r\tan\tfrac{1}{2}[90^\circ-(\epsilon+\chi)]$$

The points thus obtained are those in which the parallel to the ecliptic cuts the diameter DE. The distance between these points is consequently the diameter 2 R of the projection, and middle point of the distance is the center.

Call δ the distance from the center of the chart B to the center of this projection, then

$$(57) \qquad R=\frac{x_2-x_1}{2} \qquad \delta=\frac{x_2+x_1}{2}$$

or, substituting the values of x_0 and x_1

$$(57') \qquad R=\frac{r\sin\chi}{\sin\epsilon+\cos\chi} \qquad\qquad \delta=\frac{r\cos\epsilon}{\sin\epsilon+\cos\chi}$$

By means of these equations we can draw all the parallels to the ecliptic by merely giving χ the proper values. For the ecliptic itself $\chi=90^\circ$, and

$$R=r\,\mathrm{cosec}\,\epsilon \qquad\qquad \delta=r\cot\epsilon$$

For its pole $\chi=0$, and

$$\delta=\tan\tfrac{1}{2}\,(90^\circ-\epsilon)$$

this being, of course, the distance of its pole from the origin of co-ordinates, along the line DE.

THE ANGLE AT WHICH THE PROJECTIONS OF TWO GREAT CIRCLES CUT IS EQUAL TO THE ANGLE AT WHICH THE CIRCLES THEMSELVES CUT.

Let D (Fig. 4) be the point of sight, and P the point of intersection of two great circles, as, for example, the circles making with each other an angle equal to ω. The plane of projection passes through the diameter AB of the sphere, and is perpendicular to DC. Let m be the projection of a point M on the surface of the sphere. Now, from our general equation for the projection of meridians by the stereographic method, we have

$$On=\xi=r\tan\alpha \qquad\qquad mn=\eta=r\frac{\cot\omega}{\cos\alpha}$$

Also

$$\tan mpn=\frac{mn}{On+Op}=\cot\omega$$

since

$$Op=\cot\frac{x}{2}$$

From this it follows that $mpn=90^\circ-\omega$. If, now, pg is a circular arc whose center is at m, or the same thing, if pg is the projection of a great circle through P, the angle $npg=\omega$. Likewise a second great circle, also passing through P and making an angle ω' with the same meridian from which ω was measured, would have for its projection a circle cutting the line AB at an angle ω'. The two projections therefore would make the same angle $\omega-\omega'$ that the circles upon the sphere make with each other.

THE SAME PROPOSITION ALSO HOLDS FOR SMALL CIRCLES.

Let η (Fig. 4) be the center of the projection of a parallel of latitude θ and m the center of the projection of a meridian of longitude ω. These circles intersect at right angles on the surface of the sphere. Further let t be one of the two points of intersection of these circles. Then for the parallel

$$O\eta=\frac{r\cos\alpha}{\sin\alpha+\sin\theta} \qquad\qquad \eta t=\frac{r\cos\theta}{\sin\alpha+\sin\theta}$$

For the meridian

$$On=r\tan\alpha \qquad\qquad mn=r\frac{\cot\omega}{\cos\alpha} \qquad\qquad mt=\frac{r}{\cos\alpha\sin\omega}$$

hence follows:

$$n\eta=\frac{r\,(1+\sin\alpha\sin\theta)}{\cos\alpha\,(\sin\alpha+\sin\theta)}$$

and

$$n\eta^2-\eta t^2=\frac{r^2[1+2\sin\alpha\sin\theta+\sin^2\alpha-\cos^2\theta]}{\cos^2\alpha\,(\sin\alpha+\sin\theta)^2}=\frac{r^2}{\cos^2\alpha\sin^2\omega}-\frac{\cot^2\omega}{\cos^2\alpha}=mt^2-mn^2$$

or finally:

$$mt^2+\eta t^2=n\eta^2+mn^2=m\eta^2$$

and consequently the angle $mt\mu$ is a right angle, as is also the angle of the two circles on these diameters passing through the point t. Hence the projections of the meridians and parallels cut at right angles.

In stereographic projections we see, then, that all circles on the chart intersect at the same angle that they do on the sphere, and also that all angles on the sphere are projected in equal angles on the chart. It follows from this that the projection of any infinitely small portion of the sphere is similar to the infinitesimal itself—the only difference being in the relative sizes. This property is one which lies at the foundation of some of the most interesting and elegant investigations of the problem of projection; for the present we shall say no more concerning it, but will take it up in another place and fully develop it. The fact that circles are projected in circles, and that the infinitesimal element of surface and its projection are similar, are the reasons why the stereographic projection is the one most commonly employed for celestial and terrestrial charts. It is, moreover, evident that not only whole hemispheres but also any part of them may be projected in this way, as, for example, any single country or continent. The point of sight should be chosen as nearly as possible opposite the middle of the part to be projected, because the further the part lies from the normal upon the plane of projection from the point of sight the greater is the distortion of the projection.

THE DISTANCE BETWEEN TWO POINTS ON THE SPHERE AND ON THE PROJECTION.

Let δ be the distance between two points A and B on the sphere, and δ' the distance between A' and B' their projections. Suppose a given point M such that

$$MA=x \qquad MB=y$$

$$M'A'=x' \qquad M'B'=y'$$

We have thus a spherical triangle MAB and a plane triangle M'A'B', its projection, with the angles M and M', equal. Now, in the spherical triangle ABM we have

$$\cos\delta=\cos x\cos y+\sin x\sin y\cos M$$

and from the plane triangle

$$\delta'^2=x'^2+y'^2-2x'y'\cos M$$

Observe that

$$x'=r\tan\frac{x}{2} \qquad y'=r\tan$$

Now eliminate M, and after simple reductions we have

$$\delta'=\frac{r\sin\frac{1}{2}\delta}{\cos\frac{1}{2}x\cos\frac{1}{2}y}$$

From this it follows that if x and y are constant, for example, if they are assumed to remain upon the same parallels of latitude, then is δ' proportional to $2\sin\frac{1}{2}\delta$ or to the chord of the arc AB upon the sphere, whatever be the angle M. If $M=0$, then

$$\delta'=x'-y' \qquad \delta=x-y$$

and consequently the chord of δ

$$=\delta'\frac{\textit{chord}\ (x-y)}{x'-y'}$$

from which for every value of δ' on the chart the corresponding value of δ on the sphere can be found. This expression, of course, cannot be used when $x'=y'$ or when x' is very nearly $=y'$. For this case we must make $M=180°$, then $\delta'=x'+y'$, $\delta=x+y$, and chord of δ

$$=\delta'\frac{\textit{chord of}\ (x+y)}{x'+y'}$$

from which the value of δ can always be exactly obtained.

On perspective charts the scale of miles is different at different points. In order to measure small distances and when great accuracy is not required, it will be sufficient to take the length of a degree of longitude or latitude in any part of the chart and consider that as equal to 60 geographical miles. For greater distances, or where accuracy is important, it will be necessary to take from the chart the latitudes θ and θ' and the longitudes ω and ω' of the places and find the distance δ (radius unity) by the known formula

$$\cos \delta = \sin \theta \sin \theta' + \cos \theta \cos \theta' \cos (\omega - \omega')$$

For convenience of logarithmic computation make here

$$\cot \theta' \cos (\omega - \omega') = \cot \Omega$$

then

$$\cos \delta = \sin \theta \sin \theta' + \cos \theta \cot \Omega \sin \theta' = \frac{\sin \theta' \sin \theta \sin \Omega + \cos \theta \cos \Omega \sin \theta'}{\sin \Omega}$$

or finally

$$\cos \delta = \frac{\sin \theta' \cos (\theta - \Omega)}{\sin \Omega}$$

To find the longitude and latitude of a place from its position on the chart.

The equation of meridians (33) is

$$x^2 + y^2 - 2\,xr \tan \alpha - 2\,yr \frac{\cot \omega}{\cos \alpha} - r^2 = 0$$

That of parallels (34) is

$$x^2 + y^2 + 2\,rx \frac{\cos \alpha}{\sin \alpha + \sin \theta} + \frac{r^2 (\sin \theta - \sin \alpha)}{\sin \theta + \sin \alpha} = 0$$

Make for convenience $x^2 + y^2 = \rho^2$; then the first of these equations gives

$$\cot \omega = \frac{r^2 - \rho^2}{2\,yr} \cos \alpha + \frac{x}{y} \sin \alpha$$

the second becomes

$$\sin \theta = \frac{r^2 - \rho^2}{r^2 + \rho^2} \sin \alpha - \frac{2\,x}{1 + \rho^2} \cos \alpha$$

These equations give us the means of finding θ and ω if we know x and y.

For the stereographic equatorial projection

$$\alpha = 90^\circ \qquad \cot \omega = \frac{x}{y} \qquad \sin \theta = \frac{r^2 - \rho^2}{r^2 + \rho^2}$$

For the stereographic meridian projection

$$\alpha = 0 \qquad \sin \theta = -\frac{2\,x}{1 + \rho^2} \qquad \cot \omega = \frac{r^2 - \rho^2}{2\,yr}$$

GNOMONIC PROJECTION.

This is a perspective projection made upon a plane tangent to the sphere, the point of sight being at the center. It is clear that every great circle will here be projected in straight lines. A complete hemisphere can obviously not be constructed on this plan, as the points of intersection of the projecting lines with the plane of projection will, for the points in the circumference of the complete great circle of the hemisphere, lie at an infinite distance. For gnomonic projection we must have $e = 0$, and in consequence

$$x = \frac{r (\sin \alpha \cos \theta \cos \omega - \cos \alpha \sin \theta)}{\cos \alpha \cos \theta \cos \omega + \sin \alpha \sin \theta} \qquad y = \frac{r \cos \theta \sin \omega}{\cos \alpha \cos \theta \cos \omega + \sin \alpha \sin \theta}$$

Take first the simple cases of gnomonic equatorial and gnomonic meridian projection. For the former of these cases

$$\alpha = 90^\circ \qquad x = r \cot \theta \cos \omega \qquad y = r \cot \theta \sin \omega$$

The equation of meridians is thus

$$y = x \tan \omega$$

The meridians are thus projected in straight lines, making the same angles on the projection with the first meridian as the lines themselves do on the sphere. The equation of the parallels is

$$x^2 + y^2 = r^2 \cot^2 \theta$$

concentric circles having radii proportional to the cotangents of their latitudes.

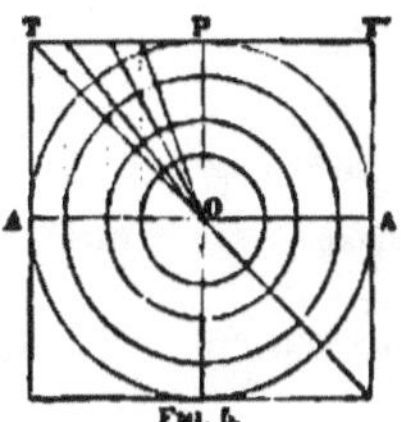

Fig. 5.

The construction is extremely simple (Fig. 5). Divide the limiting circle of the chart into any convenient number of parts and join the center to the points which express the latitudes counted from the diameter AA' perpendicular to the first meridian; these radii prolonged meet the tangent TT' parallel to this diameter and cut off on it distances equal to the radii of the parallels.

GNOMONIC MERIDIAN PROJECTION.

For this case

$$\alpha = 0 \qquad x = -r \frac{\tan \theta}{\cos \omega} \qquad y = r \tan \omega$$

The equation of the meridians is then

$$y = r \tan \omega$$

that of the parallels is

$$x^2 \cot^2 \theta - y^2 - r^2 = 0$$

The meridians are then straight lines parallel to the axis of x and simply constructed. The parallels are hyperbolas, whose major axis is in the direction of x and equals $2r \tan \theta$; whose minor axis is equal to $2r$ and is perpendicular to the first meridian.

The most convenient method of construction by points will be to employ the co-ordinate x given by

$$x = r \frac{\tan \theta}{\cos \omega}$$

and calculate the intersections of the parallels with the meridians already drawn, by giving θ a certain value and ω a series of values, 5°, 10°, 15°, &c.

We shall now take up the general case where the plane of projection is tangent at any point of latitude α.

The equation of the meridians is now

$$y \cos \omega - x \sin \alpha \sin \omega = r \cos \alpha \sin \omega$$

which is the equation of a right line making an angle with the first meridian

$$\tan^{-1} (\sin \alpha \tan \omega)$$

and cutting this meridian in a point whose distance from the center is

$$Op = r \cot \varepsilon$$

The equation of the parallels is

$$x^2 (\sin^2 \theta - \cos^2 \varepsilon) + y^2 \sin^2 \theta + 2\, rx \cos \varepsilon \sin \varepsilon - r^2 \cos^2 \theta - \cos^2 \varepsilon = 0$$

This is a conic section and is

$$\begin{gathered} \text{an ellipse if } \sin \theta > \cos \varepsilon \text{ or } \theta > 90^\circ - \varepsilon \\ \text{an hyperbola if } \sin \theta < \cos \varepsilon \text{ or } \theta < 90^\circ - \varepsilon \\ \text{a parabola if } \sin \theta = \cos \varepsilon \text{ or } \theta = 90^\circ - \varepsilon \end{gathered}$$

We will consider briefly these three cases. If $\theta > 90^\circ - \theta$, we have an ellipse whose semi-axes are

$$a = \frac{r}{2} \frac{\sin^2 \theta}{(\sin^2 \theta - \cos^2 \varepsilon)} \qquad b = \frac{r \cos \theta}{\sqrt{\sin^2 \theta - \cos^2 \varepsilon}}$$

The center of the ellipse is at

$$\xi = -\frac{r}{2} \frac{\sin 2\varepsilon}{(\sin^2 \theta - \cos^2 \theta)} \qquad \eta = 0$$

For $\theta < 90^\circ - \varepsilon$, we have an hyperbola whose semi-axes are

$$a = \frac{r \sin 2\theta}{\cos (\theta - \varepsilon) \cos (\theta + \varepsilon)} \qquad b = \frac{r \sin \varepsilon}{\sqrt{\cos (\theta - \varepsilon) \cos (\theta + \varepsilon)}}$$

and for the center

$$\xi = -\frac{r}{2} \frac{\sin 2\varepsilon}{2 \cos (\theta - \varepsilon) \cos (\theta + \varepsilon)} \qquad \eta = 0$$

For $\theta = 90^\circ - \varepsilon$: This gives for the equation of the parallels

$$y^2 = \frac{r^2 \cos 2\varepsilon - r \sin 2\varepsilon . x}{\sin^2 \varepsilon}$$

This is a parabola whose semi-parameter is

$$p = 2\, r \cot \varepsilon$$

and whose vertex is at the point

$$\xi = r \cot 2\varepsilon \qquad \eta = 0$$

For the equator we have $\theta = 0$, and its equation becomes

$$x = r \tan \varepsilon$$

the equation of a right line perpendicular to the first meridian and at a distance from the center $= O\varepsilon$ (Fig. 6).

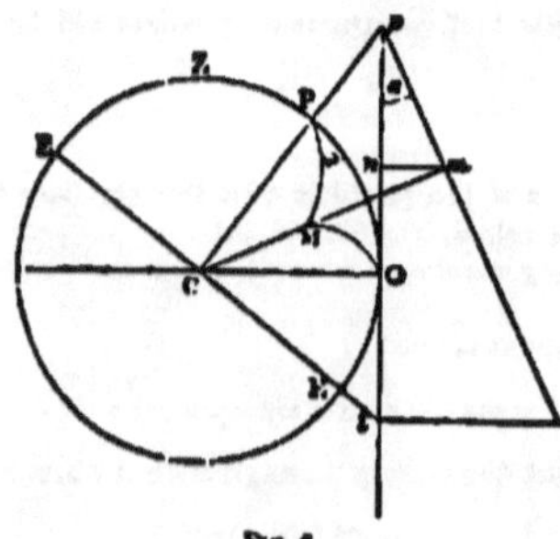

FIG. 6.

Since

$$Op = r \cot \alpha,\ p\varepsilon = pO + O\varepsilon = \frac{r}{\cos \alpha \sin \alpha}$$

Then

$$\varepsilon e = p\varepsilon \tan q, \text{ or } \varepsilon e = \frac{r \tan \omega}{\cos \alpha}$$

when $q = \tan^{-1} (\sin \alpha \tan \omega)$.

Instead of tracing the parallels directly, it will be convenient to determine in the meridians pm, $p\mu'$, &c., conceived as already drawn, points of latitude θ, and then join these points by a curve.

First, to find the projection m of the point M whose longitude is $\omega = \mathrm{OPM}$ and whose latitude is $\theta = 90° - \mathrm{PM}$. This problem reduces itself to the finding of the distance pm.

In the triangle CpO we have

$$Cp = \frac{CO}{\sin \alpha} = \frac{r}{\sin \alpha}$$

In the triangle Cpm

$$Cmp + Cpm = 90° + \theta$$

or

$$Cmp = 90° + \theta - Cpm$$

Calling K the angle Cpm and $q = \mathrm{OPM}$, we have readily

$$\cos K = \cos q \cos \alpha$$

and from the triangle Cpm

$$pm = \frac{r \cos \theta}{\sin \alpha \cos (\theta - K)}$$

For the determination of K we have

$$\cos K = \frac{\cos \alpha}{\sqrt{1 + \sin^2 \alpha \tan^2 \omega}}$$

which shows that K is constant for all points of the same meridian. For each value of θ we have, then, for the determination of pm

$$pm = \frac{r \cos \theta}{\sin \alpha \cos (\theta - K)}$$

For the construction of this projection we may proceed as follows: Lay off from the center O (Fig. 7) upon the first meridian Op a length $Op = r \cot \alpha$. This is easily constructed by erecting

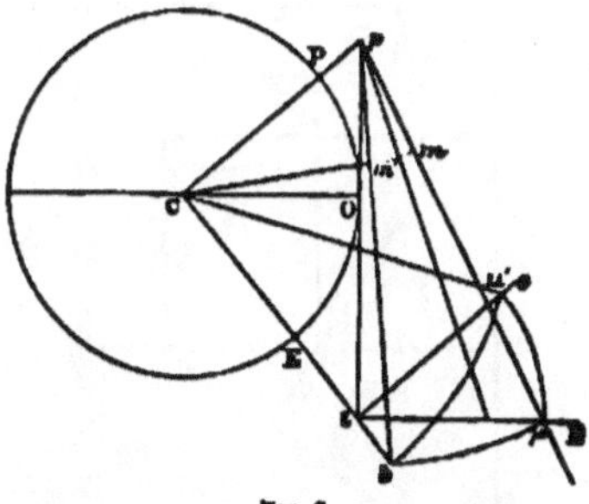

FIG. 7.

at O a perpendicular OC to Op, making $OC = r$, and at C laying off the angle $pCO = 90° - \alpha$. Similarly construct $O\varepsilon = r \tan \alpha$ by erecting $C\varepsilon$ perpendicular to Cp. Now draw εe perpendicular to $C\varepsilon$, and then draw lines from C, making angles with $C\varepsilon$ equal to the longitudes of the meridians whose projections are required. By this means we find upon εe the lengths $\varepsilon\mu'$, $\varepsilon\nu'$, of the intercepts of the meridians upon the equator $E\varepsilon$ perpendicular to $p\varepsilon$. Then, with ε as a center describe arcs of circles passing through the points μ', ν', &c., and cutting Ez in the points μ, ν, &c.; joining these points to p and we have constructed the projections of the meridians.

We will now determine the point of each meridian of which the latitude θ is given (Fig. 6). In the triangle Cpe (a triangle in space) the side Ce is in the equator, so that the figure is right-angled at C. Its intersection pm with the sphere is a meridian PM whose projection is in pm, and in which $PM=90°-\theta$, the distance of M from the pole P. The right line CM prolonged intersects the line pe in the projection m. Now, in (Fig. 7) lay off on Cz the distance $Cb=C\mu$ corresponding to the line Ce of the preceding figure; pb will be equal to the distance of the pole from the equator of the map and in consequence to $p\mu$, and might be constructed by drawing from p as a center an arc of radius $p\mu$ intersecting Cz in b. It is now only necessary to draw a line C in making with Cz the angle $m'Cz=\theta$; its intersection m' with pb gives the distance $pm'=pm$; the latter distance pm being laid off on the line $p\mu$, already drawn. It may be readily verified that

$$p\mu=\frac{r\tan\omega}{\cos\alpha}$$

and

$$pm=\frac{r\cos\theta}{\sin\alpha\cos(\theta-K)}$$

The gnomonic projection is not much employed in the construction of geographical charts, but is frequently used for celestial projections. Suppose that we take the plane of projection perpendicular to the horizon of a point C, whose geographical latitude is φ, and suppose that this plane meet the horizon at the point O, whose azimuth is ω; let Z denote the zenith of C, and P the pole of the earth. Draw the meridian PZR of the point C, and the vertical ZO, making $RZO=\omega$. The arc PZ is equal to the colatitude of C, *i. e.*, $90°-\varphi$. In order to make the preceding formulas applicable to this case, call $PCO=\alpha$. In the spherical triangle PZO, we know $PZ=90°-\varphi$; the spherical angle $PZO=180°-\omega$, and the side $ZO=90°$; we can now calculate the angle $ZOP=\Psi$, and the side $PO=90°-\alpha$ by means of the formulas

$$\sin\alpha=-\sin\varphi\cos\omega \qquad \tan\Psi=\sin\omega\cot\varphi$$

If in the plane of projection Op represent the projection of the meridian OP, and Oz that of the arc of a great circle Oz, then $zOp=ZOP=\Psi$; in like manner, Oh drawn perpendicular to Oz will represent the projection of the horizon ORH, and ie, perpendicular to pz, the projection of the equator

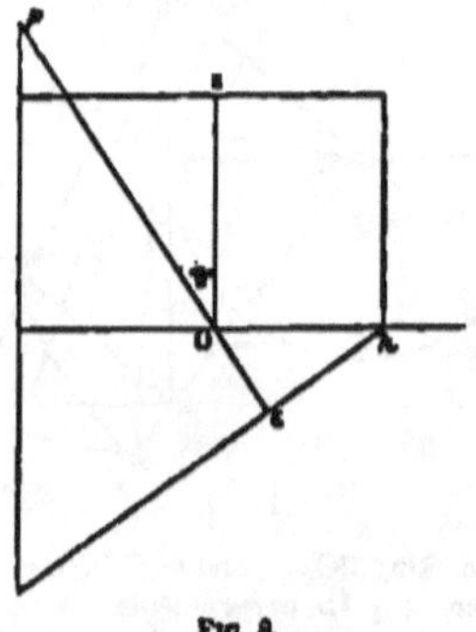

Fig. 8.

Ex. Draw the line Oh (Fig. 8), the projection of the horizon, and at O erect the perpendicular Oz, making the angle $zOp=\Psi$, Ψ being calculated by the formula

$$\tan\Psi=\cot\varphi\sin\omega;$$

op will denote the first meridian. In order to find *p* we have

$$p = r \cot \varepsilon \qquad \sin \varepsilon = -\sin \varphi \cos \omega$$

Then finally lay off on Op, $O\varepsilon = r \tan \varepsilon$, and the perpendicular εh to $p\varepsilon$ will give the projection of the equator.

DISTANCE BETWEEN TWO POINTS.

Since in this projection all great circles are projected in straight lines, it is easy to find the distance between any two points. If we apply here the general solution for all perspective projections, it is obvious that it is only necessary to draw from *o* (Fig. 9), two perpendiculars to *oa*

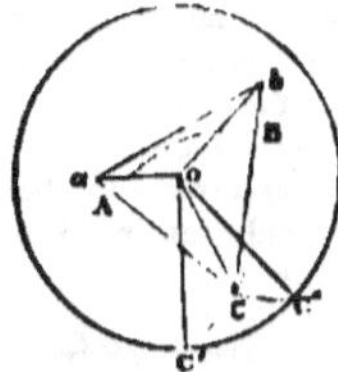

FIG. 9.

and *ob*, the radii of the two points, and make them equal to the radius of the sphere. Then, with the three sides *ab* (known) *a*C′, and *b*C″ thus determined, construct a triangle *a*C*b*, and finally, from the point C, with a radius equal to the radius of the sphere, describe an arc AB of a great circle, which will give the required distance in degrees and fractions of a degree.

These are the principal perspective projections which have been used for celestial and terrestrial charts. Any number of modifications might be given, depending upon the position of the point of sight, as *c* may range anywhere from 0 to ∞. It would be difficult, however, by this process, to simplify very much either the construction or use of the projections by such means. The stereographic projection is, from the fact that both meridians and parallels are projected in circles, the most convenient to use. The common fault of all of these projections, and one which is indeed incident to the nature of projection, is that only those portions of the sphere opposite the eye are projected in approximately their true dimensions, those near the boundaries of the map being very much distorted.

---

§ II.

ORTHOMORPHIC PROJECTION.

From the most general point of view a projection may be defined as the representation of any given surface upon any other surface, whether plane or curved, in such a way as to satisfy certain prescribed conditions. In the representation of any non-developable surface (*e. g.*, the sphere) upon a plane certain errors are of course unavoidable, but any of these errors may be diminished, or even made to disappear altogether, at the cost of increasing some other. In the particular case of projections which it is proposed now to study, we will assume that the elements of the sphere are similar to the corresponding elements of the projection, or we shall so construct the projection that corresponding infinitesimal areas upon the sphere and upon the map shall be similar. It will be convenient to use the term given by Germain to such projections, and so we shall call them orthomorphic.

The nature of a curved surface is determined by an equation between the three co-ordinates x, y, z of any one point of the same. By means of this equation any one of these co-ordinates can be expressed as a function of the other two, or, more generally, each of the quantities x, y, z may be given as a function of two new independent variables, u and v, and in consequence each point of the surface will correspond to definite values of u and v. The general consideration of this case will be reserved for another chapter. As we are here to confine ourselves to the projection of the sphere, it is obvious that the two parameters u and v correspond to the spherical co-ordinates φ and ω, or to the geographical colatitude and longitude (since $\varphi=90°-\theta$) of a point on the surface. If, as usual, r denote the radius of the sphere, then

$$x^2+y^2+z^2=r^2$$

is its equation, and the known formulas of transformation to spherical co-ordinates are

$$x=r\cos\omega\sin\varphi \qquad y=r\sin\omega\sin\varphi \qquad z=r\cos\varphi$$

Let, now, ξ, η, ζ denote the co-ordinates of a point upon any other surface on which it is desired to project the sphere. We make this general assumption here, as it is as easy to obtain the results at present sought for *any* surface as it is for the plane. The ξ, η, ζ are of course dependent upon one another, and, as in the former case, may each be given as functions of two independent parameters, u' and v'. If the points (x, y, z) and (ξ, η, ζ) correspond, then the co-ordinates ξ, η, ζ are dependent upon x, y, z, and in consequence upon u and v, or, in the case under consideration, upon φ and ω. Now, introducing Gauss' notation, we have

$$dx=a\,d\omega+a'd\varphi \qquad dy=b\,d\omega+b'd\varphi \qquad dz=c\,d\omega+c'd\varphi$$

and likewise

$$d\xi=\alpha\,d\omega+\alpha'd\varphi \qquad d\eta=\beta\,d\omega+\beta'd\varphi \qquad d\zeta=\gamma\,d\omega+\gamma'd\varphi$$

the a, b, c, α, β, γ, evidently denoting the first differential coefficients of x, y, z, ξ, η, ζ, with respect to ω, and similarly these same symbols accented denote the derivatives of x, y, &c., with respect to φ. Imagine, now, these points upon the surface to be projected, which we shall call S, infinitely near to each other; these can then be considered as the vertices of an infinitely small plane triangle. To these three points upon S there will correspond three points upon Σ (the second surface), likewise infinitely near each other and forming an infinitesimal triangle. As the condition of orthomorphic projection is that the corresponding infinitesimal areas shall be similar, it is obvious that the sides of these two triangles must be proportional. Denoting by ds and $d\sigma$ corresponding linear elements of S and Σ we have

$$d\sigma=m\,ds;$$

m, denoting the ratio of the linear elements of the two surfaces, is in general a function of ω and φ and varies from point to point of the surface. In our case m is a constant, and consequently the corresponding elements of area upon S and Σ are similar.

The ordinary expression for the element of length ds is

$$ds^2=dx^2+dy^2+dz^2$$

which becomes, on substituting the new values of dx, dy, dz,

$$ds^2=(a^2+b^2+c^2)d\omega^2+(a'^2+b'^2+c'^2)d\varphi^2+2(aa'+bb'+cc')d\omega d\varphi$$

Similarly

$$d\sigma^2=(\alpha^2+\beta^2+\gamma^2)d\omega^2+(\alpha'^2+\beta'^2+\gamma'^2)d\varphi^2+2(\alpha\alpha'+\beta\beta'+\gamma\gamma')d\omega d\varphi$$

Now, since m is constant, the equation

$$d\sigma^2=m^2ds^2$$

gives

$$\alpha^2+\beta^2+\gamma^2=m^2(a^2+b^2+c^2)=m^2\mathrm{E}$$

$$\alpha'^2+\beta'^2+\gamma'^2=m^2(a'^2+b'^2+c'^2)=m^2\mathrm{G}$$

$$\alpha\alpha'+\beta\beta'+\gamma\gamma'=m^2(aa'+bb'+cc')=m^2\mathrm{F}$$

We can thus write

$$ds=\mathrm{E}d\omega^2+2\,\mathrm{F}d\omega d\varphi+\mathrm{G}d\varphi^2$$

If $ds=0$, we find by solution of the resulting quadratic

$$d\omega=-\frac{d\varphi}{E}\left\{F\mp\sqrt{F^2-EG}\right\}$$

from this we derive immediately, i as usual denoting $\sqrt{-1}$,

$$E\,d\omega+F\,d\varphi\pm i\,d\varphi\sqrt{EG-F^2}=0$$

or

$$E\,d\omega+F\,d\varphi+i\,d\varphi\sqrt{EG-F^2}=0$$

$$E\,d\omega+F\,d\varphi-i\,d\varphi\sqrt{EG-F^2}=0$$

Call R and R′ the integrating factors of these two differential equations and assume for the integral of the first

$$p+iq=\text{const.}$$

and for that of the second

$$p-iq=\text{const.}$$

and there follows

$$E\,d\omega+F\,d\varphi+i\,d\varphi\sqrt{EG-F^2}=R^{-1}(dp+idq)$$

$$E\,d\omega+F\,d\varphi-i\,d\varphi\sqrt{EG-F^2}=R'^{-1}(dp-idq)$$

Multiplying these two equations together,

$$E\,ds^2=[RR']^{-1}(dp^2+dq^2)$$

or, making

$$[RR'E]^{-1}=n$$

then

$$ds^2=n(dp^2+dq^2)$$

In precisely the same way we can find for the surface Σ the integrals

$$P+iQ=\text{const.}\qquad\qquad P-iQ=\text{const.}$$

and for the element of length

$$d\sigma^2=N(dP^2+dQ^2)$$

These two expressions for ds^2 and $d\sigma^2$ can be written in the forms

$$ds^2=n(dp+idq)(dp-idq)\qquad\qquad d\sigma^2=N(dP+idQ)(dP-idQ)$$

and from these, by virtue of the condition $d\sigma=m\,ds$, we have

$$\frac{m^2n}{N}=\frac{(dP+idQ)(dP-idQ)}{(dp+idq)(dp-idq)}$$

It is evident that the numerator of the right-hand side of this equation is only divisible by the denominator when $dP+idQ$ is divisible by $dp+idq$, and $dP-idQ$ is divisible by $dp-idq$; or when $dP+idQ$ is divisible by $dp-idq$, and $dP-idQ$ is divisible by $dp+idq$.

In the first case $dP+idQ$ will vanish when $dp+idq=0$, or $P+iQ$ will be constant for $p+iq$ constant; *i. e.*, $P+iQ$ will be merely a function $p+iq$ and $P-iQ$ similarly will be a function of $p-iq$. Placing then

$$P+iQ=f_1(p+iq)\qquad\qquad P+iQ=f_1(p-iq)$$

$$P-iQ=f_2(p-iq)\qquad\qquad P-iQ=f_2(p+iq)$$

It is easy to see that both assumptions give results which differ only with respect to their signs. The functions f_1 and f_2 must also be of the same form since $P+iQ$ and $P-iQ$ differ only in the sign of i. All conditions will then be satisfied if we take one of the functions, say f_1, and write

$$P+iQ=f(p+iq)$$

replacing for convenience f_1 by f; P will be the real and iQ the imaginary part of $f(p+iq)$.

Assume that in general

$$\phi(u)=\frac{df(u)}{du}$$

Now we have

$$dP+idQ=df(p+iq)$$

and

$$\frac{dP+idQ}{dp+idq}=\frac{df(p+iq)}{dp+idq}$$

or, according to the above convention,

$$\frac{dP+idQ}{dp+idq}=\phi(p+iq)$$

also

$$\frac{dP-idQ}{dp-idq}=\phi(p-iq)$$

The expression for m^2 becomes now

$$m^2=\frac{N}{n}\phi(p+iq)\phi(p-iq)$$

which gives the ratio of the original element to its projection.

The results of the foregoing discussion may be briefly summarized as follows: First, find from the assumed equation $ds^2=0$ the two integrals

$$p+iq=\text{const.} \qquad p-iq=\text{const.}$$

Then, denoting by F any arbitrary function such that P shall be the real part and iQ the imaginary part of $F(p+iq)$, we find at once the two equations which give P and Q in terms of p and q, or we have the sought elements of the projection in terms of the elements of the surface to be projected.

Finally, if

$$\phi(u)=\frac{dF(u)}{du}$$

then

$$m=\sqrt{\frac{N}{n}\phi(p+iq)\phi(p-iq)}$$

which gives the ratio of the length of the linear elements of the surfaces S and Σ, where

$$N=\frac{d\sigma^2}{dP^2+dQ^2} \qquad n=\frac{ds^2}{dp^2+dq^2}$$

ORTHOMORPHIC PROJECTION OF THE SPHERE.

Suppose we have a sphere given by the equation

$$x^2+y^2+z^2=r^2$$

The formulas already given for transformation to spherical co-ordinates are

$$x=r\cos\omega\sin\varphi \qquad y=r\sin\omega\sin\varphi \qquad z=r\cos\varphi$$

Differentiating these

$$dx=-r d\omega\sin\omega\sin\varphi+r\cos\omega\cos\varphi d\varphi$$

$$dy=r d\omega\cos\omega\sin\varphi+r\sin\omega\cos\varphi d\varphi$$

$$dz=-r\sin\varphi d\varphi$$

Squaring and adding

$$ds^2=r^2\sin^2\varphi d\omega^2+r^2d\varphi^2$$

If we then make $ds=0$, we have

$$E=r^2\sin^2\varphi \qquad G=r^2 \qquad F=0$$

and

$$r^2 \sin^2 \varphi d\omega \pm i r^2 \sin \varphi d\varphi = 0$$

or

$$d\omega \pm \frac{i d\varphi}{\sin \varphi} = 0$$

The integral of this is

$$\omega \pm i \log \cot \frac{\varphi}{2} = \text{const.}$$

Now, if F denote any function whatever, we will have ξ the real and η the imaginary part of the function

$$F\left(\omega + i \log \cot \frac{\varphi}{2}\right)$$

and these values of ξ and η will be the rectangular co-ordinates of the projection of the point on the sphere whose longitude is ω and whose colatitude is φ.

MERCATOR'S PROJECTION.

The simplest supposition that we can make is that the function F (υ) is linear, or that we have $F(\upsilon) = K\upsilon$ where K is an arbitrary constant. We have then

$$\xi + i\eta = K\left(\omega + i \log \cot \frac{\varphi}{2}\right)$$

from which

$$\xi = K\omega \qquad \eta = K \log \cot \frac{\varphi}{2}$$

the known equations for the Mercator projection.

In order to find the ratio m of the corresponding elements of the sphere and plane, make

$$p + iq = \omega + i \log \cot \frac{\varphi}{2}$$

from which derive as usual

$$dp = d\omega \qquad dq = \frac{d\varphi}{\sin \varphi}$$

We had, however,

$$n = \frac{ds^2}{dp^2 + dq^2}$$

and substituting these values of dp and dq,

$$n = r^2 \sin^2 \varphi \qquad N = 1$$

Further,

$$\phi = \frac{dF(\upsilon)}{d\upsilon} = \frac{K d\upsilon}{d\upsilon} = K$$

and consequently

$$m = \sqrt{\frac{N}{n} \phi(p+iq)\, \phi(p-iq)} = \sqrt{\frac{NK^2}{n}} = \frac{K}{r \sin \varphi}$$

HARDING'S PROJECTION.

Suppose we make the supposition that $F(\upsilon) = Ke^{il\upsilon}$ where K and l are constants. As before, ξ is the real and η the imaginary part of

$$F(\omega + i \log \cot \tfrac{1}{2}\varphi)$$

or in the assumed case

$$Ke^{il\left(\omega + i \log \cot \frac{\varphi}{2}\right)} = Ke^{il\omega - l \log \cot \frac{\varphi}{2}} = Ke^{l \log \tan \frac{\varphi}{2} + il\omega} = Ke^{il\omega} \tan^l \frac{\varphi}{2}$$

Since

$$e^{il\omega} = \cos l\omega + i \sin l\omega$$

we have

$$\xi = K \tan^l \frac{\varphi}{2} \cos l\omega \qquad \eta = K \tan^l \frac{\varphi}{2} \sin l\omega$$

and in consequence, for the equation of the parallels,

$$\xi^2+\eta^2=K^2\tan^{2l}\frac{\varphi}{2}$$

a series of concentric circles with radii given by $K\tan^l\frac{\varphi}{2}$. In like manner eliminating φ we have for the meridians

$$\eta=\xi\tan l\omega$$

the equation of straight lines passing through the origin.

For the determination of m, we observe that

$$p=\omega \qquad q=\log\cot\frac{\varphi}{2} \qquad n=r^2\sin^2\varphi \qquad N=1$$

Further,

$$\phi(u)=\frac{dF(u)}{du}=ilKe^{ilu}$$

from which it is clear that

$$\phi(p+iq)=ilKe^{il(\omega+i\log\cot\frac{\varphi}{2})} \qquad \phi(p-iq)=ilKe^{-il(\omega-i\log\cot\frac{\varphi}{2})}$$

and consequently

$$m^2=\frac{l^2K^2e^{2l\log\tan\frac{\varphi}{2}}}{r^2\sin^2\varphi}$$

or simply

$$m=\frac{lK\tan^l\frac{\varphi}{2}}{r\sin\varphi}$$

For the case of $l=1$ we find

$$\xi=K\tan\frac{\varphi}{2}\cos\omega \qquad \eta=K\tan\frac{\varphi}{2}\sin\omega \qquad m=\frac{K(1-\cos\varphi)}{r\sin^2\varphi}$$

the formulas that occur in the case of stereographic equatorial projection; and thus we see that this projection has the great advantage of preserving the similarity of infinitesimal areas. Leaving any further application of this method for another place, we will now revert to the beginning of the subject again and develop the necessary formulas for the orthomorphic projection of a spheroid.

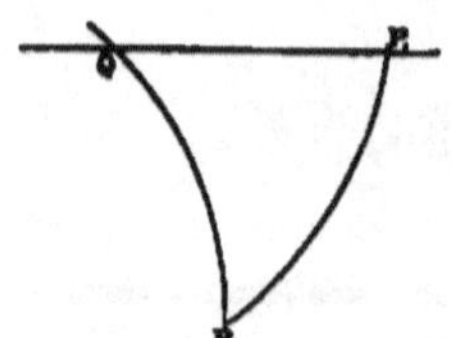

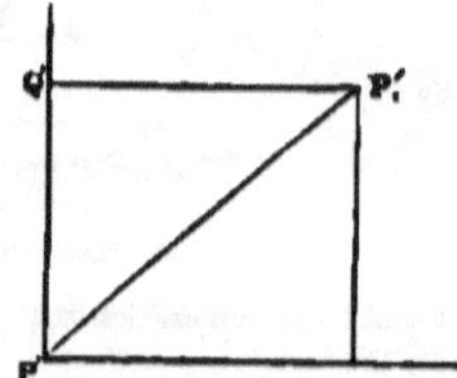

FIG. 10.

Suppose in Fig. 10 that PQ denote an element of a meridian upon the spheroid and QP_1 the element of a parallel through Q, the same letters accented to denote the representations of these quantities upon the projection. The condition of equality of angles gives for these infinitesimal areas

$$\frac{Q'P'_1}{P'Q'}=\frac{QP_1}{PQ}$$

Squaring and adding unity to both sides of this equation

$$P'Q'^2+Q'P_1'^2=\frac{P'Q'^2}{PQ^2}(PQ^2+P'Q_1^2)$$

Observe that the factor $\frac{P'Q'^2}{PQ^2}$ depends only on the latitude and longitude θ and ω of the point P; denote this factor by t^2 and we have from the figure

$$P'P_1'^2=t^2PP_1^2$$

The ratio of the corresponding linear elements PP_1 and $P'P'_1$ upon the spheroid and upon the projection depends only upon the co-ordinates ω and θ of P and not upon the direction of the element.

By the usual convention we have

$$P'P_1'^2=d\xi^2+d\eta^2$$

and we know that, denoting by ds the element PQ of the meridian

$$ds=\frac{a\,(1-\epsilon^2)\,d\theta}{\sqrt{1-\epsilon^2\sin^2\theta}}$$

ϵ denoting the eccentricity of the spheroid; if ρ denote the radius of the parallell QP, we have $QP_1=\rho d\omega$, and consequently

$$d\xi^2+d\eta^2=t^2\,(ds^2+\rho^2d\omega^2)$$

or

$$d\xi^2+d\eta^2=t^2\rho^2\left[\left(\frac{ds}{\rho}\right)^2+d\omega^2\right]$$

Denoting as before the colatitude (*i. e.*, the angle which the normal makes with the axis of the spheroid) by φ, we have $d\varphi=-d\theta$, and the quantity $\frac{ds}{\rho}$ has for its value

$$\frac{ds}{\rho}=\frac{(1-\epsilon^2)d\varphi}{(1-\epsilon^2\cos^2\varphi)\sin\varphi}$$

$\frac{ds}{\rho}$ is an exact differential and we may denote it by du; consequently

$$u=\int\frac{d\varphi}{\sin\varphi}-\epsilon^2\int\frac{\sin\varphi d\varphi}{1-\epsilon^2\cos^2\varphi}=\frac{1}{2}\log\frac{1-\cos\varphi}{1+\cos\varphi}+\log\left(\frac{1-\epsilon\cos\varphi}{1+\epsilon\cos\varphi}\right)^{\frac{\epsilon}{2}}+\log G$$

The first term of this with the constant is the value of u for the earth supposed spherical; collecting the terms this is

$$u=\log\left[G\tan\frac{\varphi}{2}\left(\frac{1-\epsilon\cos\varphi}{1+\epsilon\cos\varphi}\right)^{\frac{\epsilon}{2}}\right]$$

If we suppose such an angle ζ that

$$\tan\frac{\zeta}{2}=\tan\frac{\varphi}{2}\left(\frac{1-\epsilon\cos\varphi}{1+\epsilon\cos\varphi}\right)^{\frac{\epsilon}{2}}$$

then

$$u=\log G\tan\frac{\zeta}{2}$$

which is the same form of u that we have for the earth supposed spherical; ζ may thus be regarded as the polar distance corrected to allow for the ellipticity of the earth, and, the eccentricity being very small, we have, nearly enough, for ζ

$$\zeta=\varphi-\frac{\epsilon^2}{2}\sin 2\varphi$$

Returning now to the equation giving the ratio of the linear elements upon the spheroid, which we will again denote by S, and the plane, denoted by Σ, write

$$l\rho = m_0$$

then

$$d\xi^2+d\eta^2=m_0^2\,(du^2+d\omega^2)$$

We have now to determine ξ and η in terms of u and ω. Write for brevity

$$u+i\omega=\alpha \qquad \xi+i\eta=\alpha^1$$
$$u-i\omega=\beta \qquad \xi-i\eta=\beta^1$$

Call μ_0 the value of m_0 when the quantities u and ω, on which it depends, are replaced by their values in α and β, and we have

$$d\alpha^1 d\beta^1=\mu_0^2\,d\alpha\, d\beta.$$

Now

$$d\alpha^1=\frac{d\alpha^1}{d\alpha}d\alpha+\frac{d\alpha^1}{d\beta}d\beta \qquad d\beta^1=\frac{d\beta^1}{d\alpha}d\alpha+\frac{d\beta^1}{d\beta}d\beta$$

and, consequently,

$$\frac{d\alpha^1}{d\alpha}\frac{d\beta^1}{d\alpha}d\alpha^2+\frac{d\alpha^1}{d\beta}\frac{d\beta^1}{d\beta}d\beta^2+\left(\frac{d\alpha^1}{d\alpha}\frac{d\beta^1}{d\beta}+\frac{d\alpha^1}{d\beta}\frac{d\beta^1}{d\alpha}\right)d\alpha d\beta=\mu_0^2\,d\alpha\, d\beta$$

from which follows

$$\frac{d\alpha^1}{d\alpha}\frac{d\beta^1}{d\alpha}=0 \qquad \frac{d\alpha^1}{d\beta}\frac{d\beta^1}{d\beta}=0$$

and also

$$\frac{d\alpha^1}{d\beta}=0 \qquad \frac{d\beta^1}{d\alpha}=0$$

or

$$\frac{d\alpha^1}{d\alpha}=0 \qquad \frac{d\beta^1}{d\beta}=0$$

and by integration

$$\alpha^1=F(\alpha) \qquad \beta^1=F_1(\beta)$$

or

$$\alpha^1=\Phi(\beta) \qquad \beta^1=\Phi(\alpha)$$

These are equivalent to the results already obtained on the supposition that the angle between any two elements of the projection is equal to the angle between the two corresponding elements of the spheroid.

The functions F and Φ are quite indeterminate and may denote any arbitrary functions of $u+i\omega$ and $u-i\omega$. But since the variables ξ and η are real, as are u and ω, the functions F and Φ are not perfectly arbitrary, but have determinate values as soon as values have been assigned to F and Φ. It is obvious from very simple considerations that if the function F is real, then F_1 must denote the *same* function, and if F is imaginary F_1 will denote the conjugate function obtained by the change of i into $-i$ in F; of course the same remarks apply to the functions Φ and Φ_1; and so it is clear that each of the two solutions obtained contains but one arbitrary function, either real or imaginary. We will consider merely the first of these solutions, viz:

$$\xi+i\eta=F\,(u+i\omega) \qquad \xi-i\eta=F_1\,(u-i\omega)$$

Direct solution gives

$$\xi=\frac{1}{2}[F\,(u+i\omega)+F_1\,(u-i\omega)] \qquad \eta=\frac{1}{2i}[F(u+i\omega)-F_1\,(u-i\omega)].$$

It is easy now to find the value of m or the ratio of the lengths of two corresponding elements upon S and Σ. We had

$$l=\frac{m_0}{\rho}$$

and in order to preserve as much as possible similarity of the notation in this part of the chapter with that employed in the first part, we will make $t=m$. Now

$$d\xi^2+d\eta^2=m_0^2\,(du^2+d\omega^2)$$

Denoting the derivatives of F and F_1 with respect to $u+i\omega$ and $u-i\omega$, respectively, by F' and F_1', we find

$$d\xi+id\eta=F'\,(u+i\omega)\,(du+id\omega) \qquad d\xi-id\eta=F_1'\,(u-i\omega)\,(du-id\omega)$$

then,

$$d\xi^2+d\eta^2=F'\,(u+i\omega)\,F'_1\,(u-i\omega)\,(du^2+d\omega^2)$$

and so

$$m_0^2=F'\,(u+i\omega)\,F'_1\,(u-i\omega)$$

which gives at once

$$m=\frac{1}{\rho}\sqrt{F'\,(u+i\omega)\,F'_1\,(u-i\omega)}=\frac{1}{\sin p}\sqrt{(1-\epsilon^2\cos^2\varphi)\,F'\,(u+i\omega)\,F'_1\,(u-i\omega)}$$

Making for brevity

$$\sqrt{F'\,(u+i\omega)\,F'_1\,(u-i\omega)}=\frac{1}{\Omega}$$

we have finally

$$m=\frac{1}{\rho\Omega}$$

Linear elements in projection are altered in the ratio of $1:m$, and elementary areas in the ratio $1:m^2$.

LAGRANGE'S PROJECTION.

Observe that from the equation

$$\xi+i\eta=F\,(u+i\omega)$$

we have

$$d\xi=\alpha du \qquad d\eta=\beta du$$

ω being for the time constant, and denoting by α the real part and by β the coefficient of i in the derivative, with respect to u of $F\,(u+i\omega)$; further

$$d^2\xi=\frac{d\alpha}{du}du^2 \qquad d^2\eta=\frac{d\beta}{du}du^2$$

From these we find for the radius of curvature of the meridians

$$\rho_m=\frac{(\alpha^2+\beta^2)^{\frac{3}{2}}}{\beta\frac{d\alpha}{du}-\alpha\frac{d\beta}{du}}$$

Now, for any point whatever of the surface, we know that

$$d\xi=\alpha du-\beta d\omega \qquad d\eta=\beta du+\alpha d\omega$$

Consequently

$$\frac{d\alpha}{d\omega}=-\frac{d\beta}{du} \qquad \frac{d\beta}{d\omega}=\frac{d\alpha}{du}$$

From these results

$$\rho_m=\frac{(\alpha^2+\beta_2)^{\frac{3}{2}}}{\left(\frac{\beta d\beta}{d\omega}+\frac{\alpha d\alpha}{d\omega}\right)}$$

or, finally,

$$\frac{1}{\rho_m}=-\frac{d}{d\omega}\left(\frac{1}{\sqrt{\alpha^2+\beta^2}}\right)$$

For the parallels we find, regarding u as constant,

$$\frac{1}{\rho_p}=\frac{d}{du}\left(\frac{1}{\sqrt{\alpha^2+\beta^2}}\right)$$

Remembering the definition of the quantity Ω, viz:

$$\Omega=\sqrt{F'(u+i\omega)\,F'_1(u-i\omega)}$$

these two equations become

$$\frac{1}{\rho_m}=-\frac{d\Omega}{d\omega}\qquad\frac{1}{\rho_p}=\frac{d\Omega}{du}$$

Since $\frac{1}{\rho_m}$ is independent of u, we have that $\frac{d^2\Omega}{du\,d\omega}=0$. If, then, the meridians are represented by circles it is clear that the parallels are also circles, for $\frac{d^2\Omega}{du\,d\omega}=0$ is the condition that $\frac{d\Omega}{du}$ shall be constant with respect to ω, as well as the condition that $\frac{d\Omega}{d\omega}$ shall be constant with respect to u. The projections in these cases are of course circles, as the circle is the only curve which will satisfy the condition that the radius of curvature shall be constant.

Write

$$\frac{1}{\sqrt{F'(v)}}=\phi(v)\qquad\frac{1}{\sqrt{F'_1(v)}}=\psi(v)$$

then the condition

$$\frac{d^2\Omega}{du\,d\omega}=0$$

obviously becomes

$$\frac{\phi''(u+i\omega)}{\phi\,(u+i\omega)}=\frac{\psi''(u-i\omega)}{\psi\,(u-i\omega)}$$

the double accents denoting the second derivatives of the corresponding functions. The second number of this equation must, by virtue of the nature of the functions ϕ and ψ, be deduced from the first by the simple interchange of i with $-i$, which equality can only exist when each member of the equation is equal to the same constant k; then

$$\phi(u+i\omega)=A_0e^{\sqrt{k}(u+i\omega)}+B_0e^{-\sqrt{k}(u+i\omega)}$$

A_0 and B_0 being constants, real or imaginary, and consequently

$$F\,(u+i\omega)=H+\frac{e^{-\sqrt{k}(u+i\omega)}}{Ae^{\sqrt{k}(u+i\omega)}+Be^{-\sqrt{k}(u+i\omega)}}$$

A, B, and H being constants of the same nature as A_0 and B_0.

The constant k may be clearly either positive or negative; but if we suppose k negative, say $=-l^2$, we will evidently arrive at the same result as that which would be obtained by changing u into ω and ω into $-u$, so we shall only consider a positive value of k, $=l^2$, and the above equation becomes, on making $H=0$,

$$F\,(u+i\omega)=\xi+i\eta=\frac{e^{-l(u+i\omega)}}{Ae^{l(u+i\omega)}+Be^{-l(u+i\omega)}}$$

Retaining H is only equivalent to a transformation to parallel axes through a new origin, so, of course, nothing is lost by making $H=0$. Multiply the third term of this equation, numerator and denominator, by

$$Ae^{l(u-i\omega)}+Be^{-l(u-i\omega)}$$

This gives

$$\xi+i\eta=\frac{Be^{-2lu}+Ae^{-2il\omega}}{A^2e^{2lu}+AB\,(e^{2il\omega}+e^{-2il\omega})+Be^{-2lu}}=\frac{Be^{-2lu}+A\,(\cos 2l\omega-i\sin 2l\omega)}{A^2e^{2lu}+2\,AB\cos 2l\omega+B^2e^{-2lu}}$$

Regarding A and B as real and positive, this gives, by equating the real parts and the coefficient of the imaginary parts separately,

$$\xi=\frac{A\cos 2\,l\omega+B\epsilon^{-2u}}{A^2\epsilon^{2u}+2\,AB\cos 2\,l\omega+B^2\epsilon^{-2u}}\qquad\qquad \eta=\frac{-A\sin 2\,l\omega}{A^2\epsilon^{2u}+2\,AB\cos 2\,l\omega+B^2\epsilon^{-2u}}$$

Eliminating u from these equations and we find a relation between ξ, η, and ω, which will be the equation of the circles representing the meridians, and eliminating ω we find the relation between ξ, η, and u, which represents the parallels, also circles, as we have already seen. Square these quantities ξ and η and add the results;

$$\xi^2+\eta^2=\frac{\epsilon^{-2u}}{A^2\epsilon^{2u}+2\,AB\cos 2\,l\omega+B^2\epsilon^{-2u}}$$

and

$$\frac{\xi}{\xi^2+\eta^2}=A\epsilon^{2u}\cos 2\,l\omega+B\qquad\qquad \frac{\eta}{\xi^2+\eta^2}=-A\epsilon^{2u}\sin 2\,l\omega$$

The elimination of u from these gives

$$\xi^2+\eta^2-\frac{\eta}{B}\cot 2\,l\omega-\frac{\xi}{B}=0$$

or

$$\left(\xi-\frac{1}{2\,B}\right)^2+\left(\eta-\frac{\cot 2\,l\omega}{2\,B}\right)^2=\left(\frac{1}{2\,B\sin 2\,l\omega}\right)^2$$

a circle whose center is at

$$\xi_0=\frac{1}{2\,B}\qquad \eta_0=\frac{\cot 2\,l\omega}{2\,B}$$

and whose radius is

$$\rho_m=\frac{1}{2\,B\sin 2\,l\omega}$$

The circle obviously passes through the origin of the co-ordinates (ξ η). The axis of ξ is then cut by all the meridians in two fixed points—the origin and a point distant from the origin $=\frac{1}{2\,B}$. These two points represent the poles, and the axis of ξ is itself the first meridian.

The elimination of ω from our equations gives then the equation of the parallels as

$$\xi^2+\eta^2+\frac{2\,B\xi}{A^2\epsilon^{4u}-B^2}-\frac{1}{A^2\epsilon^{4u}-B^2}=0$$

a circle whose center is at

$$\xi_0=-\frac{B}{A^2\epsilon^{4u}-B^2}\qquad\qquad \eta_0=0$$

and whose radius is

$$\rho_p=\frac{A\epsilon^{2u}}{A^2\epsilon^{4u}-B^2}$$

Designating by O the origin of co-ordinates and by P the point distant from the origin $\frac{1}{B}$; *i. e.*, O and P are the poles of the earth in projection, and denoting by C the center of any one of the circles representing parallels; we must clearly have

$$CO=\frac{B}{A^2\epsilon^{4u}-B^2}\qquad\qquad OP=\frac{1}{B}$$

then

$$CP=CO+OP=\frac{A^2\epsilon^{4u}}{(A^2\epsilon^{4u}-B^2)B}$$

It follows from this that

$$CO\cdot CP=\rho_p^2$$

or the diameters of the projections of parallels are harmonically divided by the poles or points of intersection of the projections of meridians. By taking the origin of co-ordinates at the middle of OP we can somewhat simplify the equations of both meridians and parallels.

$$OP=\frac{1}{B},\quad =2\lambda,\ \text{say;}$$

then the equation of meridians becomes

$$\xi^2+\eta^2-2\lambda\cot 2ly-\lambda^2=0$$

that of parallels

$$\xi^2+\eta^2+2\lambda\frac{4\lambda^2A^2\epsilon^{4l\omega}+1}{4\lambda^2A^2\epsilon^{4l\omega}-1}\xi+\lambda^2=0$$

It will be convenient just here to introduce the so-called hyperbolic and Gudermannian functions. The hyperbolic functions required as given by

$$\sinh\theta=\tfrac{1}{2}(\epsilon^{\theta}-\epsilon^{-\theta})\qquad\cosh\theta=\tfrac{1}{2}(\epsilon^{\theta}+\epsilon^{-\theta})\qquad\tanh\theta=\frac{\epsilon^{\theta}-\epsilon^{-\theta}}{\epsilon^{\theta}+\epsilon^{-\theta}}$$

and

$$\coth\theta=\frac{1}{\tanh\theta}$$

The derivatives of the first two of these are

$$\frac{d\sinh\theta}{d\theta}=\cosh\theta\qquad\frac{d\cosh\theta}{d\theta}=\sinh\theta.$$

To these may be added the following expressions for Gudermannian functions:

$$\mathrm{sg}\chi=\tanh\chi=-i\tan i\chi\qquad\mathrm{cg}\chi=\mathrm{sech}\,\chi=\sec i\chi\qquad\mathrm{tg}\chi=\sinh\chi=-i\sin i\chi$$

and

$$\sin i\chi=i\sinh\chi=i\,\mathrm{tg}\chi\qquad\cos i\chi=\cosh\chi=\frac{1}{\mathrm{cg}\chi}\qquad\tan i\chi=i\tanh\chi=i\,\mathrm{sg}\chi$$

with

$$\mathrm{sg}^2\chi+\mathrm{cg}^2\chi=1$$

It frequently makes the expressions we are dealing with simpler to introduce these functions in place of the complicated exponential quantities to which they are equal. Assume now

$$4\lambda^2A^2=\epsilon^{4lh}$$

and we have for the coefficient of ξ in the above equation

$$2\lambda\frac{\epsilon^{4l(\omega+h)}+1}{\epsilon^{4l(\omega+h)}-1}=2\lambda\frac{\epsilon^{2l(h+\omega)}+\epsilon^{-2l(h+\omega)}}{\epsilon^{2l(h+\omega)}-\epsilon^{-2l(h+\omega)}}=2\lambda\coth 2l(\omega+h)=2\lambda\coth 2lv$$

giving for the equation of parallels

$$\xi^2+\eta^2+2\lambda\coth 2lv\,\xi+\lambda^2=0$$

Resume now the expressions for ξ and η, viz:

$$\xi=\frac{A\cos 2l\omega+B\epsilon^{-2l\omega}}{A^2\epsilon^{2l\omega}+2AB\cos 2l\omega+B^2\epsilon^{-2l\omega}}\qquad\eta=\frac{-A\sin 2l\omega}{A^2\epsilon^{2l\omega}+2AB\cos 2l\omega+B^2\epsilon^{-2l\omega}}$$

The change to new origin involves writing $\xi+\frac{1}{2B}$ instead of ξ, which gives

$$\xi=\frac{B^2\epsilon^{-2l\omega}-A^2\epsilon^{2l\omega}}{2B(A^2\epsilon^{2l\omega}+2AB\cos 2l\omega+B^2\epsilon^{-2l\omega})}$$

or introducing the new constants h and λ, this becomes

$$\xi = \lambda \frac{e^{-2l(u+h)} - e^{2l(u+h)}}{e^{2l(u+h)} + e^{-2l(u+h)} + 2 \cos 2 l\omega} = \frac{\lambda \tanh 2 lv}{1 + \cos 2 l\omega \operatorname{sech} 2 lv}$$

$$\eta = \frac{-2 \lambda e^{2lv} \sin 2 l\omega}{e^{4lv} + 2 e^{2lv} \cos 2 l\omega + e^{-2lv}} = \frac{-\lambda \sin 2 \omega}{\cosh 2 lv + \cos 2 l\omega}$$

Now, since

$$\sinh 2 lv = -i \sin 2 ilv$$

$$\cosh 2 lv = \cos 2 ilv$$

we have

$$\xi + i\eta = -i\lambda \frac{\sin 2 ilv + \sin 2 l\omega}{\cos 2 ilv + \cos 2 l\omega}$$

or

$$F(u + i\omega) = -i\lambda \tan l(iv + \omega)$$

and

$$F(u - i\omega) = -i\lambda \tan l(iv - \omega)$$

Of course these can be given in terms of the hyperbolic functions, but there would be no gain in so doing; and similarly the values of ξ and η might be given in terms of the Gudermannian functions by means of the preceding formulas, but the results would be interesting only from an entirely theoretical point of view. We have now for the value of Ω

$$\Omega = \frac{1}{\sqrt{F'(u + i\omega) F_1'(u - i\omega)}} = \frac{\cos l(iv + \omega) \cos l(iv - \omega)}{l\lambda}$$

The value of u, which is to be used in all these formulas, is

$$u = \log \Omega \tan \frac{\zeta}{2}$$

ζ denoting the polar distance, corrected, to allow for ellipticity, in the case of the earth.

In the construction of this projection there are two indeterminate quantities

$$\frac{1}{B} = 2\lambda = PP'$$

upon which the scale of the map will depend, and the constant l; there is also indeterminate the position of that point of the earth's surface which is to be taken as the center of the map.

Values for all these indeterminates should be so found that the alteration in magnitude consequent upon the projection of any part of the spherical or spheroidal surface shall be the least possible. The solution of this problem involves finding a point for which m is a minimum, or the neighborhood of which m is least altered. Returning for a moment to the equation of the parallels

$$\xi^2 + \eta^2 + \frac{2 B\xi}{A^2 e^{2lu} - B^2} - \frac{1}{A^2 e^{2lu} - B^2} = 0$$

we will solve the problem of finding the points upon the axis of x which will harmonically divide the diameters of these circles, or as we may state the problem, to find the two points upon the axis of x whose distances from an arbitrary part upon the circumference of any one of these circles shall be in a constant ratio.

Take the equation

$$(x - g)^2 + y^2 = K^2[(x - g')^2 + y^2]$$

this is the equation of the locus of points whose distances from two fixed points (g, o) (g', o) upon the axis of x have a constant ratio K; comparing this with the above equation we have

$$\frac{-2g + 2g'K^2}{1 - K^2} = \frac{2 B}{A^2 e^{2lu} - B^2} \qquad \frac{K^2 g'^2 - g^2}{1 - K^2} = \frac{1}{A^2 e^{2lu} - B^2}$$

These are multiplied by

$$g'=0 \qquad g=\frac{1}{B} \qquad K=\frac{A}{B}e^{2lv}$$

or we have that the circles which represent the parallels are the loci of points whose distances from the two fixed points P, P′ have a constant ratio $=\frac{A}{B}e^{2lv}$. This is rather more general than the principle already obtained, viz, that P and P′ divide the diameters harmonically. Confining ourselves now to a spherical earth, conceive that the constant l has been chosen and assume upon a horizontal line the points P and P′ for the poles, taking the north pole on the right. This line PP′ is the meridian $\omega=0$; a line QQ′ perpendicular to PP′ at the middle point is obviously the parallel corresponding to lv, or $v=0$. This meridian and parallel can of course be made to pass through any point of the earth's surface, and this point will then be the center of the projection or map. A knowledge of this place infers a knowledge of the meridian from which longitudes are reckoned and affords the means of finding the constant h. For the center of the chart $\omega=0$ and $lv=0$; calling φ_0 the co-latitude of the center (instead of ζ_0 as in the case of a spheroid) the value of u at the center will be $=\log\tan\frac{\varphi_0}{2}$ $(G=1)$ and $lv=u+h$ becomes

$$\log\tan\frac{\varphi_0}{2}+h=0$$

Now, to find the meridian of longitude ω, draw with PP′ for base a segment containing the angle $\pi-2\omega$, if ω is positive, or $\pi+2\omega$ if ω is negative. And for a parallel of latitude θ or $90°-\varphi$, and for which $u=\log\tan\frac{\varphi}{2}$, describe a circle the locus of points whose distances from P and P′ have the ratio

$$\frac{A}{B}e^{2lv}=e^{2l(u+h)}=\tan^{2l}\frac{\varphi}{2}\cot^{2l}\frac{\varphi_0}{2}=k$$

Now, take up the subject of the increase of magnitude resulting from the projection of any portion of the surface of the earth. We have for m the value

$$m=\frac{-2lA\sqrt{1-e^2\cos^2\varphi}}{a\left[A^2e^{2lv}+2AB\cos 2l\omega+B^2e^{-2lv}\right]\sin\varphi}$$

$$=\frac{-4lh\sqrt{1-e^2\cos^2\varphi}}{a\left[\dfrac{\tan^{2l}\frac{\varphi}{2}}{\tan^{2l}\frac{\varphi_0}{2}}+2\cos l\omega+\dfrac{\tan^{2l}\frac{\varphi_0}{2}}{\tan^{2l}\frac{\varphi}{2}}\right]\sin\varphi}$$

for a spherical earth of radius r,

$$m=\frac{-4lh}{r\left[\dfrac{\tan^{2l}\frac{\varphi}{2}}{\tan^{2l}\frac{\varphi_0}{2}}+2\cos l\omega+\dfrac{\tan^{2l}\frac{\varphi_0}{2}}{\tan^{2l}\frac{\varphi}{2}}\right]\sin\varphi}$$

With respect to ω, it is obvious that m is a minimum where $\omega=0$, that is, the alteration is a minimum along the central meridian. Assume, then, $\omega=0$, and confine ourselves to the sphere, we have then for the first meridian

$$m=\frac{4lh}{r\sin\varphi\left[\tan^l\frac{\varphi}{2}\cot^l\frac{\varphi_0}{2}+\tan^l\frac{\varphi_0}{2}\cot^l\frac{\varphi}{2}\right]^2}$$

This is a minimum for the denominator a maximum; write

$$Q=\sqrt{\sin\varphi}\left[\tan^l\frac{\varphi}{2}\cot^l\frac{\varphi_0}{2}+\tan^l\frac{\varphi_0}{2}\cot^l\frac{\varphi}{2}\right]$$

for a maximum,

$$\frac{dQ}{d\varphi}=0$$

this gives readily

$$\tan^n\frac{\varphi}{2}\cot^n\frac{\varphi_0}{2}=\frac{2t-\cos\varphi}{2t+\cos\varphi}$$

Thus, the distance PP' is divided in the ratio $\frac{2t-\cos\varphi}{2t+\cos\varphi}$ by a point at which the alteration is the least possible. Substitution in the above value of m gives for the minimum of this quantity

$$m_0=\frac{t l\,(4t^2-\cos^2\varphi)}{4r\sin\varphi}$$

This is dependent upon t and φ, and we can again assume that there is a value of φ such that the derivative of m_0 with respect to φ shall vanish; *i. e.*, if we give a slight increment to φ the resulting change in m_0 will be $=0$, or

$$\frac{dm_0}{d\varphi}=0$$

Now

$$\frac{dm_0}{d\varphi}=tl\,\frac{(1+\sin^2\varphi-4t^2)}{4r\sin^2\varphi}\cos\varphi=0$$

this gives

$$2t=\sqrt{1+\sin^2\varphi}$$

an equation for determining t when the colatitude φ is given. The practical construction of this projection will be given in Part II, and need not be referred to here.

The entire theory of the Lagrangian projection might have been obtained from the general considerations at the beginning of this chapter. If we assume $F(v)=\cos v$, we have

$$\left[\cos\omega+i\log\cot\frac{\varphi}{2}\right]=\frac{1}{2}\left[e^{i\left(\omega+i\log\cot\frac{\varphi}{2}\right)}+e^{-i\left(\omega+i\log\cot\frac{\varphi}{2}\right)}\right]=\frac{1}{2}\left[e^{i\omega}\tan\frac{\varphi}{2}+e^{-i\omega}\cot\frac{\varphi}{2}\right]$$

$$=\frac{1}{2}\left[(\cos\omega+i\sin\omega)\tan\frac{\varphi}{2}+(\cos\varphi-i\sin\omega)\cot\frac{\varphi}{2}\right]$$

Equating separately the real and imaginary parts—

$$\xi=\frac{\cos\omega}{\sin\varphi}\qquad\qquad \eta=-\frac{\sin\omega}{\tan\varphi}$$

Or more generally, let

$$F(v)=K\cos(\alpha+\beta v)$$

K, α, β being constants, the resulting values of ξ, η become in this case

$$\xi=\frac{K}{2}\cos(\alpha+\beta\omega)\left(\tan^\beta\frac{\varphi}{2}+\cot^\beta\frac{\varphi}{2}\right)\qquad\qquad \eta=\frac{K}{2}\sin(\alpha+\beta\omega)\left(\tan^\beta\frac{\varphi}{2}-\cot^\beta\frac{\varphi}{2}\right)$$

And again, if we write

$$F(v)=K\tan(\alpha+\tfrac{1}{2}\beta v)$$

and use the known formula

$$\tan x=\frac{e^{2ix}-1}{i(e^{2ix}+1)}$$

we will ultimately come to the formulas of Lagrange's projection. But enough has now been said on this subject.

It has been seen already that, if we assume for F a linear function, that is $F(v)=Kv$, where K is a constant, we obtain Mercator's projection. Assume now for F the value

$$F(u+i\omega)=Ke^{l(u+i\omega)}$$

and also

$$F_1(u-i\omega)=Ke^{l(u-i\omega)}$$

Since

$$e^{il\omega}=\cos l\omega+i\sin l\omega$$

and

$$e^{-il\omega}=\cos l\omega-i\sin l\omega$$

these values of F and F_1 give us

$$\xi=Ke^{lu}\cos l\omega \qquad \eta=Ke^{lu}\sin l\omega$$

In the case of a spherical earth

$$u=\log\tan\frac{\varphi}{2}$$

and

$$e^{lu}=\tan^l\frac{\varphi}{2}$$

giving

$$\xi=K\tan^l\frac{\varphi}{2}\cos l\omega \qquad \eta=K\tan^l\frac{\varphi}{2}\sin l\omega$$

The parallels are thus projected into circles given by

$$\xi^2+\eta^2=K^2\tan^{2l}\frac{\varphi}{2}$$

and the meridians into right lines, all passing through the center of the concentric circles representing the parallels, whose equation is

$$\frac{\eta}{\xi}=\tan l\omega$$

For the ratio m there results

$$m=\frac{\sqrt{F'F_1'}}{\sin\varphi}=\frac{lK\tan^l\frac{\varphi}{2}}{\sin\varphi}$$

Since l is an arbitrary constant, we are at liberty to assign to it any value that we please For $l=1$ we have the stereographic equatorial projection whose equations are

$$\text{meridians } \frac{\eta}{\xi}=\tan\omega \qquad \text{parallels } \xi^2+\eta^2=K^2\tan^2\frac{1}{2}\left(\frac{\pi}{2}-\theta\right)$$

since $\varphi=90^\circ-\theta$. Here K represents the radius of the equator. For this case also

$$m=\frac{1}{2\cos^2\frac{\varphi}{2}}$$

The general values of ξ and η may be put in the form

$$\xi=Q\cos l\omega \qquad \eta=Q\sin l\omega$$

and also

$$m=\frac{lQ}{\sin\varphi}$$

It is obvious from these formulas that the co-ordinates of any point whose longitude is ω are the same as those which correspond to a longitude $=\omega l$ in the stereographic equatorial projection for which $l=1$. If, then, l is a fraction <1 the projection of the entire sphere will lie in a sector of a circle which is that fraction of the entire circular area. This projection was proposed by Lambert, but fully elaborated and discussed by Gauss.

Since l is arbitrary, we may determine it so as to satisfy the condition that the lengths of degrees upon two given parallels of the projection shall have the same ratio as they have upon the sphere. Call φ_0 and φ_1 the colatitudes of the parallels upon the sphere, their degrees are in the ratio of

$$\sin\varphi_0 : \sin\varphi_1$$

and for the chart it is then necessary to write

$$\left(\frac{\tan\frac{\varphi_1}{2}}{\tan\frac{\varphi_0}{2}}\right)^l = \frac{\sin\varphi_1}{\sin\varphi_0}$$

giving then

$$l = \frac{\log\sin\varphi_1 - \log\sin\varphi_0}{\log\tan\frac{\varphi_1}{2} - \log\tan\frac{\varphi_0}{2}}$$

For φ_1 and φ_0 may be taken the extreme values of φ. For the construction of this projection, called *Lambert's orthomorphic conic projection*, draw an indefinite line PA for the central meridian, and with P, the pole, as a center draw circles of radii

$$\rho = K\tan^l\frac{\zeta}{2}$$

ζ again denoting the correct polar distance for the spheroid; these circles are the parallels; K is an arbitrary constant which fixes the scale of the chart; it may be determined by giving, for example, the value of the latitude, for which the radius ρ is equal to the corresponding arc of the meridian upon the spheroid. Suppose that $\theta = ok$, that is the distance from the pole to the equator on the map is equal to q the quarter meridian. Now

$$q = \frac{1}{2}\pi a\left(1 - \frac{1}{4}e^2 - \frac{3}{64}e^4 - \ldots\right) = \frac{1}{2}\pi a\left(1 - \frac{\varepsilon}{2}\right)$$

ε denoting the ellipticity and a the equatorial radius. Then, on the above supposition,

$$K = \frac{1}{2}\pi a\left(1 - \frac{\varepsilon}{2}\right)$$

The meridians which on the sphere make angles with the central meridian $= \omega$, make on the chart angles with the representation of that meridian $= l\omega$, and these are the only angles that are not preserved in their true size. If the arbitrary l be determined by the condition that the degrees of the colatitudes φ_0 and φ_1 shall have the same ratio as upon the sphere, we know that

$$l = \frac{\log\sin\varphi_1 - \log\sin\varphi_0}{\log\tan\frac{\varphi_1}{2} - \log\tan\frac{\varphi_0}{2}}$$

There yet remains one method for obtaining the values of the co-ordinates ξ and η in orthomorphic projection, which differs entirely from all that we have so far examined; this is known as the method of indeterminate coefficients. The development of the theory of this system gives rise to quite complicated formulas, and consequently from a practical point of view, the general method is useless, but there is one particular case in which the results simplify themselves to such an extent as to make it worth while to examine briefly the method. The particular case referred to is known as Lambert's orthomorphic cylindric projection.

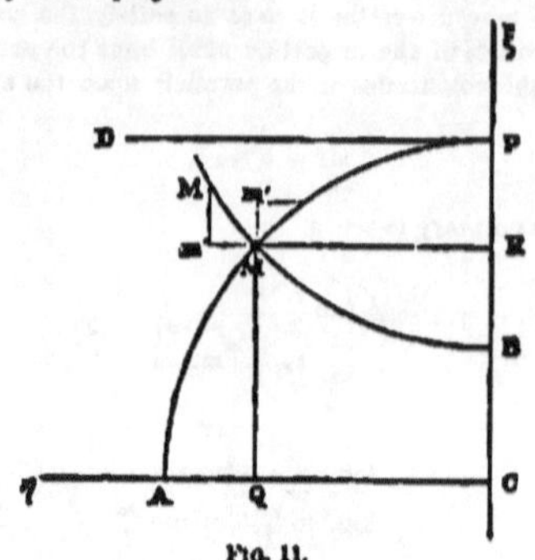

FIG. 11.

In Fig. 11, let PA represent a meridian of longitude ω, and MB a parallel of latitude θ; the longitude of M'' is $\omega + d\omega$ and the latitude of M' is $\theta - d\theta$. The first condition of this projection makes M''MM' a right angle and consequently the triangles M''m''M, M'm'M similar. Since also the degrees of longitude and latitude preserve the same ratio that they do upon the sphere we must have

$$\frac{MM'}{MM''} = \frac{d\theta}{d\omega \cos\theta}$$

and consequently,

$$\frac{Mm''}{MM'} = \frac{d\eta''}{d\xi'} = \frac{d\omega \cos\theta}{d\theta} \qquad \frac{M''m''}{M'm'} = \frac{d\xi''}{d\eta'} = \frac{d\omega \cos\theta}{d\theta}$$

Now, since ξ and η are functions of ω and θ

$$d\xi = \frac{d\xi}{d\omega}d\omega + \frac{d\xi}{d\theta}d\theta \qquad d\eta = \frac{d\eta}{d\omega}d\omega + \frac{d\eta}{d\theta}d\theta$$

For a given meridian $\omega = \text{const.}$, and $d\omega = 0$; then

$$-d\xi' = \frac{d\xi}{d\theta}d\theta \qquad d\eta' = \frac{d\eta}{d\theta}d\theta$$

For a given parallel $d\theta = 0$, and

$$d\xi'' = \frac{d\xi}{d\omega}d\omega \qquad d\eta'' = \frac{d\eta}{d\omega}d\omega$$

The above equations of condition are, then, in general

$$\frac{d\eta}{d\omega} = -\frac{d\xi}{d\theta}\cos\theta \qquad \frac{d\xi}{d\omega} = \frac{d\eta}{d\theta}\cos\theta$$

[Multiply the first of these by $\frac{d\eta}{d\theta}$ and the second by $\frac{d\xi}{d\theta}$ and add; there results

$$\frac{d\xi}{d\theta}\frac{d\xi}{d\omega} + \frac{d\eta}{d\theta}\frac{d\eta}{d\omega} = 0$$

the well-known condition of orthogonality of the lines $\xi = \text{const.}$ and $\eta = \text{const.}$]

We know that any variable quantity z a function of two independent variables ω and θ can be given in the form of the series

$$z = \Sigma A_i \theta^i + \omega \Sigma B_i \theta^i + \omega^2 \Sigma C_i \theta^i + \ldots\ldots$$

where A, B, C, &c., are constants to be determined when we know a sufficient number of values of z or its successive derivatives for the given values of θ and ω. The series may be still more compactly written if we denote the A_i in the above by $A_i^{(0)}$ the B_i by $A_i^{(1)}$ the C_i by $A_i^{(2)}$, &c. The series is then

$$z=\sum_0^{\infty}\omega^j\sum_0^{\infty}A_i^j\,\theta^i$$

The form chosen by Lambert for this series, in the two particular cases of representing ξ and η in such a form, is

$$\xi=\sum_0^{\infty}\omega^j\sum_0^{\infty}a_i^{(j)}\cos i\theta \qquad\qquad \eta=\sum_0^{\infty}\omega^j\sum_0^{\infty}A_i^{(j)}\sin i\theta$$

a term $A_0^{(0)}+A_0^{(1)}\omega+A_0^{(2)}\omega^2$ having to be added in the second formula, since $\sin 0\theta=0$. Now, the equations

$$\frac{d\eta}{d\omega}=-\frac{d\xi}{d\theta}\cos\theta \qquad\qquad \frac{d\xi}{d\omega}=\frac{d\eta}{d\theta}\cos\theta$$

give on expanding the above forms

$$a_0^{(1)}=\frac{1}{2}A_1^{(0)}$$

$$a_1^{(1)}=\frac{1}{2}\left[2A_2^{0}\right]$$

$$a_2^{(1)}=\frac{1}{2}\left[A_1^{0}+3A_3^{0}\right]$$

. . . .

$$a_i^{(1)}=\frac{1}{2}\left[(i-1)A_{i-1}^{0}+(i+1)A_{i+1}^{0}\right]$$

$$a_0^{(2)}=\frac{1}{4}A_1^{(1)}$$

$$a_1^{(2)}=\frac{1}{4}\left[2A_2^{(1)}\right]$$

$$a_2^{(2)}=\frac{1}{4}\left[A_1^{(1)}+3A_3^{(1)}\right]$$

. . . .

$$a_i^{(2)}=\frac{1}{4}\left[(i-1)A_{i-1}^{(1)}+(i+1)A_{i+1}^{(1)}\right]$$

$$a_0^{(j)}=\frac{1}{2j}A_1^{(j-1)}$$

$$a_1^{(j)}=\frac{1}{2j}\left[2A_2^{(j-1)}\right]$$

$$a_2^{(j)}=\frac{1}{2j}\left[A_1^{(j-1)}+3A_3^{(j-1)}\right]$$

. . . .

$$a_i^{(j)}=\frac{1}{2j}\left[(i-1)A_{(i-1)}^{(j-1)}+(i-1)A_{i+1}^{(j-1)}\right]$$

Similarly

$$A_i^{(1)}=\frac{1}{2}\left[(i-1)a_{i-1}^{(0)}+(i+1)a_{i+1}^{(0)}\right]$$

$$A_i^{(2)}=\frac{1}{4}\left[(i-1)a_{i-1}^{(1)}+(i+1)a_{i+1}^{(1)}\right]$$

. . . .

$$A_i^{(j)}=\frac{1}{2j}\left[(i-1)a_{(i-1)}^{(j-1)}+(i+1)a_{(i+1)}^{(j-1)}\right]$$

We find readily now

$$a_i^{(1)}=\frac{1}{4}\Big[(i-1)A_{i-1}^{(0)}+(i+1)A_{i+1}^{(0)}\Big]=\frac{1}{2.4}\Big[(i-1)(i-2)a_{i-2}^{(0)}+2i^2a_i^{(0)}+(i+1)(i+2)a_{i+2}^{(0)}\Big]$$

$$a_i^{(2)}=\frac{1}{6}\Big[(i-1)A_{i-1}^{(1)}+(i+1)A_{i+1}^{(1)}\Big]=\frac{1}{2.4.6}\Big\{(i-1)(i-2)(i-3)a_{i-3}^{(0)}+(i-1)(3i^2-3i+2)a_{i-1}^{(0)}$$

$$+(i+1)(3i^2+3i+2)a_{i+1}^{(0)}+(i+1)(i+2)(i+3)a_{i+3}^{(0)}\Big\}$$

It is not necessary to give the general case of $a_i^{(i)}$; the law being obvious, the reader can readily construct it for himself.

In the particular case mentioned above, Lambert's orthomorphic cylindric projection, take for the central meridian a straight line (*vide* figure), and upon it lay off the actual lengths of the degrees of latitude; a second straight line at right angles to the first denotes the equator; other parallels and meridians are orthogonal curves cutting in such a way that the degrees of longitude shall be represented in their true length.

Taking for axes of co-ordinates the central meridian and the equator, it is clear that, the figure being symmetrical with respect to these axes, η should contain even powers of ω and odd powers of θ, and that ξ should contain even powers of θ and odd powers of ω. Also for $\theta=0$ we should have $\eta=0$, and ξ a function of ω only; for $\omega=0$, $\eta=0$, and $\xi=0$. These series are thus,

$$\eta=K\theta+\omega^2\sum_0^\infty A_i^{(0)}\theta^{2i+1}+\omega^4\sum_0^\infty A_i^{(1)}\theta^{2i+1}+\ldots$$

$$\xi=\omega\sum_0^\infty a_i^{(0)}\theta^{2i}+\omega^3\sum_0^\infty a_i^{(1)}\theta^{2i}+\ldots$$

Satisfying, as before, the equations of condition

$$\frac{d\eta}{d\omega}=-\frac{d\xi}{d\theta}\cos\theta \qquad \frac{d\xi}{d\omega}=\frac{d\eta}{d\theta}\cos\theta$$

and these series are readily found to become

$$\xi=\omega\Big\{1-\frac{\theta^2}{2}+\frac{\theta^4}{4!}-\frac{\theta^6}{6!}+\ldots(-)^i\frac{\theta^{2i}}{i!}\Big\}+\omega^3\Big\{\frac{1}{6}-\frac{5}{12}\theta^2+\frac{41}{144}\theta^4-\frac{73}{864}\theta^6+\ldots\Big\}$$

$$+\omega^5\Big\{\frac{1}{24}-\frac{61}{240}\theta^2+\frac{191}{480}\theta^4-\Big\}+\omega^7\Big\{\frac{61}{5040}-\frac{139}{1008}\theta^2+\ldots\Big\}$$

$$\eta=\theta+\omega^2\Big\{\frac{1}{2}\theta-\frac{4}{2}\frac{\theta^3}{3!}+\frac{16}{2}\frac{\theta^5}{5!}-\frac{64}{2}\frac{\theta^7}{7!}+\ldots\Big\}+\omega^4\Big\{\frac{5}{2}\theta-\frac{7}{18}\theta^3+\frac{5}{18}\theta^5-\ldots\Big\}$$

$$+\omega^6\Big\{\frac{61}{720}\theta-\frac{443}{1440}\theta^3+\ldots\Big\}$$

By properly dividing the numerical coefficients in these series, they are readily found to assume the forms

$$\xi=\omega\cos\theta+\omega^3\frac{\cos\theta+\cos3\theta}{4.\ 1.\ 3.}+\omega^5\frac{4\cos\theta+10\cos3\theta+6\cos5\theta}{4.\ 8.\ 1.\ 3.\ 5.}$$

$$+\omega^7\frac{34\cos\theta+154\cos3\theta+210\cos5\theta+90\cos7\theta}{4.\ 8.\ 12.\ 1.\ 3.\ 5.\ 7.}+\ldots+\ldots$$

$$\eta=\theta+\omega^2\frac{\sin2\theta}{4}+\omega^4\frac{4\sin2\theta+3\sin4\theta}{4.\ 8.\ 1.\ 3.}+\omega^6\frac{34\sin2\theta+60\sin4\theta+30\sin6\theta}{4.\ 8.\ 12.\ 1.\ 3.\ 5.}$$

$$+\omega^8\frac{496\sin2\theta+1512\sin4\theta+1620\sin6\theta+680\sin8\theta}{4.\ 8.\ 12.\ 16.\ 1.\ 3.\ 5.\ 7.}+\ldots$$

Again, group the terms in sin 2θ, sin 4θ, &c., and those in cos θ, cos 3θ, &c., and these become

$$\xi=2\cos\theta\tan\frac{\omega}{2}+\frac{2}{3}\cos 3\theta\tan^3\frac{\omega}{2}+\frac{2}{5}\cos 5\theta\tan^5\frac{\omega}{2}+\frac{2}{7}\cos 7\theta\tan^7\frac{\omega}{2}+\ \&c.$$

$$\eta=\theta+\sin 2\theta\tan^2\frac{\omega}{2}+\frac{1}{2}\sin 4\theta\tan^4\frac{\omega}{2}+\frac{1}{3}\sin 6\theta\tan^6\frac{\omega}{2}+\frac{1}{4}\sin 8\theta\tan^8\frac{\omega}{2}+\ \&c.$$

The law of these last two developments is obvious, the general term of ξ being

$$=\frac{2}{2i-1}\cos(2i-1)\theta\tan^{2i-1}\frac{\omega}{2}$$

that of η

$$=\frac{1}{i-1}\sin 2(i-1)\theta\tan^{2(i-1)}\frac{\omega}{2}$$

i denoting the number of the term. A final grouping of the terms will conduct us to the formulas

$$\xi=\tfrac{1}{2}\log\left(\frac{1+2\tan\frac{\omega}{2}\cos\theta+\tan^2\frac{\omega}{2}}{1-2\tan\frac{\omega}{2}\cos\theta+\tan^2\frac{\omega}{2}}\right)\qquad \eta=\theta+\tan^{-1}\left(\frac{\sin 2\theta\tan^2\frac{\omega}{2}}{1-\cos 2\theta\tan^2\frac{\omega}{2}}\right)$$

or, by introducing the colatitude φ,

$$\xi=\tfrac{1}{2}\log\left(\frac{1+\sin\varphi\sin\omega}{1-\sin\varphi\sin\omega}\right)\qquad \eta=90^\circ-\varphi+\cot^{-1}\left(\cot 2\varphi+\cot^2\frac{\varphi}{2}\operatorname{cosec} 2\varphi\right)$$

or simply

$$\xi=\tfrac{1}{2}\log\left(\frac{\operatorname{cosec}\varphi+\sin\omega}{\operatorname{cosec}\varphi-\sin\omega}\right)\qquad \cot\eta=\cos\omega\tan\varphi.$$

Forming the differential coefficients of ξ and η with respect to φ and ω, we obtain

$$\frac{d\xi}{d\varphi}=\frac{\sin\omega\cos\varphi}{1-\sin^2\omega\sin^2\varphi}\qquad \frac{d\eta}{d\varphi}=\frac{-\cos\varphi}{1-\sin^2\omega\sin^2\varphi}$$

$$\frac{d\xi}{d\omega}=\frac{\cos\omega\sin\varphi}{1-\sin^2\omega\sin^2\varphi}\qquad \frac{d\eta}{d\omega}=\frac{\sin\omega\sin\varphi\cos\varphi}{1-\sin^2\omega\sin^2\varphi}$$

These obviously satisfy the known equations of condition which must exist between these differential coefficients, viz:

$$\frac{d\xi}{d\varphi}\sin\varphi=\frac{d\eta}{d\omega}\qquad \frac{d\eta}{d\varphi}\sin\varphi=-\frac{d\xi}{d\omega}$$

The ratio of the change in elementary areas is easily arrived at from the above values of the differential coefficients; using the formula $d\sigma=\sqrt{d\xi^2+d\eta^2}$, consider a small quadrilateral on the sphere comprised between two parallels and two meridians; its area is $=d\theta\, d\omega\cos\theta$. Now, making $\varphi=90^\circ-\theta$, we have for the arc of a parallel, when $d\theta=0$,

$$d\sigma=\frac{d\omega\cos\theta}{\sqrt{1-\cos^2\theta\sin^2\omega}}$$

and for the arc of a meridian, for which $d\omega=0$,

$$d\sigma'=\frac{d\theta}{\sqrt{1-\cos^2\theta\sin^2\omega}}$$

the area of the rectangle is then

$$\frac{d\omega\, d\theta\cos\theta}{\sqrt{1-\cos^2\theta\sin^2\omega}}$$

and for the ratio of increase

$$m^2=\frac{1}{1-\cos^2\theta\sin^2\omega}$$

If $\omega=0$, $m=1$; or, the ratio of areas on the sphere and on the projection is = unity along the central meridian. The principal advantages of this projection are thus seen to be:

(*a*) That it preserves all the angles in their true size, and consequently gives orthogonal intersections of the meridians and parallels.

(*b*) The degrees of latitude are equal upon the central rectilinear meridian, and the degrees of longitude in the neighborhood of this meridian differ but little from their true size. This projection will be obtained, as we have already seen, by passing a cylinder tangent to the sphere along a meridian, projecting the sphere upon the cylinder and developing the latter. It is on this account that the name orthomorpho-cylindric has been chosen for this projection.

The general subject of orthomorphic projection will be resumed in another place, and a fuller mathematical theory given of this most interesting problem, but before leaving the subject it is of importance to note that if either of the variables ξ or η be given, the other can be found by simple integration. For, from the equations of condition

$$\frac{d\xi}{d\omega}=\frac{d\eta}{d\theta}\cos\theta \qquad \frac{d\eta}{d\omega}=\frac{d\xi}{d\theta}\cos\theta$$

we have

$$d\xi=\frac{d\eta}{d\theta}\cos\theta\, d\omega-\frac{d\eta}{d\omega}\,\frac{d\theta}{\cos\theta} \qquad d\eta=\frac{d\xi}{d\omega}\,\frac{d\theta}{\cos\theta}-\frac{d\xi}{d\theta}\cos\theta\, d\omega$$

If, then, either ξ or η is given, forming its differential coefficients, and substituting in the corresponding one of these two equations, we have the means of obtaining the remaining co-ordinate. For example, let $\eta=\omega$; then

$$\frac{d\eta}{d\omega}=1 \qquad \frac{d\eta}{d\theta}=0$$

and

$$d\xi=-\frac{d\theta}{\cos\theta}$$

from which

$$\xi=\log\tan\tfrac{1}{2}(90^\circ-\theta) \qquad \eta=\omega$$

the equations of Mercator's projection. Littrow gives the projection of which one of the equations is

$$\xi=\tan\theta\cos\omega$$

The differential coefficients are

$$\frac{d\xi}{d\omega}=-\tan\theta\sin\omega \qquad \frac{d\xi}{d\theta}=\frac{\cos\omega}{\cos^2\theta}$$

by means of which we find

$$d\eta=-\frac{(\sin\theta\sin\omega\, d\theta+\cos\theta\cos\omega\, d\omega)}{\cos^2\omega}$$

Integrating thus

$$\eta=-\frac{\sin\omega}{\cos\theta}$$

Combining the value of ξ with this value of η in such a way as to eliminate ω, we have for the equation of the parallels

$$\xi^2+\eta^2\sin^2\theta-\tan^2\theta=0$$

and in like manner for the meridians is found

$$\xi^2\sin^2\omega-\eta^2\cos^2\omega+\sin^2\omega\cos^2\omega=0$$

Thus the parallels are projected into ellipses and the meridians into hyperbolas having their common center at the origin of the co-ordinates and their major axes in the direction of the axis η. Littrow speaks of both the meridians and parallels as being projected in hyperbolas,* which is evidently a slight error on the part of the eminent astronomer.

---

* *Chorographie*, page 113.

Finally assume

$$\xi=\frac{1}{2}\log\frac{1+\sin\omega\cos\theta}{1-\sin\omega\cos\theta}$$

Differentiating this gives

$$\frac{d\xi}{d\theta}=\frac{\sin\omega\sin\theta}{1-\sin^2\omega\cos^2\theta}\ .\qquad\qquad \frac{d\xi}{d\omega}=\frac{\cos\omega\cos\theta}{1-\sin^2\omega\cos^2\theta}$$

From these we have

$$d\eta=\frac{\cos\omega\,d\theta}{1-\sin^2\omega\cos^2\theta}+\frac{\sin\omega\sin\theta\cos\theta}{1-\sin^2\omega\cos^2\theta}d\omega$$

of which the integral is

$$\eta=\tan^{-1}\frac{\tan\theta}{\cos\omega}$$

which is identical with

$$\cot\eta=\cos\omega\tan\varphi$$

the formula already obtained.

§ III.

ORTHOMORPHIC PROJECTION—(Continued).

We will in this chapter take up two projections closely allied to each other, and very interesting in the methods of development employed by the illustrious authors. The first of these is by Sir John Herschel and is found in Volume XXX of the Journal of the Geographical Society of London for 1860. The paper is entitled "On a New Projection of the Sphere" and the author further calls it "Investigation of the conditions under which a spherical surface can be projected on a plane, so that the representation of any small portion of the surface shall be similar in form to the original;" this is, of course, merely the fundamental proposition of all orthomorphic projections.

Assume the radius of the sphere $=1$ and denote as usual latitude and longitude by θ and ω, and the plane coordinates by ξ and η. We must clearly have, since ξ and η are functions of θ and ω,

$$d\xi=\mathrm{M}d\omega+\mathrm{N}d\theta\qquad\qquad d\eta=\mathrm{P}d\omega+\mathrm{Q}d\theta$$

M, N, P, Q being, of course, functions of θ and ω. The elementary rectangle included between two meridians, whose difference of longitude is $d\omega$, and two parallels whose difference of latitude is $d\theta$, will have for its sides $d\theta$ and $d\omega\cos\theta$, having to each other the ratio

$$\frac{d\omega}{d\theta}\cos\theta$$

In passing along the projection of any one meridian ω does not vary. In passing then from the point whose projection is defined by ξ, η to the point *on the small meridian* whose projection is defined by $\xi+d\xi$, $\eta+d\eta$, ξ and η must vary by the variation of θ alone, or

$$d\xi=\mathrm{N}d\theta\qquad\qquad d\eta=\mathrm{Q}d\theta$$

and the distance between these two projected points is, on the same meridian

$$=d\theta\sqrt{\mathrm{N}^2+\mathrm{Q}^2}$$

Similarly supposing ourselves to pass along the same parallel we have for the distance between two infinitely near points, $d\theta$ being $=0$,

$$d\omega\sqrt{\mathrm{M}^2+\mathrm{P}^2}$$

These, then, are the sides of the elementary figure on the plane of projection corresponding to the infinitesimal rectangle on the surface of the sphere, and these two figures must be similar; which condition, being satisfied, obviously carries with it the similarity of an infinitesimal figure on the sphere and its projection. The sides then must be in the same ratio and the angle they include a right one. The first of these conditions gives

$$\frac{d\omega}{d\theta}\cos\theta = \frac{\sqrt{M^2+P^2}\,d\omega}{\sqrt{N^2+Q^2}\,d\theta}$$

or

$$\cos^2\theta = \frac{M^2+P^2}{N^2+Q^2}$$

The tangent of the angle made by the projected element of the meridian with the ordinate η is evidently represented by

$$\frac{\frac{d\xi}{d\omega}d\omega}{\frac{d\eta}{d\omega}d\omega} \text{ or by } \frac{M}{P}$$

and that of the projected element of the parallel by

$$\frac{\frac{d\xi}{d\theta}d\theta}{\frac{d\eta}{d\theta}\times -d\theta} \text{ or } -\frac{N}{Q}$$

because, lying on opposite sides of the ordinate η, if one tangent be taken positively the other must be taken negatively. The condition, then, of rectangularity requires that the product of these tangents shall be $=1$, which gives for the other essential equation

$$\frac{M}{P}\times -\frac{N}{Q} = 1$$

or

$$PQ = -MN$$

But

$$\cos^2\theta = \frac{M^2}{N^2}\cdot\frac{1+\left(\frac{P}{M}\right)^2}{1+\left(\frac{Q}{N}\right)^2} = \frac{M^2}{N^2}\cdot\frac{1+\left(\frac{P}{M}\right)^2}{1+\left(\frac{M}{P}\right)^2}$$

or

$$\cos^2\theta = \left(\frac{P}{N}\right)^2$$

whence we get the following

$$P = N\cos\theta \qquad M = -Q\cos\theta$$

Assume now

$$\alpha = \int\frac{d\theta}{\cos\theta} + \omega \qquad \beta = \int\frac{d\theta}{\cos\theta} - \omega$$

which gives

$$d\omega = \tfrac{1}{2}(d\alpha - d\beta) \qquad d\theta = \tfrac{1}{2}(d\alpha + d\beta)\cos\theta$$

and by substituting these in the equations

$$d\xi = M\,d\omega + N\,d\theta \qquad d\eta = P\,d\omega + Q\,d\theta$$

we find

$$d\xi = \tfrac{1}{2}(P+M)\,d\alpha + \tfrac{1}{2}(P-M)\,d\beta \qquad d\eta = \tfrac{1}{2}(P-M)\,d\alpha - \tfrac{1}{2}(P+M)\,d\beta$$

whence, adding and subtracting, we obtain

$$d(\xi+\eta) = P\,d\alpha - M\,d\beta \qquad d(\xi-\eta) = M\,d\alpha + P\,d\beta$$

The first members of these equations being exact differentials, the second must be so also. The conditions for this are

$$\frac{dP}{d\beta}=-\frac{dM}{d\alpha} \qquad \frac{dP}{d\alpha}=\frac{dM}{d\beta}$$

But universally

$$dP=\frac{dP}{d\alpha}d\alpha+\frac{dP}{d\beta}d\beta$$

whence, substituting,

$$dP=\frac{dM}{d\beta}d\alpha-\frac{dM}{d\alpha}d\beta$$

The first member of this being an exact differential, the second must be such also. This gives

$$\frac{d^2M}{d\alpha^2}+\frac{d^2M}{d\beta^2}=0$$

The known form for the integral of this partial differential equation is Φ and Ψ, denoting arbitrary functions

$$M=\Phi(\alpha+i\beta)+\Psi(\alpha-i\beta)$$

Substituting this for M in the expression for dP, reducing and integrating,

$$P=i[\Phi(\alpha+i\beta)-\Psi(\alpha-i\beta)]$$

and putting for brevity

$$A=\alpha+i\beta=\left(\int\frac{d\theta}{\cos\theta}+\omega\right)+i\left(\int\frac{d\theta}{\cos\theta}-\omega\right) \qquad B=\alpha-i\beta=\left(\int\frac{d\theta}{\cos\theta}+\omega\right)-i\left(\int\frac{d\theta}{\cos\theta}-\omega\right)$$

we find

$$d(\xi+\eta)=i[\Phi(A)dA-\Psi(B)dB] \qquad d(\xi-\eta)=\Phi(A)dA+\Psi(B)dB$$

which, by writing $2F(A)$ for $\int\Phi(A)dA$, and $2f(B)$ for $\int\Psi(B)dB$, affords the following values of ξ and η:

$$\xi=(1+i)F(A)+(1-i)f(B) \qquad -\eta=(1-i)F(A)+(1+i)f(B)$$

in which F and f are the characteristics of any two functions, both completely arbitrary and independent. Suppose, for example, we take

$$F(u)=f(u)=u$$

Then

$$\xi=(1+i)A+(1-i)B=(A+B)+i(A-B)=2(\alpha-\beta)=4\omega$$

and

$$-\eta=(1-i)A+(1+i)B=(A+B)-i(A-B)=2(\alpha+\beta)=4\int\frac{d\theta}{\cos\theta}=4\log\tan\tfrac{1}{2}(90^\circ-\theta)$$

which is the law of Mercator's projection.

The equations

$$\xi=(1+i)F(A)+(1-i)f(B) \qquad -\eta=(1-i)F(A)+(1+i)f(B)$$

being subject to no restriction, it is evident that we may superadd to the general conditions of the problem any which will suffice either to determine altogether or to limit the generality of the arbitrary functions F and f, in the view of obtaining convenient forms of projected representations. Suppose, for instance, that we assume as a condition that the projected representations of all circles about a fixed pole on the sphere shall be concentric circles about a fixed center on the plane. Since the origin of the co-ordinates ξ and η is arbitrary, we will fix it in that center; and since the condition is that when θ is given, and therefore

$$\int\frac{d\theta}{\cos\theta}=\text{a const., say } k$$

the equation between ξ and η shall be that of a circle about the center, we have

$$\xi^2+\eta^2=\rho^2=\text{a function of } \theta \text{ or } k$$

For brevity put

$$F(A)=X \qquad f(B)=Y$$

Then we have

$$\xi=(1+i)\,X+(1-i)\,Y \qquad -\eta=(1-i)\,X+(1+i)\,Y$$

and substituting and reducing

$$\xi^2+\eta^2=8XY$$

that is to say

$$F(A)\,f(B)=\frac{\rho^2}{8}$$

Since any function of an arbitrary function is itself an arbitrary function, we may without any loss of generality write $e^{F(A)}$ for $F(A)$, and $e^{f(B)}$ for $f(B)$. Now

$$F(\alpha+i\beta)+f(\alpha-i\beta)=\Psi(\alpha+\beta)$$

because

$$\alpha+\beta=2\int\frac{d\theta}{\cos\theta}=2k$$

It does not appear that this equation can be satisfied by any forms of F and f more general than the following, viz:

$$F(u)=(g+ih)\,u \qquad f(u)=(-h+ig)\,u$$

which give for the value of $\Psi(\alpha+\beta)$

$$[(g-h)+i\,(g+h)]\,(\alpha+\beta)$$

or, what comes to the same thing,

$$2\,[(g-h)+i\,(g+h)]\int\frac{d\theta}{\cos\theta}$$

Practically speaking, this expression is useless unless the imaginary term vanishes, or $g+h=0$, $g-h=2g$, in which case it reduces itself to

$$4g\int\frac{d\theta}{\cos\theta}=4gk$$

whence also

$$\rho^2=8e^{4gk} \qquad \rho=2\sqrt{2}\,e^{2g\int\frac{d\theta}{\cos\theta}}$$

which, since

$$\int\frac{d\theta}{\cos\theta}=\log\tan\tfrac{1}{2}\,(90^\circ-\theta)$$

reduces itself to

$$\rho=2\sqrt{2}\,\tan^{2g}\tfrac{1}{2}\,(90^\circ-\theta)$$

Suppose $g=\frac{1}{2}$. This is the law of the stereographic projection, and the values of ξ and η become

$$\xi=2\rho\,[\cos\omega+\sin\omega]=2\rho\sqrt{2}\sin(45^\circ+\omega)$$

$$-\eta=2\rho\,[\cos\omega-\sin\omega]=2\rho\sqrt{2}\cos(45^\circ+\omega)$$

$$\rho=2\sqrt{2}\tan\tfrac{1}{2}\,(90^\circ-\theta)$$

In the more general case of $2g=n$, we find

$$\xi=2\rho\,[\cos n\omega+\sin n\omega]=2\rho\sqrt{2}\sin(45^\circ+n\omega)$$

$$-\eta=2\rho\,[\cos n\omega-\sin n\omega]=2\rho\sqrt{2}\cos(45^\circ+n\omega)$$

$$\rho=2\sqrt{2}\tan^{n}\tfrac{1}{2}\,(90^\circ-\theta)$$

To interpret these expressions we have only to consider that when ω increases by any number of degrees $n\omega$ increases by n times that number; so that if ω increases from 0 to 360°, $n\omega$ increases n

times 360°. The co-ordinates of the projection of any point, therefore, are those corresponding to n times the longitude in the case of the stereographic projection. If, then, n be a fraction less than unity, the projection of the whole spherical surface will, instead of occupying the whole area of a circle, be comprised within a sector, the same fractional part of the whole area. Thus if $n=\frac{1}{2}$, the projection of the whole sphere in longitude will be comprised within a semicircle; if $n=\frac{1}{3}$, within a sector of 120°; if $n=\frac{2}{3}$, within a sector of 240°, &c.; and the entire parallels of latitude will in like manner be represented by the portions of concentric circles comprised between the extreme radii of these respective sectors. If φ be the polar distance of any parallel of latitude, and ρ the radius of the circular segment representing that parallel, we have, neglecting the coefficient $-2\sqrt{2}$ or taking 1 for the equatorial radius *in the projection*,

$$\rho=\tan^n\frac{\varphi}{2}$$

from which it is easy to calculate ρ for each polar distance from 0° to 180°. The values for the four cases $n=1, \frac{1}{2}, \frac{1}{3}, \frac{2}{3}$ for $\varphi=0°, 10°, 20°$, &c., are given in the table.

The second case that we take up is in its development very similar to Herschel's projection; it is an investigation of orthomorphic projection by Professor Boole. The idea was suggested to Boole by reading Herschel's paper. The investigation is contained in the supplementary volume to Boole's Differential Equations, published by Todhunter, after the author's death. The general theory as given by Boole is applicable to the projection of any surface upon a plane.

Let x, y, z denote the rectangular rectilinear co-ordinates of any point of the given surface; ξ, η the co-ordinates of the corresponding point on the plane of projection. Let the equation of the given surface be

$$F(x, y, z)=0$$

or simply

$$F=0.$$

Regarding ξ and η as ultimately functions of x, y, z we have

$$d\xi=\frac{d\xi}{dx}dx+\frac{d\xi}{dy}dy+\frac{d\xi}{dz}dz \qquad d\eta=\frac{d\eta}{dx}dx+\frac{d\eta}{dy}dy+\frac{d\eta}{dz}dz$$

dx, dy, dz being connected by the relation

$$\frac{dF}{dx}dx+\frac{dF}{dy}dy+\frac{dF}{dz}dz=0$$

Now for brevity write

$$\frac{dF}{dx}=A \qquad \frac{dF}{dy}=B \qquad \frac{dF}{dz}=C$$

$$\frac{d\xi}{dx}=a \qquad \frac{d\xi}{dy}=b \qquad \frac{d\xi}{dz}=c$$

$$\frac{d\eta}{dx}=a' \qquad \frac{d\eta}{dy}=b' \qquad \frac{d\eta}{dz}=c'$$

then

(1) $$d\xi=adx+bdy+cdz$$

(2) $$d\eta=a'dx+b'dy+c'dz$$

(3) $$0=Adx+Bdy+Cdz$$

The two conditions to be fulfilled are, as we already know, the equality of corresponding angles, and the proportionality of corresponding sides, of the element on the surface and the corresponding element on the plane of projection.

Assuming now any point ξ, η on the plane of projection, let ξ alone vary, and the infinitesimal line generated by $d\xi$; and since $d\eta=0$, we have

$$a'dx+b'dy+c'dz=0 \qquad Adx+Bdy+Cdz=0$$

Denoting by ∇ the determinant

$$\begin{vmatrix} a, & b, & c \\ a', & b', & c' \\ A, & B, & C \end{vmatrix}$$

write

$$\frac{d\nabla}{da}=L \qquad \frac{d\nabla}{db}=M \qquad \frac{d\nabla}{dc}=N$$

then the above equations give

$$\frac{dx}{L}=\frac{dy}{M}=\frac{dz}{N}$$

so that the direction-cosines of the infinitesimal line on the surface F corresponding to the line $d\xi$ on the plane are

$$\frac{L}{\sqrt{L^2+M^2+N^2}} \qquad \frac{M}{\sqrt{L^2+M^2+N^2}} \qquad \frac{N}{\sqrt{L^2+M^2+N^2}}$$

In like manner, if η alone vary we shall find for the direction-cosines of the infinitesimal line on F corresponding to $d\eta$ on the plane

$$\frac{L'}{\sqrt{L'^2+M'^2+N'^2}} \qquad \frac{M'}{\sqrt{L'^2+M'^2+N'^2}} \qquad \frac{N'}{\sqrt{L'^2+M'^2+N'^2}}$$

where

$$L'=\frac{d\nabla}{da'} \qquad M'=\frac{d\nabla}{db'} \qquad N'=\frac{d\nabla}{dc'}$$

Since the angle between ξ and η is a right angle, the angle on the surface between the lines whose direction-cosines have been found must also be right. This gives at once

$$LL'+MM'+NN'=0$$

The ratio of the element of length $d\xi$ to the corresponding element on the surface is

$$\frac{d\xi}{\sqrt{dx^2+dy^2+dz^2}}$$

or

$$\frac{adx+bdy+cdz}{\sqrt{dx^2+dy^2+dz^2}}$$

and finally

$$\frac{aL+bM+cN}{\sqrt{L^2+M^2+N^2}}$$

Equating this ratio to the corresponding ratio of the length of $d\eta$ to that of its projection on the surface, we have

$$\frac{aL+bM+cN}{\sqrt{L^2+M^2+N^2}}=\frac{a'L'+b'M'+c'N'}{\sqrt{L'^2+M'^2+N'^2}}$$

Now we know that

$$aL+bM+cN=-\nabla$$

and

$$a'L'+b'M'+c'N'=\nabla$$

the numerators of the last equation only differing in sign. But ∇, being the determinant of the system

$$adx+bdy+cdz=0 \qquad a'dx+b'dy+c'dz=0 \qquad Adx+Bdy+Cdz=0$$

expresses, when equated to zero, the condition that if $d\xi$ vanishes $d\eta$ must also vanish; and $d\xi$ and $d\eta$ being independent, this condition cannot be satisfied; so that the above equation reduces to

$$\frac{1}{\sqrt{L^2+M^2+N^2}}-\frac{1}{\sqrt{L'^2+M'^2+N'^2}}=0$$

or

$$(\alpha)\qquad L'^2+M'^2+N'^2-(L^2+M^2+N^2)=0$$

and this, with the already-found condition of orthogonality

$$(\beta)\qquad LL'+MM'+NN'=0$$

will fully express the conditions of similarity. If we multiply β by $2i$, and add and subtract the result from α, we obtain the equivalent system

$$(L'+iL)^2+(M'+iM)^2+(N'+iN)^2=0 \qquad (L'-iL)^2+(M'-iM)^2+(N'-iN)^2=0$$

Now

$$L'\pm iL=\frac{dF}{dy}\frac{d\xi}{dz}-\frac{dF}{dz}\frac{d\xi}{dy}\pm i\left(\frac{dF}{dy}\frac{d\eta}{dz}-\frac{dF}{dz}\frac{d\eta}{dy}\right)=\frac{dF}{dy}\frac{d(\xi\pm i\eta)}{dz}-\frac{dF}{dz}\frac{d(\xi\pm i\eta)}{dy}$$

Writing then

$$\xi+i\eta=u \qquad \xi-i\eta=v$$

we have

$$L'+iL=\frac{dF}{dy}\frac{du}{dz}-\frac{dF}{dz}\frac{du}{dy} \qquad L'-iL=\frac{dF}{dy}\frac{dv}{dz}-\frac{dF}{dz}\frac{dv}{dy}$$

Similarly

$$M'+iM=\frac{dF}{dz}\frac{du}{dx}-\frac{dF}{dx}\frac{du}{dz} \qquad M'-iM=\frac{dF}{dz}\frac{dv}{dx}-\frac{dF}{dx}\frac{dv}{dz}$$

$$N'+iN=\frac{dF}{dx}\frac{du}{dy}-\frac{dF}{dy}\frac{du}{dx} \qquad N'-iN=\frac{dF}{dx}\frac{dv}{dy}-\frac{dF}{dy}\frac{dv}{dx}$$

Substituting these in the last equations, there result

$$\left(\frac{dF}{dy}\frac{du}{dz}-\frac{dF}{dz}\frac{du}{dy}\right)^2+\left(\frac{dF}{dz}\frac{du}{dx}-\frac{dF}{dx}\frac{du}{dz}\right)^2+\left(\frac{dF}{dx}\frac{du}{dy}-\frac{dF}{dy}\frac{du}{dx}\right)^2=0$$

$$\left(\frac{dF}{dy}\frac{dv}{dz}-\frac{dF}{dz}\frac{dv}{dy}\right)^2+\left(\frac{dF}{dz}\frac{dv}{dx}-\frac{dF}{dx}\frac{dv}{dz}\right)^2+\left(\frac{dF}{dx}\frac{dv}{dy}-\frac{dF}{dy}\frac{dv}{dx}\right)^2=0$$

Each of these can be written in the form of a symmetrical determinant of the second order. Designating by θ_1, θ_2, θ_3, any three quantities whatever, make U equal to the determinant

$$\begin{vmatrix} \theta_1, & \theta_2, & \theta_3 \\ \frac{dF}{dx}, & \frac{dF}{dy}, & \frac{dF}{dz} \\ \frac{du}{dx}, & \frac{du}{dy}, & \frac{du}{dz} \end{vmatrix}$$

Then the first of these two differential equations is simply

$$\left(\frac{dU}{d\theta_1}\right)^2+\left(\frac{dU}{d\theta_2}\right)^2+\left(\frac{dU}{d\theta_3}\right)^2=0$$

and, as is well known, can be written

$$(\mathrm{I})\qquad \begin{vmatrix} \left(\frac{dF}{dx}\right)^2+\left(\frac{dF}{dy}\right)^2+\left(\frac{dF}{dz}\right)^2, & \frac{dF}{dx}\frac{du}{dx}+\frac{dF}{dy}\frac{du}{dy}+\frac{dF}{dz}\frac{du}{dz} \\ \frac{dF}{dx}\frac{du}{dx}+\frac{dF}{dy}\frac{du}{dy}+\frac{dF}{dz}\frac{du}{dz}, & \left(\frac{du}{dx}\right)^2+\left(\frac{du}{dy}\right)^2+\left(\frac{du}{dz}\right)^2 \end{vmatrix}=0$$

and in like manner for the second

$$\text{(II)}\qquad \begin{vmatrix} \left(\frac{dF}{dx}\right)^2+\left(\frac{dF}{dy}\right)^2+\left(\frac{dF}{dz}\right)^2, & \frac{dF}{dx}\frac{dv}{dx}+\frac{dF}{dy}\frac{dv}{dy}+\frac{dF}{dz}\frac{dv}{dz} \\ \frac{dF}{dx}\frac{dv}{dx}+\frac{dF}{dy}\frac{dv}{dy}+\frac{dF}{dz}\frac{dv}{dz}, & \left(\frac{dv}{dx}\right)^2+\left(\frac{dv}{dy}\right)^2+\left(\frac{dv}{dz}\right)^2 \end{vmatrix}=0$$

These are the partial differential equations of the first order, serving to determine u and v as functions of x, y, and z.

But it is not necessary to solve these equations in their general form. It is well known that the co-ordinates x, y, and z of any point on a surface, being connected by the equation of the surface, can always be expressed as functions of two independent variable parameters, and these parameters, when fixed upon, become the independent variables of the problem. Let θ and ω represent such parameters, and let their expressions in terms of x, y, z give

$$\theta=\Phi_1(x, y, z) \qquad \omega=\Phi_2(x, y, z)$$

which equations, combined with that of the given surface, will reciprocally determine x, y, z as functions of θ and ω. The differential coefficients

$$\frac{dF}{dx} \qquad \frac{dF}{dy} \qquad \frac{dF}{dz}$$

which are functions of x, y, z, now become functions of θ and ω; and further

$$\frac{du}{dx}=\frac{du}{d\theta}\frac{d\theta}{dx}+\frac{du}{d\omega}\frac{d\omega}{dx} \qquad \frac{du}{dy}=\frac{du}{d\theta}\frac{d\theta}{dy}+\frac{du}{d\omega}\frac{d\omega}{dy} \qquad \frac{du}{dz}=\frac{du}{d\theta}\frac{d\theta}{dz}+\frac{du}{d\omega}\frac{d\omega}{dz}$$

and as $\frac{d\theta}{dx}$, $\frac{d\omega}{dx}$, &c., are known functions of x, y, and z, they are also expressible in terms of θ and ω. The result of these substitutions will then be to convert (I) into a partial differential equation in which u is the dependent and θ and ω the independent variables, and this equation being, like (I), of the first order and second degree in the differential coefficients of u, will be of the form

$$P\left(\frac{du}{d\theta}\right)^2+Q\frac{du}{d\theta}\frac{du}{d\omega}+R\left(\frac{du}{d\omega}\right)^2=0$$

For v we have an exactly similar equation, with the same coefficients.

The above equation is, by the solution of a quadratic, resolvable into two equations of the form

$$\frac{du}{d\theta}-\lambda_1\frac{du}{d\omega}=0 \qquad \frac{du}{d\theta}-\lambda_2\frac{du}{d\omega}=0$$

To these correspond the respective auxiliary equations

$$d\omega+\lambda_1 d\theta=0 \qquad d\omega+\lambda_2 d\theta=0$$

If the integrals of these are

$$S=c_1 \qquad T=c_2$$

respectively, then we have

$$u=\Phi(S) \qquad v=\Psi(T)$$

Now, v being determinable by an equation of the same form as u, it follows that of the above two values of u one must be assigned to v, so that the solution of the problem will be contained in the system

$$u=\Phi(S) \qquad v=\Psi(T)$$

or in the system

$$u=\Phi(T) \qquad v=\Psi(S)$$

The particular forms of the arbitrary functions Φ and Ψ will depend solely upon the nature of the problem under consideration.

The first members of (I) and (II) are obviously essentially positive; and so, if the intermediate transformations are real, is the first member of the equation whose coefficients are P, Q, R. Hence the quadratic determining λ_1 and λ_2 will have imaginary roots of the form $\alpha \pm i\beta$. Ultimately, therefore, it will suffice to integrate one equation of the auxiliary system

$$d\omega+\lambda_1 d\theta=0 \qquad d\omega+\lambda_2 d\theta=0$$

and then to deduce the solution of the other by changing i into $-i$.

Suppose, now, that the surface to be represented is an oblate spheroid, such as the earth; take the plane of the equator for that of projection and the center for origin. Let the co ordinates x, y pass through the meridians of 0° and 90°, respectively, and z through the poles. The equation of the surface will be

$$\frac{x^2+y^2}{a^2}+\frac{z^2}{b^2}=1$$

when a is the earth's equatorial, and b its polar radius. Let also the latitude of the points (x, y, z) be represented by θ and its longitude by ω. We have

$$F=\frac{x^2+y^2}{a^2}+\frac{z^2}{b^2}-1=0$$

$$\frac{dF}{dx}=\frac{2x}{a^2} \qquad \frac{dF}{dy}=\frac{2y}{a^2} \qquad \frac{dF}{dz}=\frac{2z}{b^2}$$

and substituting in (I), we have

$$\begin{vmatrix} \frac{x^2+y^2}{a^4}+\frac{z^2}{b^4} & \frac{x}{a^2}\frac{du}{dx}+\frac{y}{a^2}\frac{du}{dy}+\frac{z}{b^2}\frac{du}{dz} \\ \frac{x}{a^2}\frac{du}{dx}+\frac{y}{a^2}\frac{du}{dy}+\frac{z}{b^2}\frac{du}{dz} & \left(\frac{du}{dx}\right)^2+\left(\frac{du}{dy}\right)^2+\left(\frac{du}{dz}\right)^2 \end{vmatrix}$$

or, expanding this and writing $\frac{a^2}{b^2}=k^2$

$$\text{(III)} \qquad (x^2+y^2+k^4z^2)\left\{\left(\frac{du}{dx}\right)^2+\left(\frac{du}{dy}\right)^2+\left(\frac{du}{dz}\right)^2\right\}-\left(x\times\frac{du}{dx}+y\frac{du}{dy}+k^2z\frac{du}{dz}=0\right)$$

Now as x, y are co-ordinates in the plane of the equator, and x passes through the first meridian we have

$$\frac{y}{x}=\tan\omega$$

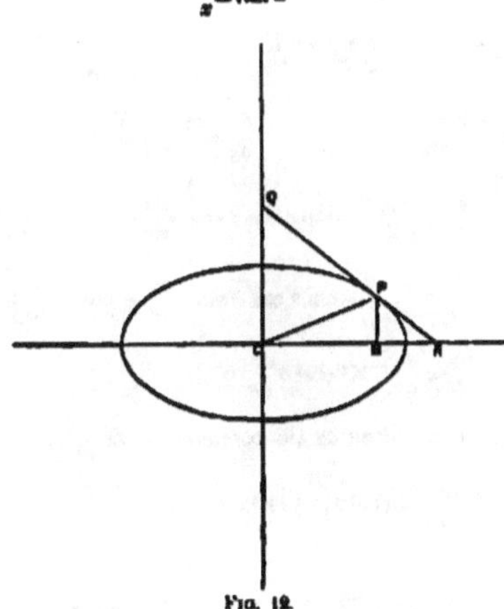

FIG. 12.

Again, representing in Fig. 13 the meridian of the point P, or (x, y, z) touched by the straight line QR in the same plane, we have $CM=\sqrt{x^2+y^2}$ and $MP=z$. Therefore if $\sqrt{x^2+y^2}=\rho$, the equation of the meridian is

$$\frac{\rho^2}{a^2}+\frac{z^2}{b^2}=1$$

and that of the tangent is

$$\frac{\rho\rho'}{a^2}+\frac{zz'}{b^2}=1$$

ρ', z' being current rectangular co-ordinates of the tangent. Hence

$$\tan CQR=\frac{a^2z}{b^2\rho}=\frac{k^2z}{\sqrt{x^2+y^2}}$$

But $CQR=\theta$. Therefore finally

$$\theta=\tan^{-1}\frac{k^2z}{\sqrt{x^2+y^2}} \qquad \omega=\tan^{-1}\frac{y}{x}$$

and we must now transform (III) so as to make θ and ω the independent variables. From the last equations combined with that of the surface we have readily

$$x=\frac{ak\cos\omega}{\sqrt{k^2+\tan^2\theta}} \qquad y=\frac{ak\sin\omega}{\sqrt{k^2+\tan^2\theta}} \qquad z=\frac{a\tan\theta}{k\sqrt{k^2+\tan^2\theta}}$$

and substituting in (III) we obtain

$$\text{(IV)}\qquad 0=\sec^2\theta\left\{\left(\frac{du}{dx}\right)^2+\left(\frac{du}{dy}\right)^2+\left(\frac{du}{dz}\right)^2\right\}-\left(\cos\omega\frac{du}{dx}+\sin\omega\frac{du}{dy}+\tan\theta\frac{du}{dz}\right)^2$$

Again

$$\frac{du}{dx}=\frac{du}{d\theta}\frac{d\theta}{dx}+\frac{du}{d\omega}\frac{d\omega}{dx} \qquad \frac{du}{dy}=\frac{du}{d\theta}\frac{d\theta}{dy}+\frac{du}{d\omega}\frac{d\omega}{dy} \qquad \frac{du}{dz}=\frac{du}{d\theta}\frac{d\theta}{dz}+\frac{du}{d\omega}\frac{d\omega}{dz}$$

Now

$$\frac{d\theta}{dx}=\frac{-k^2zx}{\sqrt{x^2+y^2}\,(x^2+y^2+k^4z^2)}=\frac{-\sin\theta\cos\theta\cos\omega\sqrt{K}}{aK}$$

when $K=k^2+\tan^2\theta$. In like manner

$$\frac{d\theta}{dy}=\frac{-\sin\theta\cos\theta\sin\omega\sqrt{K}}{aK} \qquad \frac{d\theta}{dz}=\frac{K\cos^2\theta\sqrt{K}}{a}$$

$$\frac{d\omega}{dx}=\frac{-\sin\omega\sqrt{K}}{aK} \qquad \frac{d\omega}{dy}=\frac{\cos\omega\sqrt{K}}{aK} \qquad \frac{d\omega}{dz}=0$$

Hence

$$\frac{du}{dx}=\frac{\sqrt{K}}{aK}\left[-\sin\theta\cos\theta\cos\omega\frac{du}{d\theta}-\sin\omega\frac{du}{d\omega}\right]$$

$$\frac{du}{dy}=\frac{\sqrt{K}}{aK}\left[-\sin\theta\cos\theta\sin\omega\frac{du}{d\theta}+\cos\omega\frac{du}{d\omega}\right]$$

$$\frac{du}{dz}=\frac{\sqrt{K}}{aK}\,K^2\cos^2\theta\frac{du}{d\theta}$$

Substituting these in (IV), and dividing by the common factor $\frac{K}{a^2K^2}$, we have, on reduction,

$$\left(\frac{du}{d\omega}\right)^2+\cos^2\theta\,[1+(K^2-1)\cos^2\theta]^2\left(\frac{du}{d\theta}\right)^2=0$$

which is resolvable into

$$\frac{du}{d\omega}-i\cos\theta\,[1+(K^2-1)\cos^2\theta]\frac{du}{d\theta}=0 \qquad \frac{du}{d\omega}+i\cos\theta\,[1+(K^2-1)\cos^2\theta]\frac{du}{d\theta}=0$$

partial differential equations, of which the integrals are included in the common formula

$$u=\phi\left(\int\frac{d\theta}{\cos\theta[1+(K^2-1)\cos^2\theta]}\pm i\omega\right)$$

Now

$$\int\frac{d\theta}{\cos\theta[1+(K^2-1)\cos^2\theta]}=\int\frac{d\theta}{\cos\theta}+(1-K^2)\int\frac{\cos\theta\, d\theta}{1+(K^2-1)\cos^2\theta}$$

$$=\int\frac{d\theta}{\cos\theta}+(1-K^2)\int\frac{\cos\theta\, d\theta}{K^2-(K^2-1)\sin^2\theta}$$

$$=\int\frac{d\theta}{\cos\theta}+\epsilon^2\int\frac{\cos\theta\, d\theta}{1-\epsilon^2\sin^2\theta}\left(\text{since } \epsilon^2=\frac{a^2-b^2}{a^2}\right)$$

$$=\log\tan\tfrac{1}{2}\left(\frac{\pi}{2}+\theta\right)+\frac{\epsilon}{2}\log\frac{1-\epsilon\sin\theta}{1+\epsilon\sin\theta}$$

$$=\log\left\{\left(\frac{1+\epsilon\sin\theta}{1+\epsilon\sin\theta}\right)^{\frac{\epsilon}{2}}\tan\tfrac{1}{2}\left(\frac{\pi}{2}+\theta\right)\right\}$$

Hence

$$u=\phi\left[\log\left\{\left(\frac{1-\epsilon\sin\theta}{1+\epsilon\sin\theta}\right)^{\frac{\epsilon}{2}}\tan\tfrac{1}{2}\left(\frac{\pi}{2}+\theta\right)\right\}+i\omega\right]$$

or changing $\phi(\omega)$ into $\phi(e^{\omega})$, since ϕ is an arbitrary function

$$u=\phi\left\{\left(\frac{1-\epsilon\sin\theta}{1+\epsilon\sin\theta}\right)^{\frac{\epsilon}{2}}\tan\tfrac{1}{2}\left(\frac{\pi}{2}+\theta\right)e^{+i\omega}\right\}$$

$$v=\psi\left\{\left(\frac{1-\epsilon\sin\theta}{1+\epsilon\sin\theta}\right)^{\frac{\epsilon}{2}}\tan\tfrac{1}{2}\left(\frac{\pi}{2}+\theta\right)e^{-i\omega}\right\}$$

Let ρ and α denote the polar co-ordinates of that point in the plane of projection which corresponds to the point whose latitude and longitude on the surface are θ and ω; and let

$$S=\left(\frac{1-\epsilon\sin\theta}{1+\epsilon\sin\theta}\right)^{\frac{\epsilon}{2}}\tan\tfrac{1}{2}\left(\frac{\pi}{2}+\theta\right)$$

then the complete solution assumes the very simple form

$$\text{(V)}\qquad \rho e^{i\alpha}=\phi(Se^{i\omega})\qquad \rho e^{-i\alpha}=\psi(Se^{-i\omega})$$

Assume that the parallels of latitude are projected into circles round the pole. This requires that ρ be independent of ω, a condition which is satisfied in the most general manner by assuming

$$\phi(w)=Cw^n\qquad \psi(w)=C'w^n$$

We then find

$$\rho e^{i\alpha}=CS^ne^{in\omega}\qquad \rho e^{-i\alpha}=C'S^ne^{-in\omega}$$

whence, on multiplication and division,

$$\rho^2=CC'S^{2n}\qquad e^{2i\alpha}=\frac{C}{C'}e^{2in\omega}$$

whence, A and B being now arbitrary constants derived from C and C',

$$\rho=AS^n\qquad \alpha=\pm n\omega+B$$

Observing that α and ω should vanish together, we have $B=0$, and the equation $\alpha=\pm n\omega$ shows that the surface of the sphere will be projected into a sector of a circle, the arc of which is to the circumference of the circle $n:1$. Thus if $n=\frac{1}{4}, \frac{1}{2}$, &c., the sphere is projected into a quadrant, a semicircle, &c. These are of course the results that we have already obtained in several different places.

The other equation for ρ gives

$$\rho=A\left\{\tan\tfrac{1}{2}\left(\frac{\pi}{2}+\theta\right)\right\}^{n}\left(\frac{1-e\sin\theta}{1+e\sin\theta}\right)^{\frac{ne}{2}}$$

If $\theta=0$, we find

$$\rho=A$$

whence A is the distance of the equator from the pole in the plane of projection; and if that distance, which is arbitrary, be assumed as the unit, we have

$$\rho=\left\{\tan\tfrac{1}{2}\left(\frac{\pi}{2}+\theta\right)^{n}\left(\frac{1-e\sin\theta}{1+e\sin\theta}\right)^{\frac{ne}{2}}\right\}$$

for the distance from the pole of that parallel whose latitude is θ. We can throw this expression into a more familiar form by assuming $p=\frac{\pi}{2}+\theta$, and introducing an auxiliary quantity q defined by the relation

$$e\cos p=\cos q$$

We have then

$$\rho=\left(\tan\frac{p}{2}\right)^{n}\left(\cot\frac{q}{2}\right)^{ne}$$

Table IV gives the values of ρ for the sphere and for the spheroid whose eccentricity is .08 (about that of the earth), for each ten degrees of polar distance, for the values $n=1$ and $n=\frac{1}{2}$.

The preceding investigation by Boole is seen to be much more general than that of Herschel, the latter confining himself merely to the projection of a sphere upon a plane.

---

§ IV.

PROJECTIONS BY DEVELOPMENT.

In order that a surface may be represented upon a plane without any change of angles or areas, it must be such an one as can by actual development be rolled out upon a plane—all parts of it coming by a continuous motion to coincide with the plane—as, for example, all cones and cylinders. If we desire to make a projection of a comparatively small region, the operation will be rendered quite simple if we can substitute for the actual surface to be projected a certain portion of some developable surface upon which are drawn the meridians and parallels. The construction of these lines upon the developable must of course be such as to make the new elements correspond as closely as possible with the actual elements of the sphere. The attempt to make projections of this kind has naturally given rise to two methods: (1) Conical Projections, (2) Cylindric Projections. We will first consider the former of these.

Conceive a cone passed tangent to the sphere along the parallel of latitude which is at the middle of the region to be projected. Also imagine the planes of the different parallels and meridians to be produced until they cut the cone. We will then have upon the surface of the cone small quadrilaterals corresponding to those of the sphere; the magnitudes are different, but the angles are obviously the same. Now develop the cone upon a plane; the meridians will clearly become right lines from the vertex of the cone to the different points of the developed parallel of tangency (or any other), and the parallels will be concentric circles, the vertex of the cone being the common center. The parallel of tangency is obviously the only one unaltered by the development. The quadrilaterals upon the sphere are reproduced upon the plane still as rectangular, but the magnitudes are different, as equal distances of latitude upon the sphere are represented by distances which diminish towards the pole and increase towards the equator. The differences of longitude are all greater upon the surface of the cone than upon the sphere, except for the parallel of tangency. The error in latitude may be completely, and that in longitude partially, eliminated by laying off along the middle meridian of the development the rectified lengths of the distances between the parallels, and through the points thus obtained, with the vertex of the cone as a center, describing arcs of circles. By this means we obtain for the differences in latitude their true values, and for the differences in longitude values which are more nearly correct than those given by the first method. Fig. 13 shows both methods, the dotted lines corresponding to the second method.

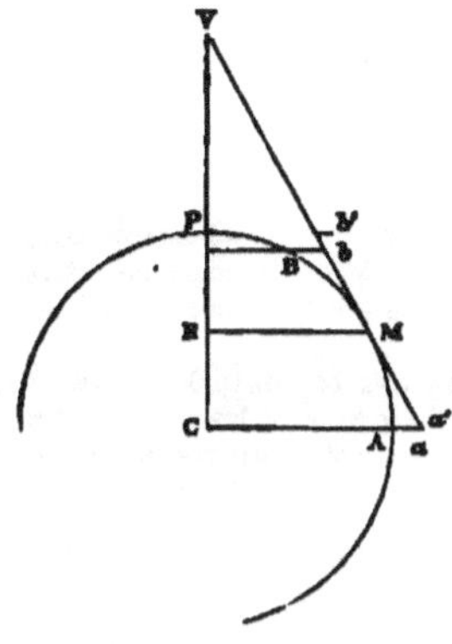

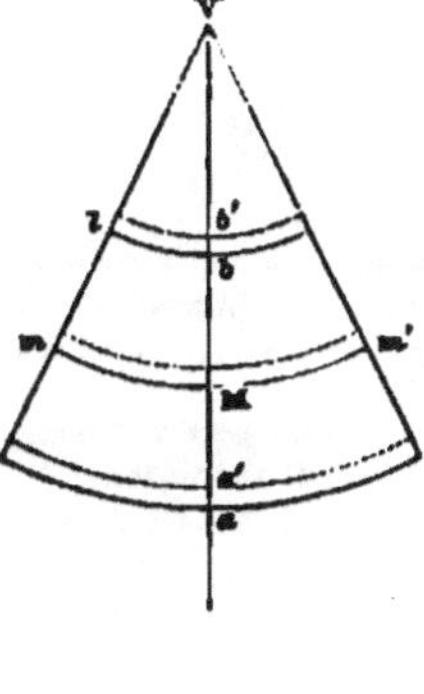

FIG. 13.

We have clearly, from the first figure,

$$\frac{180^\circ}{\pi r \cos\theta_0}=\frac{x}{mm'}$$

when θ_0 is the latitude of the middle parallel RM, and x is the difference of longitude of the extreme meridians which are to be projected. Let also V denote the angle of the extreme elements of the cone which appear in the development. The radius VM of the middle parallel is given by

$$VM=r\cot\theta_0$$

and from figure (2) follows

$$\frac{180^\circ}{\pi r \cot\theta_0}=\frac{V}{mm'}$$

Combination of these two values for mm' gives

$$V=x\sin\theta_0$$

It is obvious now how to construct the projection: The angle V being determined, we have for the radius of the middle parallel, $VM=r\cot\theta_0$. Lay off from M the distances Ma' and Mb' as obtained by actual rectification. If the distance ab contains n degrees,

$$ab=\frac{\pi r n}{180^\circ}$$

and Mb', Ma' each

$$=\frac{\pi r n}{360^\circ}$$

sponding to ω upon the sphere, we have clearly

$$\frac{\epsilon}{\omega}=\frac{V}{\pi}=\sin\theta_0$$

The radius of the parallel at latitude θ will be

$$=r\left[\cot\theta_0-(\theta-\theta_0)\right]$$

and the corresponding arc of longitude ω will be

$$=r\omega\sin\theta_0\left[\cot\theta_0-(\theta-\theta_0)\right]$$

The error for each degree of the parallel will then be

$$=r(\theta-\theta_0)\sin\theta_0$$

Euler investigated at some length the theory of conic projection and determined a cone fulfilling the following conditions: (1) That the errors at the northern and southern extremities of the chart should be equal. (2) That they shall be equal to the greatest error which occurs near the mean parallel.

The cone in this case is obviously a secant and not a tangent cone to the sphere. Let θ_a denote the least latitude of the region to be projected, and θ_b the greatest value of the latitude. Let AB denote the portion of the middle meridian comprised between these extreme latitudes. Designate

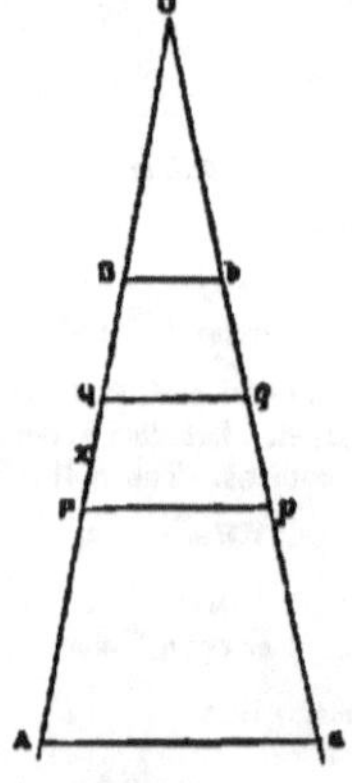

FIG. 14.

by δ the length of 1° of the meridian, and let P and Q be the intersections of the central meridian with the parallels along which the degrees shall preserve upon the map their exact ratio with the actual degrees of latitude; also call θ_p and θ_q the latitudes of these two parallels, upon each of which a degree of longitude has respectively the values $\delta\cos\theta_p$ and $\delta\cos\theta_q$. Lay off these two values of 1° along the lines Pp and Qq perpendicular to AB, and join pq; this line will represent the meridian removed one degree from AB. The point of intersection O will obviously be the common point of meeting of all the meridians and the center of all the parallels. The distance from O to any parallel is readily found; we have, since OPp is a right angle

$$\frac{Pp-Qq}{PQ}=\frac{Pp}{PO}$$

or

$$\frac{\delta(\cos\theta_p-\cos\theta_q)}{\theta_p-\theta_q}=\frac{\delta\cos\theta_p}{PO}$$

from which

$$PO=\frac{\cos\theta_p\,(\theta_p-\theta_q)}{\cos\theta_p-\cos\theta_q}$$

Having determined the center O, it is only necessary to draw an arc of radius OP and upon it lay off lengths $=\delta\cos\theta_p$; these will give the points through which the meridians pass; then laying off, along the middle meridian, distances equal to the number of degrees of latitude of the different parallels to be constructed, draw through the points thus found circles having their centers at O, and the projections of the parallels will be constructed.

We will now determine the errors resulting from this construction upon the extreme parallels through A and B. Calling ω the angle POp, we find

$$\omega=\frac{Pp}{PO}=\frac{\delta(\cos\theta_p-\cos\theta_q)}{\theta_p-\theta_q}$$

which becomes

$$\omega=\frac{\cos\theta_p-\cos\theta_q}{(\theta_p-\theta_q)\upsilon},$$

if we take $\delta=1°$, and express the denominator in parts of radius, which is done by making

$$\upsilon=0.01745329$$

the value of 1° in a circle of radius unity. Call z the distance in degrees from the center O to the pole. The distance from P to the pole will be $90°-\theta_p$, from P to O will be $90°-\theta_p+z$; the value of this in parts of radius will be

$$=\upsilon\,(90°-\theta_p+z)$$

It is easy to see now that we must have

$$z=\frac{(\theta_q-\theta_p)\cos\theta_p}{\cos\theta_p-\cos\theta_q}-90°+\theta_p$$

The distance of the extreme parallel A from O will be, in parts of radius,

$$AO=\upsilon\,(90°-\theta_a+z)$$

Multiplying this by the value of ω, we have for the value of the degree upon this parallel

$$Aa=\delta\frac{(90°-\theta_a+z)(\cos\theta_p-\cos\theta_q)}{\theta_q-\theta_p}$$

instead of $\delta\cos\theta_a$. The difference of these two values gives the error along the parallel through A. For B the error is the difference of

$$\delta\frac{(90°-\theta_b+z)(\cos\theta_p-\cos\theta_q)}{\theta_q-\theta_p}\ \text{and}\ \delta\cos\theta_b$$

Euler's proposition was to determine the parallels P and Q in such a manner as to make the extreme errors at A and B equal. Equating these two errors and reducing, we have

$$(\theta_a-\theta_b)(\cos\theta_p-\cos\theta_q)+(\theta_q-\theta_p)(\cos\theta_a-\cos\theta_b)=0$$

For the length of one degree upon the parallels of A and B we have

$$\upsilon(90°-\theta_a+z)\,\omega\ \text{and}\ \upsilon(90°-\theta_b+z)\,\omega$$

We have from these

$$\upsilon(90°-\theta_a+z)\,\omega-\cos\theta_a=\upsilon(90°-\theta_b+z)-\cos\theta$$

from which follows

$$\omega = \frac{\cos\theta_a - \cos\theta_b}{\upsilon(\theta_b - \theta_a)}$$

Further, equate both of these errors to the greatest error which occurs between A and B, supposing in the first instance that it occurs at the point X half way from A to B. The latitude of X is

$$= \frac{\theta_a + \theta_b}{2}$$

The error there is

$$= -\left[\upsilon\left(90^\circ - \frac{\theta_a + \theta_b}{2} - z\right)\omega - \cos\frac{\theta_a + \theta_b}{2} \right]$$

its sign being opposite to the signs of the errors at A and B. The condition is now expressed by the two equations

$$\upsilon(90^\circ - \theta_a + z)\,\omega - \cos\theta_a = \cos\frac{\theta_a + \theta_b}{2} - \upsilon\left(90^\circ - \frac{\theta_a + \theta_b}{2} - z\right)\omega$$

$$\upsilon(90^\circ - \theta_b + z)\,\omega - \cos\theta_b = \cos\frac{\theta_a + \theta_b}{2} - \upsilon\left(90^\circ - \frac{\theta_a + \theta_b}{2} - z\right)\omega \, .$$

Giving ω its value

$$\frac{\cos\theta_a - \cos\theta_b}{\upsilon(\theta_b - \theta_a)}$$

we find readily

$$\frac{(180^\circ - \frac{3}{2}\theta_a - \frac{1}{2}\theta_b + 2z)(\cos\theta_a - \cos\theta_b)}{\theta_b - \theta_a} = \cos\theta_a + \cos\frac{\theta_a + \theta_b}{2}$$

which reduces to

$$(180^\circ - \tfrac{3}{2}\theta_a - \tfrac{1}{2}\theta_b + 2z) = \frac{\theta_b - \theta_a}{\cos\theta_a - \cos\theta_b}\left[\cos\theta_a + \cos\frac{\theta_a + \theta_b}{2}\right]$$

from which z is readily found. Applying this to the construction of a map of Russia, it is only necessary to write

$$\theta_a = 40^\circ \qquad \theta_b = 70^\circ \qquad \frac{\theta_a + \theta_b}{2} = 55^\circ$$

The formula for ω gives now at once

$$\omega = \frac{\cos 40^\circ - \cos 70^\circ}{30\,\upsilon} = 48' \, 44''$$

The equation

$$(180^\circ - \tfrac{3}{2}\theta_a - \tfrac{1}{2}\theta_b + 2z)\,\upsilon\omega = \cos\theta_a + \cos\frac{\theta_a + \theta_b}{2}$$

gives now

$$(85^\circ - 2z)\,\upsilon\omega = 1.33962$$

Now

$$\upsilon\omega = 0.0141$$

therefore

$$2z = \frac{1.33962}{0.0141} - 85^\circ = 10^\circ$$

or

$$z = 5^\circ$$

So far we have assumed that the maximum error lay at the middle of AB, but we will now find the correct point, and assume that for this place the latitude is θ; the error will now be

$$\upsilon(90^\circ - \theta + z)\,\omega - \cos\theta$$

Differentiating this with respect to θ and equating to zero we find for the position of maximum error

$$\sin\theta = \omega = 0.8098270$$

or

$$\theta = 54^\circ \, 4'$$

Equating the error at θ to those of A and B

$$v(180^\circ-\theta_a-\theta+2z)=\cos\theta_a+\cos\theta$$

from which

$$z=5^\circ\ 0'\ 30''$$

The values of z and θ differ very little from their assumed values of 5° and 55° respectively. The errors at A and B are then equal to

$$v(90^\circ-\theta_a+z)-\cos\theta_a=0.00946$$

A degree on the parallel of 40° is then expressed by 0.77550 instead of 0.76604, its true value upon the sphere. This degree is, then, about $\frac{1}{81}$ greater than the true degree on the parallel of 40°; and the degree on the parallel of 70° is about $\frac{1}{34}$ too great, its true value being 0.34202.

MURDOCH'S PROJECTION.

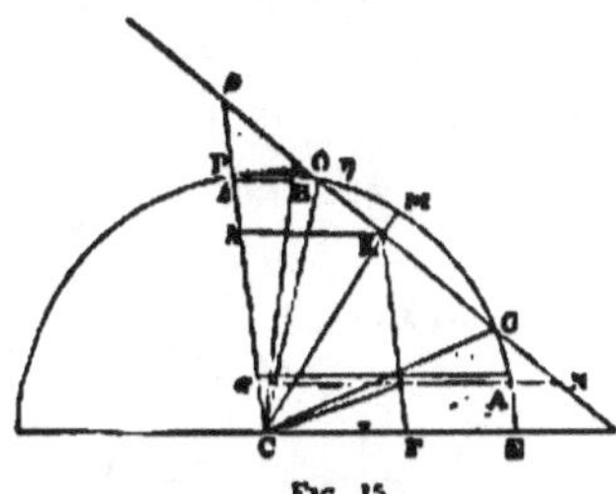

FIG. 15.

In Fig. 15, let θ_a and θ_b denote the latitude of two extreme parallels Aa and Bb, which limit a spherical zone whose projection is to be determined. The latitude of M half way between A and B is $\frac{\theta_a+\theta_b}{2}$. Murdoch's projection consists in making the entire area of the chart equal to the entire area of the zone to be projected. In order to effect this it will be necessary, supposing PN and PO the radii of the extreme parallels of the chart (obtained by rectification), that the surface generated by the revolution of ON (=AB) about PC shall be $=2\pi r(ab)$, when r = radius of the sphere expressed in degrees. Let δ denote the equal angles yCM, zCM; we must then have

$$2\pi Kk\,AB=2\pi r(ab)$$

From the similar triangles Kck and MFC, give

$$\frac{Kk}{FO}=\frac{KC}{MC}$$

Consequently

$$Kk=r\cos\frac{\theta_a+\theta_b}{2}\cos\delta$$

and substituting this in the above equation

$$\frac{\theta_b-\theta_a}{2}\cos\frac{\theta_a+\theta_b}{2}\cos\delta=\sin\theta_b-\sin\theta_a=2\sin\frac{\theta_b-\theta_a}{2}\cos\frac{\theta_a+\theta_b}{2}$$

This gives for $\cos\delta$ the value

$$\cos\delta=\frac{\sin\frac{\theta_b-\theta_a}{2}}{\frac{\theta_b-\theta_a}{2}}$$

It is easy to see that for the radius $Kp=R$ of the middle parallel we have

$$Kp=r\cos\frac{\theta_s+\theta_n}{2}\frac{\cos\delta}{\sin\frac{\theta_s+\theta_n}{2}} \qquad R=r\cot\frac{\theta_s+\theta_n}{2}\cos\delta$$

The quantities which we have already denoted by z and V are here connected by the relation

$$V=z\sin\frac{\theta_s+\theta_n}{2}$$

Murdoch, in order to draw the intermediate parallels, divided the right line dn into equal parts, giving for the radius of any parallel θ

$$R+\frac{\theta_s+\theta_n}{2}=\theta$$

This method, although perfectly arbitrary, had the effect of diminishing the errors in the chart.

Mayer, who resumed the problem proposed by Murdoch, gave the radii $p\zeta$ and $p\eta$ as

$$p\eta=pk-K\eta \qquad p\zeta=pK+K\zeta$$

and since

$$K\eta=K\zeta=r\sin\delta$$

$$p\eta=R-r\sin\delta=r\frac{\cos\left(\frac{\theta_s+\theta_n}{2}+\delta\right)}{\sin\frac{\theta_s+\theta_n}{2}}$$

$$p\zeta=R+r\sin\delta=r,\frac{\cos\left(\frac{\theta_s+\theta_n}{2}-\delta\right)}{\sin\frac{\theta_s+\theta_n}{2}}$$

A second method of projection was given by Murdoch, in which the eye is placed at the center of the sphere, as in gnomonic projection, and a perspective is made which is subject to the condition of preserving the entire surface of the zone which is to be represented. Lambert was the first to indicate a method of conic development which should preserve all the angles except the one at the vertex of the cone, when the 360° having upon the sphere the pole for center will obviously be represented in different manners, according to the different conditions to be fulfilled. A full account of this method is given in the chapter on orthomorphic projections.

BONNE'S PROJECTION.

This method of projection is that which has been almost universally employed for the detailed topographical maps based on the detailed trigonometrical surveys of the several states of Europe. It was originated by Bonne, was thoroughly investigated by Henry and Puissant in connection with the map of France, and tables for France were computed by Plessen. In constructing a map on this projection a central meridian and a central parallel are first assumed. A cone tangent along the central parallel is then assumed, and the central meridian developed along that element of the cone which is tangent to it, and the cone is then developed on a tangent plane. The parallel falls into an arc of a circle with its center at the vertex, and the meridian becomes a graduated right line. Concentric circles are then conceived to be traced through points of this meridian at elementary distances along its length. The zones of the sphere lying between the parallels through these points are next conceived to be developed, each between its corresponding arcs. Thus all the parallel zones of the sphere are rolled out on a plane in their true relations to each other and to the central meridian, each having in projection the same width, length, and relation to the neighboring zones as on the spheroidal surface. As there are no openings between consecutive developed elements, the total area is unaltered by the development. Each meridian of the projection is so traced as to cut each parallel in the same point in which it intersected it on the sphere.

If the case in hand be that involving the greatest extension of the method, or that of the projection of the entire spheroidal surface, a prime or central meridian must first be chosen, one-half of which gives the central straight line of the development, and the other half cuts the zones apart and becomes the outer boundary of the total developed figure. Next the latitude of the governing parallel must be assumed, thus fixing the center of all the concentric circles of development. Having then drawn a straight line and graduated it from 90° north latitude to 90° south latitude, and having fixed the vertex or center of development on it, concentric arcs are drawn from the center through the different graduations. There results from this process an oblong kidney-shaped figure, which represents the entire earth's surface, and the boundary of which is the double-developed lower half of the meridian first assumed. This projection preserves in all cases the areas developed, without any change. The meridians intersect the central parallel at right angles, and along this, as along the central meridian, the map is strictly correct. For moderate areas the intersections approach tolerably to being rectangular. All distances along parallels are correct, but distances along the meridians are increased in projection in the same ratio as the cosines of the angle between the radius of the parallel and the tangent to the meridian at the point of intersection are diminished. Thus, in a full earth projection the bounding meridian is elongated to about twice its original length. While each quadrilateral of the map preserves its area unchanged, its two diagonals become unequal; one increasing and the other decreasing in receding towards the corners of the map, the greatest inequality being towards the east and west polar corners.

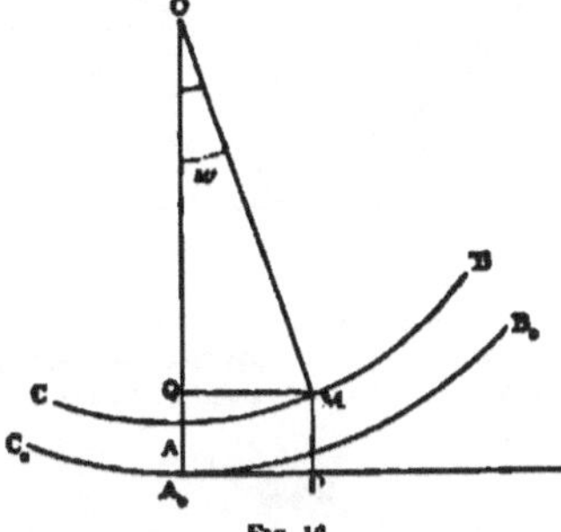

Fig. 16.

Denote the radius of the central parallel by ρ_0; then

$$OA_0=\rho_0=r\cot\theta_0$$

Denote by s the length of the arc AA_0, and the arc passing through a given point M; θ_0 of course denotes the latitude of the central parallel, and θ that of the parallel BC. The latitude of M is $=\theta_0+\frac{s}{r}$, and thus

$$MA=\rho\omega=\omega r\cos\left(\theta_0+\frac{s}{r}\right) \qquad \rho=\rho_0-s=r\cot\theta_0-s$$

$$x=MQ=\rho\sin\omega \qquad y=MP=r\cot\theta_0-\rho\cos\omega$$

It is not difficult in this projection to take account of the spheroidal form of the earth. It is only necessary to multiply cot θ_0 by the principal normal n_0, and replace the spherical arc s by the elliptic arc S, given by

$$S=a(1-e^2)[A(\theta-\theta_0)-B\sin(\theta-\theta_0)\cos(\theta+\theta_0)+\tfrac{1}{2}C\sin 2(\theta-\theta_0)\cos 2(\theta+\theta_0)]$$

Then

$$\rho_0 = \frac{a \cot \theta_0}{(1-e^2 \sin^2 \theta_0)^{\frac{1}{2}}} = n_0 \cot \theta_0 \qquad \rho = n_0 \cot \theta - S \qquad \omega = \frac{n\theta_0}{\rho} \cos\left(\theta_0 + \frac{\delta}{a}\right)$$

These give the radii of the projections of the parallels, which are then readily constructed. Lay off from the central meridian, upon the parallels now constructed, lengths equal to one degree upon each different parallel, and through these points pass a curve, which will be the projection of the meridians. The lengths are given by the formula

$$\delta = \frac{2\pi}{360} n \cos \theta = \frac{\pi a \cos \theta}{180\,(1-e^2 \sin^2 \theta)^{\frac{1}{2}}}$$

The concave parts of these curves are all turned towards the central meridian.

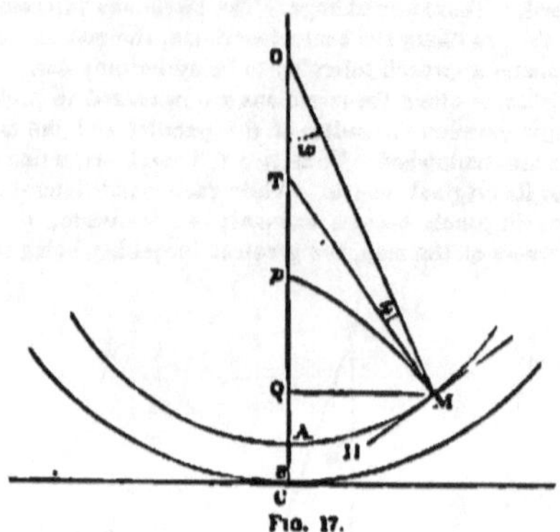

FIG. 17.

The angle x, in Fig. 17, is the angle which the tangent to the meridian at M makes with the radius OM of the parallel through that point. This angle is also the difference between the angle that the meridian makes with the parallel at this point and 90°.

We have obviously

$$\tan x = \frac{\rho d\omega}{d\rho}$$

but

$$\rho = \rho_0 + S$$

therefore

$$d\rho = ds \qquad \tan x = \frac{\rho d\omega}{ds}$$

Now

$$\rho\omega = n\omega \cos \theta = \frac{a\omega \cos \theta}{(1-e^2 \sin^2 \theta)^{\frac{1}{2}}}$$

Differentiating this gives

$$\rho d\omega + \omega d\rho = -\frac{a\omega \sin \theta\,(1-e^2)\, d\theta}{(1-e^2 \sin^2 \theta)^{\frac{3}{2}}}.$$

and we have

$$\rho d\omega + \omega d\rho = \rho d\omega + \omega ds$$

But we know that

$$ds = \frac{a(1-e^2)\, d\theta}{(1-e^2 \sin^2 \theta)^{\frac{3}{2}}}$$

Consequently we may write

$$\frac{\rho d\omega}{ds} + \omega = -\omega \sin \theta$$

and

$$\tan x = \omega \sin \theta - \omega$$

For $\theta = \theta_0$

$$\omega = \frac{a\omega \cos \theta_0}{\rho_1(1-e^2 \sin^2 \theta_0)^{\frac{1}{2}}} \qquad \rho_1 = \frac{a \cot \theta_0}{(1-e^2 \sin^2 \theta_0)^{\frac{1}{2}}}$$

Combining these

$$\omega = \omega \sin \theta_0$$

and for this case $\tan x = 0$ or $x = 0$, which only shows what we already know, viz: that the meridians and central parallel cut at right angles.

If for the central parallel we assume the equator, the vertex of the tangent cone is removed to an infinite distance, the parallels all fall into straight lines, and we have the so-called Flamsteed's projection. The kidney-shaped Bonne becomes an elongated oval with the half meridian for one axis and the whole equator for the other. The co-ordinates for any point in this projection are readily found to be

$$\zeta = \frac{\pi}{180} a\theta \qquad x = \frac{\omega\pi}{360} a \cos \theta = K \cos \frac{\zeta}{a}$$

The form of the equation giving x has induced M. d'Avezac to give this projection the name *sinusoidal.*

This projection, which should really be called Sanson's projection, is evidently only a particular case of Bonne's method; it is based upon a resolution of the earth's surface into zones or rings by parallels of latitude taken at successive elementary distances laid off along the central meridian of the area to be projected. Having developed this center meridian on a straight line of the plane of projection, a series of perpendiculars is conceived to be erected at the elementary distances along this line. Between these perpendiculars the elementary zones are conceived to be developed in the correct relations to each other and the center meridian. Each zone being of uniform width occupies a constant length along its entire developed length, and consequently the area of the plane projection is exactly equal to that of the spheroidal surface thus developed. The meridians of the developed spheroid are traced through the same points of the parallels in which they before intersected them. They all cut the parallels obliquely and are concave towards the centre meridian. Thus while each quadrilateral between parallels and meridians contains the same area and points after development as before, the form of the configuration is considerably distorted in receding from the central meridian, and the obliquity of the intersections between parallels and meridians grows to be highly unnatural.

WERNER'S EQUIVALENT PROJECTION.

If the vertex of the cone approaches the sphere instead of receding from it, as in the preceding case, we have finally, when the tangent cone becomes a tangent plane, the projection known as Werner's Equivalent Projection. The parallels are now arcs of circles described about the pole as a center and with radii equal to their actual distances from the pole, *i. e.*, equal to the rectified arc of the colatitudes. The meridians are drawn by laying off on the parallels the actual distances between the meridians as they intersect the parallels on the sphere. This projection is not of enough importance to spend any time in obtaining any of the formulas connected with it.

POLYCONIC PROJECTIONS.

In all the cases of conic projection that we have treated so far we have supposed that a narrow zone of the earth was to be projected and that for the zone was substituted a developable surface upon which the parallels and meridians were constructed according to any manner that may be desirable. We have seen that this kind of projection is only available when but a small portion of the earth is to be represented and that to make a projection of a country of great extent in latitude some modification would be necessary.

The system which is used in America and in England replaces each narrow zone of the earth's surface by the corresponding conic zone in such a way as to preserve the orthogonality of the meridians and parallels. This is the projection of which we have already spoken at length in the introduction, under the title of Polyconic Projection. As a very full account of this system has been already given, and comparisons made with the other ordinary methods of projection, we will not say anything on the subject here, but will proceed to develop the theory of the system.

The name *rectangular* polyconic projection is applied to the method in which each parallel of the spheroid is developed symmetrically from an assumed central meridian by means of the cone tangent along its circumference. Supposing each element thus developed relative to the common central meridian, it is evident that a projection results in which all parallels and meridians intersect at right angles. The parallels will be projected in circles, and the meridians in curves which cut those circles at right angles. The radii of the parallels are equal to the cotangents of their latitudes (to radius supposed unity), and the centers are upon the line which has been chosen as the central meridian. Along this meridian the parallels preserve the same distance as they do upon the sphere.

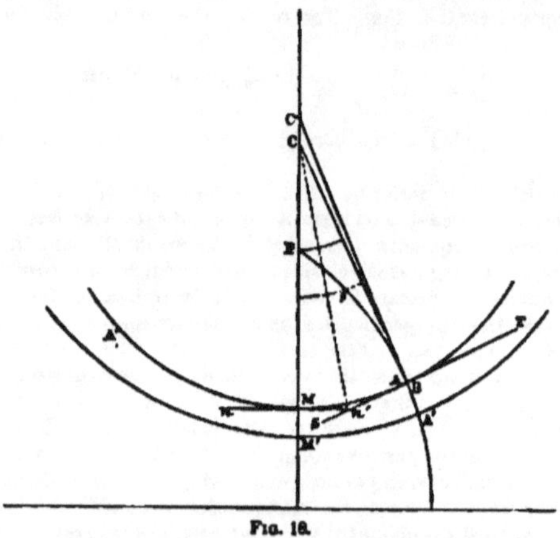

FIG. 18.

In Fig. 18 let M be any point of the central meridian of which the latitude is $\theta = 90^\circ - u$; P the pole, the arc $PM = ru$. The center of the parallel through the point M is given by $CM = r \tan u$. If M' be a point infinitely near to M (*i. e.*, $MM' = r du$) and C' the center of the corresponding circle, we have $C'M' = r \tan (u + du)$, or

$$\rho = r \tan u \qquad \rho + d\rho = r \tan (u + du)$$

Expanding the second of these, we have

$$d\rho = r \sec^2 u du$$

but

$$d\rho = CC' + MM' = CC' + r du$$

Therefore

$$CC' = r \tan^2 u du$$

We have from the triangle CC'B

$$\frac{\sin \varphi}{\sin B} = \frac{C'B}{CC'}$$

or

$$-\frac{d\varphi}{\sin \varphi} = \tan u du$$

and integrating

$$\log \cos u = \log \tan \frac{\varphi}{2} + \text{const.}$$

or passing to exponentials

$$\tan\frac{\varphi}{2}=c\cos u$$

Now

$$\tan u=\frac{\rho}{r}$$

therefore

$$\cos u=\frac{r}{\sqrt{r^2+\rho^2}}$$

Substituting this in the equation for the meridians we have

$$\frac{r\sqrt{c^2-\tan^2\frac{\varphi}{2}}}{\tan\frac{\varphi}{2}}=\rho$$

or

$$\rho=r\sqrt{c^2\cot^2\frac{\varphi}{2}-1}=\frac{r}{\sin\frac{\varphi}{2}}\sqrt{c^2-\sin^2\frac{\varphi}{2}(1+c^2)}$$

The distance from any point A to the central meridian is $=\rho\sin\varphi$ or $=r\tan u\sin\varphi$; but

$$r\tan u\sin\varphi=2c\frac{r\sin u}{1+c^2\cos^2 u}$$

For $u=90°$, or, at the equator, this becomes

$$=2cr$$

The constant c must then represent one-half the longitude of the given meridian, the equator being developed in its true length and divided into equal parts in the same manner as the central meridian. The following construction for this projection is due to Mr. O'Farrell, of the topographical department of the War Office, England. All data being as already given, draw at M the tangent nn' perpendicular to PM. In order to determine the point A, whose longitude is given as ω, lay off from M the lengths $Mn=Mn'$ equal to the true length of the required arc on the parallel θ, *i. e.*, = the arc $\frac{\omega}{2}$ described with a radius $=r\sin u$. With n and n' as centers and $n'C$ and nC as radii, draw arcs cutting the given parallel in the points A and A',

$$Mn=\frac{\omega}{2}\sin u=c\sin u$$

and, since

$$CM=r\tan u$$

we have

$$\tan MCn=c\cos u=\tan\frac{\varphi}{2}$$

or, finally,

$$ACM=\varphi$$

and the distance from A to the central meridian is

$$=r\tan u\sin\varphi$$

The radius of the curvature of the meridian whose longitude is ω is readily obtained. We have

$$AA'=ds \qquad CC'=r\tan^2 u\,du$$

Now

$$ds=r(\sec^2 u-\tan^2 u\cos\varphi)du$$

also

$$\sec^2\frac{\varphi}{2}\frac{d\varphi}{2}=-c\sin u\,du$$

therefore, if ρ denote the radius of curvature of the meridian, we have by easy reductions

$$\rho=r\frac{1+c^2+c^2\sin^2 u}{2c\sin u}$$

Now consider the distortion in this case, and for this purpose imagine a small square described on the sphere, having its sides parallel and perpendicular to the meridian. Let u and ω $(=2c)$ define its position, and let s be the length of the side. If we differentiate the equation $\tan\frac{\varphi}{2}=c\cos u$ on the supposition that u is constant, we have

$$\sec^2\varphi\, d\varphi=\cos u\, dc$$

also, the length of the representation of $2dc$ is $\tan u\, d\varphi$, or

$$\sin^2 u\cos^2\frac{\varphi}{2}\, d2c$$

Hence that side of the square which is parallel to the equator will be represented by a line equal to

$$s\cos^2\frac{\varphi}{2}$$

Similarly the meridian side will be represented by

$$s\cos^2\frac{\varphi}{2}(1+c^2+c^2\sin^2 u)$$

The square is therefore represented by a rectangle whose sides have the ratio

$$1+c^2+c^2\sin^2 u : 1$$

and its area is increased in the ratio

$$\frac{1+c^2+c^2\sin^2 u}{(1+c^2\cos^2 u)^2} : 1$$

If we make this ratio = unity, then results the equation

$$c^4\cos^4 u+3c^2\cos^2 u-2c^2=0$$

which is satisfied either by $c=0$, *i. e.*, $\omega=0$, or by

$$c^2\cos^4 u+3\cos^2 u-2=0$$

We see from this that there is no exaggeration of area along the meridian or along the curve given by the last equation. This curve crosses the central meridian at right angles in the latitude of about 54° 44′; it thence slowly inclines southward, and at 90° of longitude from the central meridian reaches 50° 20′ of longitude; at 180°, or the opposite meridian, it has reached 43° 40′. The areas of all tracts of countries lying on the north side of this curve will be diminished in the representation, and for all tracts of countries south of this curve the areas will be increased in the representation.

If we represent the whole surface of the globe continuously, the area of the representation is

$$r^2\left[(4+\pi^2)\tan^{-1}\frac{\pi}{2}+2\pi\right]$$

which is greater than the true surface of the globe in the ratio 8 : 5.

The perimeter of the representation is equal to the perimeter of the globe multiplied by $\sqrt{4+\pi^2}-1$, or 2.72.

It is desirable in certain cases to retain the lengths of the degrees on all the parallels at the sacrifice of their perpendicularity to the meridians. We thus obtain what is known as the ordinary polyconic projection, which applied to the representation of the entire surface of the globe gives a figure with two rectangular axes and from equal quadrants as in the rectangular polyconic projection. The central meridian alone is perpendicular to the parallels and is developed in its true length; upon each parallel described with the cotangent of its latitude as a radius we lay off the true lengths of the degrees of longitude and draw through the corresponding points so obtained curves which will be the projections of the meridians. The ordinary polyconic method has been adopted by the United States Coast Survey because its operations being in great part limited to a narrow belt along the seaboard, and not being intended to furnish a map of the country in regular uniform sheets, it is preferred to make an independent projection for each plane table and hydrographic sheet, by means of its own central meridian.

The method of projection in common use in the Coast Survey Office for small areas, such as those of plane-table and hydrographic sheets, is called the equidistant polyconic projection. This is to be regarded as a convenient graphic approximation, admissible within certain limits, rather than as a distinct projection, though it is capable of being extended to the largest areas and with results quite peculiar to itself. In constructing such a projection a central meridian and a central parallel are chosen, and they are constructed as in the rectangular polyconic method. The top or bottom parallel and a sufficient number of intermediate parallels are constructed by means of the tables prepared for the purpose, and the points of intersection of the different meridians with these parallels are then found and the meridians drawn.

Then starting from the central parallel the distance to the next parallel is taken from the central meridian and laid off on each other meridian. A parallel is traced through the points thus found. Each parallel is constructed by laying off equal distances on the meridians in like manner, and the tabular auxiliary parallels are, all except the central one, erased. In fact, as only the points of intersection are required, the auxiliary parallels should not be actually drawn. From this process of construction results a projection in which equal meridian distances are intercepted everywhere between the same parallels.

CYLINDRIC PROJECTION.

So far, in treating of projection by development of some auxiliary surface, we have confined ourselves to the case of intersecting or tangent cones. The next most natural case to consider is when the developable surface is a cylinder. We cannot obviously, as in the case of the cone, pass one cylinder through the upper and lower parallels of a spherical zone, so that we cannot here have more than one of these parallels developed in its true size; if the zone is above the equator, the lower parallel may be developed in its true size by circumscribing a cylinder, and the upper parallel may be represented in its true size by inscribing a cylinder. The better plan, however, and one which in general reduces distortion, is to pass the cylinder through some parallel intermediate between the extreme parallels of the zone to be projected. We will, however, first consider the case where the cylinder is tangent to the sphere either along the equator or along a meridian.

THE SQUARE PROJECTION.

The simplest, but rude, method is one in which the cylinder being tangent along the equator, the meridians and parallels appear as equidistant parallel straight lines, forming squares. Degrees of latitude and longitude are here all supposed equal in length. Distances and areas, especially in an east-and-west direction, are grossly exaggerated, though for an elementary surface the true proportions of a figure are preserved. This method is occasionally used for representing small surfaces near the equator.

PROJECTIONS WITH CONVERGING MERIDIANS.

This is a modification of the square projection designed to conform nearly to the condition that arcs of longitude shall appear proportional to the cosines of their respective latitudes. The straight line representing the central meridian being properly graduated, that is, the true length of an arc of a degree of latitude (or of a minute or multiple thereof, as the case may be) having been laid off according to the scale adopted, two straight lines are drawn at right angles to the meridian to represent parallels, one near the bottom and the other near the top of the chart. These parallels are next graduated, the arcs representing degrees (multiples or subdivisions) of longitude on each having, by scale, the true length belonging to the latitude. The corresponding points of equal nominal angular distance from the middle meridian thus marked on the parallels, when connected by straight lines, will produce the system of convergent meridians. The disadvantages of this projection are in the facts that but two of the parallels exhibit the lengths of arcs of longitude in their true proportion and that the central meridian is the only one which cuts the parallels at right angles. The projection is suitable for the projection of tolerably large areas, the above defects not being of a serious nature within ordinary limits; it also recommends itself by the ease with which points can be projected or taken off the chart by means of latitude and longitude.

THE RECTANGULAR PROJECTION.

A less defective method of delineation than the square projection consists in presenting the lengths of degrees of longitude along the *middle* parallel of the chart in their true relation to the corresponding degrees on the sphere; they will therefore appear smaller than the degrees of latitude in the proportion $1 : \cos \theta_1$. In an east-and-west direction the chart is unduly expanded above and unduly contracted below the middle parallel.

THE RECTANGULAR EQUAL-SURFACE PROJECTION.

This differs from the last in that the distances of the parallels, instead of being equal, are now drawn parallel to the equator at distances proportional to the sine of the latitude. This gives it the distinctive property of having the areas of rectangles or zones on the projection proportional to the areas of the corresponding figures on the sphere. The distortion, however, becomes quite excessive in the higher latitudes.

CASSINI'S PROJECTION.

This projection makes no use of the parallels of latitude, but substitutes for them a second system of co-ordinates, viz, one at right angles to the principal or central meridian; it is consequently convenient in connection with rectangular spherical co-ordinates having their origin in the middle of the chart; the projection of Cassini's chart of France consisted of squares and had neither meridians (excepting one) nor parallels. This simple form is, however, not the one which is generally known under Cassini's name. In the projection commonly called Cassini's the cylinder is tangent along a meridian; through the different points of division of the equator, planes are passed parallel to the plane of this meridian; and through the points of division of the meridian planes are passed intersecting the plane of the equator in a common diameter of the sphere. The first system of planes, of course, cuts small circles, and the second great circles, from the sphere. The cylinder is now developed, the generatives passing through the points of division of the meridian representing the great circles perpendicular to this meridian, while the small circles which are parallel to it have for their projection the development of these intersections of the cylinder with their planes.

This projection is not now employed, as it offers no facilities for platting positions by latitude and longitude; moreover, the distortion rapidly increases with the distance from the central meridian of the chart.

We will now obtain formulas which will enable us to find the forms of the projections of the parallels and meridians. In Fig. 10, M denotes the center of the sphere of radius $MA=r$; P is an arbitrary point in the surface, for which we have

$$EP=AD=\theta \qquad DP=AE=\omega$$

AB denotes a quadrant of the equator and AQ a quadrant of the first meridian. The determination of the position of P in the case of Cassini's projection is effected by means of the great circle passing through B and P, and the circle GH whose plane is parallel to that of the first meridian AQ; write

$$FP=AG=\theta_1 \qquad GP=AF=\omega_1$$

We have now, in the right triangle MPN,

$$PN = r \sin \theta$$

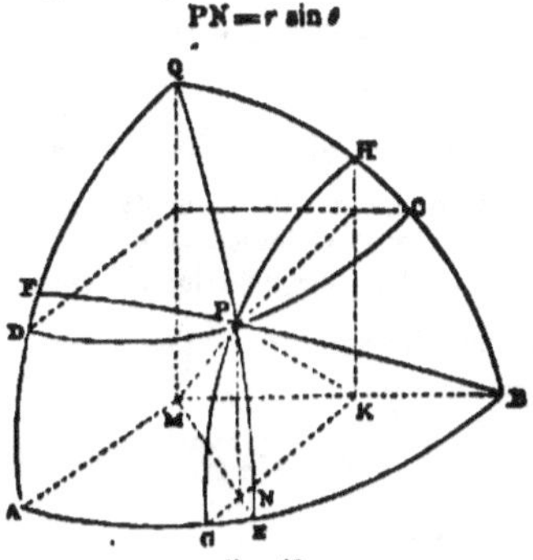

Fig. 19.

and from the right triangle PKN

$$PN = PK \sin \omega_1$$

but from the triangle MPK we have

$$KP = MP \sin BP = r \cos \theta_1$$

and consequently

$$PN = r \cos \theta_1 \sin \omega_1$$

Equating the two values of PN

$$\sin \theta = \cos \theta_1 \sin \omega_1$$

and in like manner

$$\sin \theta_1 = \cos \theta \sin \omega$$

From these two equations we obtain readily

$$\cot \omega = \cos \omega_1 \cot \theta_1 \qquad \cot \omega_1 = \cos \omega \cot \theta$$

and we also have the formulas

$$\sin \theta_1 = \cos \theta \sin \omega \qquad \sin \theta = \cos \theta_1 \sin \omega_1$$

$$\cot \omega_1 = \cot \theta \cos \omega \qquad \cot \omega = \cot \theta_1 \cos \omega$$

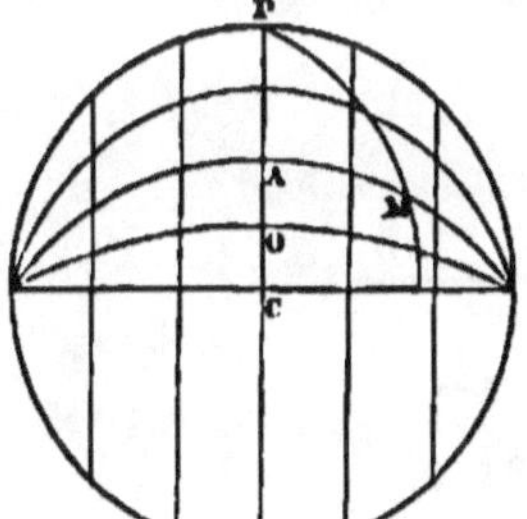

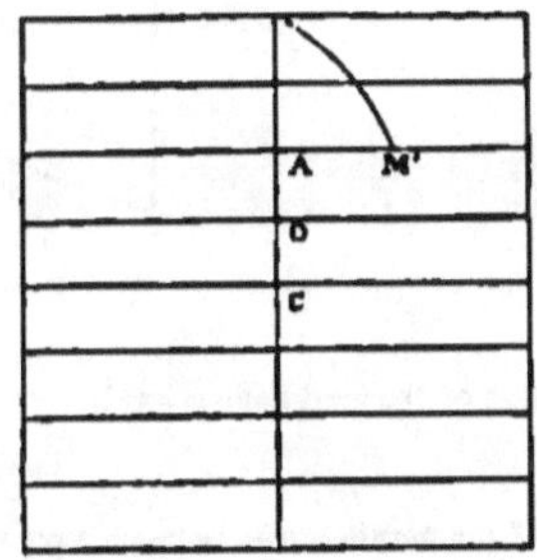

Fig. 20.

Let now, in Cassini's projection (Fig. 20), O, O' denote the center of co-ordinates, $\eta = OA = O'A'$, $\xi = AM = A'M'$; also call θ_0 the latitude of O; then $\theta_0 + \eta$ is the quantity denoted by ω, in the preceding formulas, and ξ is identical with θ, so we have

$$\sin \xi = \cos \theta \sin \omega \qquad \cot (\theta_0 + \eta) = \cos \omega \cot \theta$$

By elimination of ω from these equations we have for the equation of the projections of the parallels

$$\sin^2\xi+\cot^2(\theta_0+\eta)\sin^2\theta=\cos^2\theta$$

and by elimination of θ have for the meridians

$$[\cos^2\omega+\cot^2(\theta_0+\eta)][\sin^2\omega-\sin^2\xi]=\sin^2\omega\cos^2\omega$$

Both meridians and parallels are thus given in this projection as transcendental curves. In these last equations we have regarded the radius of the sphere as $=1$; now make the radius $=r$; then for ξ and η we must write $\frac{\xi}{r}$ and $\frac{\eta}{r}$. When the projection only represents a narrow region included between two meridians and two parallels very near together, the ratios $\frac{\xi}{r}$ and $\frac{\eta}{r}$ are very small and so is the difference $\theta-\theta_0$; so we can write

$$\sin\xi=\frac{\xi}{r}\qquad\qquad\cos\xi=1-\frac{1}{2}\frac{\xi^2}{r^2}\qquad\qquad\tan\eta=\frac{\eta}{r}$$

and the equation of the parallels becomes

$$(\eta-r\cot\theta_0)^2+\xi^2=r^2\cot^2\theta_0[\cot\theta_0+4\sin\tfrac{1}{2}(\theta-\theta_0)]$$

and that of the meridians

$$(\eta+r\tan\theta_0)^2+\frac{2r}{\cos\theta_0\tan\omega}\xi=r^2(2+\tan^2\theta_0)$$

We see that in this case the meridians are projected in parabola and the parallels in circles.

Write for convenience $\theta_0+\eta=\lambda$, and let the angle χ, in Fig. 21, denote the angle which the tangent PM to the projection of a meridian makes with the axis η; also, let χ' denote the angle which the tangent PL to the projection of a parallel makes with the same axis. Then we have

$$\tan\chi=\frac{d\xi}{d\lambda}$$

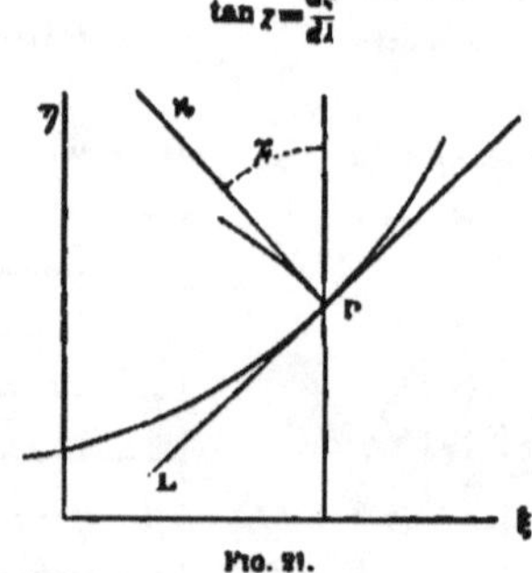

FIG. 21.

The equation of the meridians is easily thrown into the form

$$\tan\xi=\cos\lambda\tan\omega$$

and that of the parallels also becomes very readily

$$\sin\theta=\sin\lambda\cos\xi$$

For these we may substitute in practice the group

$$\cot\lambda=\cot\theta\cos\omega\qquad\qquad\sin\xi=\cos\theta\sin\omega$$

We now have

$$\frac{d\xi}{d\lambda}=-\frac{\tan\omega\sin\lambda}{1+\cos^2\lambda\tan^2\omega}=\tan\chi$$

and also

$$\tan\chi=-\tan\omega\cos\xi\sin\theta$$

Now, from the equation of the parallels, we have

$$\frac{d\xi}{d\lambda}=\frac{1}{\tan\lambda\tan\xi}=\tan\chi'$$

or, since $\tan\xi=\cos\lambda\tan\omega$

$$\tan\chi'=\frac{1}{\sin\lambda\tan\omega}=\frac{\cos\xi\cot\omega}{\sin\theta}$$

Combining these values of $\tan\chi$ and $\tan\chi'$ by multiplication, there results

$$\tan\chi\tan\chi'=\cos^2\xi$$

The condition that the projections of the meridians and parallels should cut at right angles is

$$\tan\chi\tan\chi'=-1$$

So it is clear that in general in Cassini's projection the meridians and parallels are not represented by orthogonal curves. For $\xi=0$ we have

$$\tan\chi\tan\chi'=1$$

or the projections of all parallels are perpendicular to the central meridian. If $\lambda=90^\circ$ we have $\chi=\omega$, or the projections of the meridians make the same angles with each other as the meridians themselves. From the equations for the meridians and parallels obtain the values of $\frac{d\eta}{d\xi}$ and $\frac{d^2\eta}{d\xi^2}$, and substitute each set of values in the formula

$$\rho=\frac{\left(1+\left(\frac{d\eta}{d\xi}\right)^2\right)^{\frac{3}{2}}}{\frac{d^2\eta}{d\xi^2}}$$

and by very simple reductions we find for the radius of curvature of the projections of the meridians

$$\rho_m=\frac{r\sec\chi}{\sin\chi\left[\cot\lambda\cos\chi+2\tan\xi\sin\chi\right]}\,.$$

and for the parallels

$$\rho_p=\frac{r\sin^3\lambda\sin^2\xi}{\cos^3\chi'\cot\xi(\sin^2\xi+\cot\lambda)}$$

For the case of $\chi=0$ and $\lambda=0$, or the point where a meridian cuts the equator, the expression for ρ_m becomes indeterminate, but by the ordinary means for finding the value of indeterminate quantities ρ_m is found to be for this point

$$=\frac{2r}{\sin 2\omega}$$

The radius ρ_p becomes infinite for the same latitude, *i. e.*, for $\lambda=0$, but is indeterminate for the points at which $\xi=0$; one has for these points

$$\rho_p=\frac{r}{\tan\lambda}$$

or

$$\rho_p=\frac{r}{\tan\theta}$$

MERCATOR'S PROJECTION.

If a cylinder be passed tangent to the sphere at the equator, and the planes of the meridians and parallels be produced to cut the surface of the cylinder, the meridians will be represented upon the cylinder by right lines and the parallels by circles, the right sections of the cylinder. If the cylinder be developed, we obtain a projection in which both meridians and parallels are represented by right lines, the angles between the lines being right angles.

It will be convenient here to define a loxodromic curve, or simply a loxodromic; this is a line drawn upon the surface of the sphere in such a manner as to cut all the meridians at the same angle. Any straight line drawn up on a chart constructed as above will of course represent a loxodromic upon the sphere. This projection has already been alluded to under the head of the square projection; its disadvantages are obviously very great, only east-and-west and north-and-south directions being preserved, and degrees of longitude only preserving their true length upon the line of contact of the cylinder and sphere, *i. e.*, upon the equator.

Reduced charts, or Mercator's charts, are charts whose construction is such that not only are the meridians given as right lines, a necessary condition that the loxodromic curve may be represented by a right line, but so that the angle between any two curvilinear elements upon the sphere is represented upon the chart by an equal angle between the representatives of these elements. This is effected by a proper spacing of the distances between the parallels, which are also represented as right lines upon the chart. We will now give the means of determining the proper position of any parallel upon the chart by observing the condition that the angles formed by two curvilinear elements upon the sphere shall be preserved upon the chart. Let ab, Fig. 22, denote

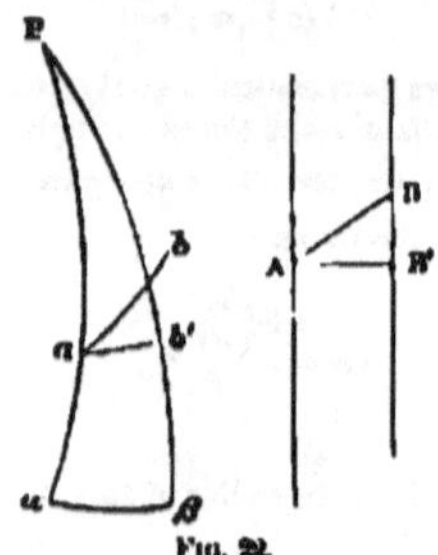

FIG. 22.

an element of a loxodromic cutting the two infinitely near meridians Pa and Pb; draw the parallel ab'. Now, in order that the angle ABB′ upon the chart shall $=abb'$ upon the sphere, we must have

$$\frac{\mathrm{BB}'}{bb'}=\frac{\mathrm{AB}'}{ab'}$$

Now, as the distance between any two meridians is everywhere equal to the distance between the same meridians at the equator, AB is $=\alpha\beta$, the element of the equator. Let ds represent an element of a meridian upon the earth (hereafter we will write earth instead of sphere, as the intention is to take account of the true shape of spheroid) and ds the element corresponding to this upon the chart. Let θ represent the latitude of the extremity of the elliptic arc ds (which is measured, of course, from the equator), ρ the radius of the parallel of latitude θ, e the eccentricity, and, as before, a the radius of the equator. The condition

$$\frac{\mathrm{BB}'}{bb'}=\frac{\mathrm{AB}}{ab'}$$

becomes now

$$\frac{ds}{ds}=\frac{a}{\rho}$$

but

$$ds=\frac{a(1-\epsilon^2)d\theta}{(1-\epsilon^2\sin^2\theta)^{\frac{3}{2}}}$$

and

$$\rho=\frac{a\cos\theta}{(1-\epsilon^2\sin^2\theta)^{\frac{1}{2}}}$$

therefore

$$d\sigma=\frac{a(1-\epsilon^2)d\theta}{\cos\theta(1-\epsilon^2\sin^2\theta)}$$

Multiplying ϵ^2 in the numerator of this expression by $\sin^2\theta+\cos^2\theta$, this becomes

$$d\sigma=\frac{a d\theta}{\cos\theta}-a d\theta\frac{\epsilon^2\cos\theta}{1-\epsilon^2\sin^2\theta}$$

Integrating from $\theta=0$ to θ, *i. e.*, from the equator to latitude θ

$$\sigma=a\int_0^\theta\frac{\cos\theta d\theta}{\cos^2\theta}-a\epsilon\int_0^\theta\frac{\epsilon\cos\theta d\theta}{1-\epsilon^2\sin^2\theta}$$

$$=a\int_0^\theta\frac{d(\sin\theta)}{1-\sin^2\theta}-a\epsilon\int_0^\theta\frac{d(\epsilon\sin\theta)}{1-\epsilon^2\sin^2\theta}$$

$$=\frac{a}{M}\left[\tfrac{1}{2}\log\frac{1+\sin\theta}{1-\sin\theta}-\tfrac{1}{2}\epsilon\log\frac{1+\epsilon\sin\theta}{1-\epsilon\sin\theta}\right]$$

where $M=0.4342945$ is the modulus of the common logarithm. Since

$$\log\frac{1+t}{1-t}=2\left(t+\frac{t^3}{3}+\frac{t^5}{5}+\ .\ .\ .\right)$$

this formula can be written

$$\sigma=\frac{a}{M}\left[\log\tan\left(45^\circ+\frac{\theta}{2}\right)-\left(\epsilon\sin\theta+\frac{(\epsilon\sin\theta)^3}{3}+\ .\ .\ .\ .\right)\right]$$

Again expressing σ in minutes of arc, and writing for $\frac{10800}{\pi}\cdot\frac{1}{M}$, and $\frac{10800}{\pi}$ their values, we have finally for σ

$$\sigma=7915'.704674\log\tan\left(45^\circ+\frac{\theta}{2}\right)-3437'.7\left(\epsilon^2\sin\theta+\frac{\epsilon^4\sin^3\theta}{3}+\ .\ .\ .\ .\ .\right)$$

where powers of ϵ above the fourth may be neglected. The further consideration of this projection is reserved for Part II.

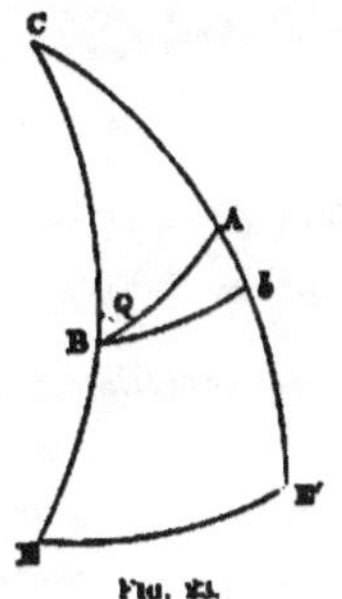

FIG. 23.

clear that a loxodromic curve upon the sphere is a species of spiral which winds around the sphere approaching indefinitely near but never passing through the poles, which are consequently assymptotic points to the curve. This is obvious when we consider that the loxodromic making equal angles with all the meridians, at the pole it would have to make the same angle with all the meridians, which would be impossible.

Consider two points A and B (Fig. 23) on consecutive meridians, the geographical co-ordinates of A being θ and ω, and those of B being θ_1 and ω_1. The angle CBA, measured from the north, is the constant angle that the loxodromic makes with the meridians; call this angle Q. The arc EE′ of the equator measures the angle between the meridians. Let $d\theta$ represent the change of latitude in passing from B to A, and $d\omega$ the infinitesimal arc EE′. Draw the parallel Bb; then, calling the radius of the equator r, we have, since $Bb=d\omega\cos\theta$,

$$d\omega\cot Q=\frac{r\,d\theta}{\cos\theta}=r\frac{d\sin\theta}{1-\sin^2\theta}$$

Integrating this, we have

$$\text{(A)}\qquad (\omega-\omega_1)\cot Q=r\log\frac{\tan\left(45^\circ+\frac{\theta}{2}\right)}{\tan\left(45^\circ+\frac{\theta_1}{2}\right)}.$$

For Q=0, or 180°, this reduces to $\omega-\omega_1=0$, which gives the curve as the meridian of B. For Q=90° the equation is satisfied only by $\theta=\theta_1$, or the curve is the parallel passing through B; for the particular value $\theta_1=0$, the loxodromic is the equator itself. For the computation of the arc $\omega-\omega_1$ in minutes, we have

$$\omega-\omega_1=7915'.705\tan Q\log\frac{\tan\left(45^\circ+\frac{\theta}{2}\right)}{\tan\left(45^\circ+\frac{\theta_1}{2}\right)}$$

In the above equation (A) we have, by passing to exponentials,

$$^{*}\text{exp.}\frac{(\omega-\omega_1)}{r}\cot Q=\frac{\tan\left(45^\circ+\frac{\theta}{2}\right)}{\tan\left(45^\circ+\frac{\theta_1}{2}\right)}$$

Now

$$\tan\left(45^\circ+\frac{\theta}{2}\right)=\frac{1+\tan\frac{\theta}{2}}{1-\tan\frac{\theta}{2}},\text{ or say, }=\epsilon$$

then

$$\tan\frac{\theta}{2}=\frac{\epsilon-1}{\epsilon+1}$$

Substituting the value of ϵ as obtained from the above equation, we have finally

$$\text{(B)}\qquad \tan\frac{\theta}{2}=\frac{\tan\left(45^\circ+\frac{\theta_1}{2}\right)\text{exp.}\frac{\omega-\omega_1}{r}\cot Q-1}{\tan\left(45^\circ+\frac{\theta_1}{2}\right)\text{exp.}\frac{\omega-\omega_1}{r}\cot Q+1}$$

The introduction of rectangular rectilinear co-ordinates in this equation by means of the formulas

$$x=r\cos\omega\cos\theta\qquad x_1=r\cos\omega_1\cos\theta_1$$

$$y=r\sin\omega\cos\theta\qquad y_1=r\sin\omega_1\cos\theta_1$$

$$z=r\sin\theta\qquad z_1=r\sin\theta_1$$

* Exp. is used as an abbreviation of "*e to the power of*."

gives us for the general equation of a loxodromic passing through the point x_1, y_1, z_1 on the surface of the sphere $x^2+y^2+z^2=r^2$

$$r-\sqrt{x^2+y^2}=z\;\frac{(z_1-\sqrt{x_1^2+y_1^2}+r)\,\text{exp.}\,\frac{\cot Q}{r}\left[\cos^{-1}\frac{x}{\sqrt{r^2-z^2}}-\cos^{-1}\frac{x_1}{\sqrt{r_1^2-z_1^2}}\right]-[z_1-\sqrt{x_1^2+y_1^2}-r]}{(z_1-\sqrt{x_1^2+y_1^2}+r)\,\text{exp.}\,\frac{\cot Q}{r}\left[\cos^{-1}\frac{x}{\sqrt{r^2-z^2}}-\cos^{-1}\frac{x_1}{\sqrt{r_1^2-z_1^2}}\right]+[z_1-\sqrt{x_1^2+y_1^2}-r]}$$

EQUATION OF A GREAT CIRCLE.

All the data remaining as before, conceive a great circle to pass through B; its azimuth $z=$ CBG.

FIG. 24.

In the triangle CBG, Fig. 24, we have

$$\cot CG \sin CB=\cos CB \cos C+\sin C \cot B$$

or

$$\tan\theta \cos\theta_1=\sin\theta_1 \cos(\omega-\omega_1)+\sin(\omega-\omega_1)\cot Z$$

and, since

$$\tan\theta=\frac{2\tan\frac{\theta}{2}}{1-\tan^2\frac{\theta}{2}}$$

it follows that

$$\text{(O)}\qquad \tan\frac{\theta}{2}=\tfrac{1}{2}\left(1-\tan^2\frac{\theta}{2}\right)\left\{\tan\theta_1\cos(\omega-\omega_1)+\frac{\cot Z}{\cos\theta_1}\sin(\omega-\omega_1)\right\}$$

which is the equation of the great circle passing through B, and making with meridian of B an angle $=$ Z.

In Fig. 25, $pADp'$ represents a circle which is parallel to the meridian PBP' and distant

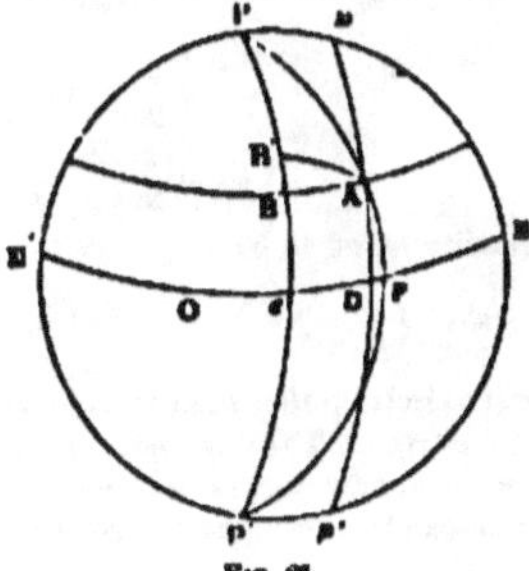

FIG. 25.

from $oD = OD - Oo = \omega_2 - \omega_1$, if the longitude of D be taken $= \omega_2$. Draw AB′ perpendicular to PP″; AB′ $= oD$, and the right-angled triangle PAB′ gives us for the relation between the latitude FA $= \theta$ and the longitude $oF = \omega - \omega_1$ of any point of this circle

$$\text{(D)} \qquad \sin(\omega - \omega_1) = \frac{\sin(\omega_2 - \omega_1)}{\cos\theta}$$

EQUATIONS OF THE PROJECTIONS OF THE LOXODROMIC, THE GREAT CIRCLE, AND A PARALLEL TO A MERIDIAN.

Since the equator is developed into its actual length, taking the rectification as the axis of ξ, we can replace $\omega - \omega_1$, expressed in the same unit as the radius r of the equator, by ξ, and equation A becomes

$$\text{(E)} \qquad \xi \cot Q = r \log \tan\left(45° + \frac{\theta}{2}\right) - r \log \tan\left(45° + \frac{\theta_1}{2}\right)$$

Now, in order to find out upon the chart how many of these curve units are contained in the abscissa ξ of a point of the curve corresponding to any latitude θ upon the sphere, it will be necessary to introduce in this equation the already-found values for Q, θ_1, and θ. The ordinate η of the same point will be found by making E=0 in the value already found for the length s. This gives

$$\text{(F)} \qquad \eta = r \log \tan\left(45° + \frac{\theta}{2}\right)$$

Eliminating θ between E and F, we have, for the equation of the projection of the loxodromic,

$$\eta = \xi \cot Q + r \log \tan\left(45° + \frac{\theta_1}{2}\right)$$

a straight line making an angle Q with the meridian, which is here taken for the axis of η, and passing through the point of the chart which corresponds to the point on the sphere of latitude θ_1.

The equations of the great circle and parallel before spoken of are obtained by eliminating θ between equation F and equations C and E respectively. Write equation F in the form

$$\tan\left(45° + \frac{\theta}{2}\right) = e^{\frac{\eta}{r}}$$

then we deduce immediately

$$\tan\frac{\theta}{2} = \frac{e^{\frac{\eta}{r}} - 1}{e^{\frac{\eta}{r}} + 1}$$

Substituting this in equation C, and writing $\omega - \omega_1 = \frac{\xi}{r}$, we have for the equation of the projection of a great circle

$$\text{(G)} \qquad e^{\frac{\eta}{r}} - e^{-\frac{\eta}{r}} = 2\left[\tan\theta_1 \cos\frac{\xi}{r} + \frac{\cot z}{\cos\theta_1} \sin\frac{\xi}{r}\right]$$

Since

$$\cos\theta = \frac{1 - \tan^2\frac{\theta}{2}}{1 + \tan^2\frac{\theta}{2}}$$

the projection of the parallel is readily found to be

$$\text{(H)} \qquad \sin\frac{\xi}{r} = \frac{1}{2} \sin\frac{\xi_1}{r}\left(e^{\frac{\eta}{r}} + e^{-\frac{\eta}{r}}\right) \qquad \left(\omega_2 - \omega_1 = \frac{\xi_1}{r}\right)$$

Thus we see that both the great circles (other than meridians) and the parallels to any meridian are projected in transcendental curves. The discussion of these equations is very simple, and need not be given here; the subject will, however, be resumed in another place, and more general forms of equations obtained for the loxodromics upon the general ellipsoid and upon the ellipsoids of revolution.

§ V.

ZENITHAL PROJECTIONS.

A projection is said to be zenithal when all points upon the earth's surface that are equidistant from a certain assumed central point are represented upon the chart in the circumference of a circle whose center is the projection of the assumed point upon the sphere.

The new co-ordinates to which the position of a point is referred are the almucantars and azimuthal circles of the assumed central point, and, as in the case of perspective projections, these circles of the sphere are given upon the chart as concentric circles, the almucantars and their diameters the azimuthal circles. The angles between the azimuthal circles are conserved, as in the case of perspective projection already alluded to. The name *zenithal projections* is obviously derived from the fact that they can always be considered as the representation of the hemisphere situated above the horizon of the given point, and having the zenith for pole. For the determination of the new co-ordinates in terms of the latitudes and longitudes, supposed known, we have only to solve a very simple spherical triangle.

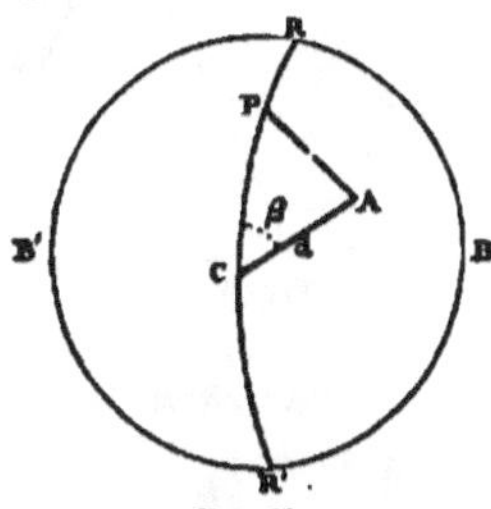

FIG. 26.

Let C (Fig. 26) be the point upon the horizon of which the projection is to be made, P the pole; then RPCR' is the meridian, and RBR'B' the horizon of C. Let A denote any point whose latitude and longitude θ and ω are known, and let λ represent the latitude of C. Now, in the spherical triangle PCA (the circles of PC and PA being, of course, great circles) we have given the side $PC = 90° - \lambda$, the side $PA = 90° - \theta$, and the angle at P = the longitude ω counted from the meridian RPCR', to find the side $CA = \alpha$ and $PCA = \beta$. We have at once

$$\cos \alpha = \sin \theta \sin \lambda + \cos \theta \cos \lambda \cos \omega$$

or, introducing an auxiliary angle ζ given by $\tan \zeta = \cos \omega \cot \lambda$

$$\cos \alpha = \frac{\sin (\theta + \zeta)}{\cos \zeta} \sin \lambda$$

and

$$\sin \beta = \frac{\sin \omega}{\sin \alpha} \cos \theta$$

In the case of $\lambda = 0$, or the center of the chart assumed upon the equator, these become

$$\cos \alpha = \cos \omega \cos \theta \qquad \tan \beta = \tan \omega \cot \theta$$

The construction of the projection is now, of course, quite simple, if we know the law which is to connect the radius of the representation of each almucantar upon the chart with the corresponding angle α; *i. e.*, if we know some such relation as $\rho = F(\alpha)$. It would only be necessary in using this projection to have tables giving the values of ρ for each value of α, and to lay off these values of ρ upon the diameters making with the meridian PC the corresponding angles β as determined by another table. The projection will evidently be symmetrical with respect to the meridian PC and also with respect to the equator if the center be chosen upon that line. The simplest form that we can give to ρ is evidently $\rho = ra$. This is equivalent to saying that the radius of each almucantar is equal to the arc of the great circle upon the sphere which joins the center of the region to be projected to the small circle under consideration. If the pole is taken for center, the almucantars become the parallels and they are given by

$$\rho = r(90^\circ - a)\frac{\pi}{180}$$

This projection was first employed by Guillaume Postel in 1581; afterwards by Lambert in 1772, he regarding it as a zenithal projection; then again in 1790 Antonio Cagnoli believed that he had invented it for the first time, and gave it his name. We shall adopt the name apparently given for the first time by Germain, and speak of it as the equidistant zenithal projection.

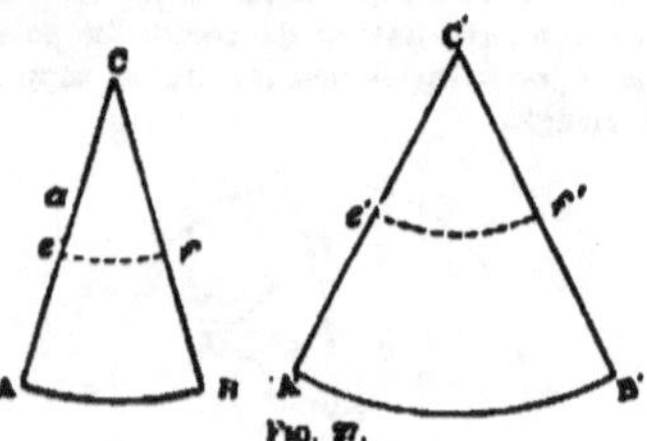

Fig. 27.

Expressing in units of angular measure, or degrees, the radii of the sphere and of the different circles that we wish to project, we have obviously, from Fig. 27,

$$\frac{A'B'}{AB} = \frac{90^\circ \pi}{180^\circ} = \frac{90}{57.29}$$

Now

$$ef = AB \sin a$$

and

$$\frac{e'f'}{A'B'} = \frac{a}{90}$$

therefore

$$\frac{A'B'}{AB} = \frac{90}{57.29} = \frac{90}{a}\frac{e'f'}{\cdot} \cdot \frac{\sin a}{ef}$$

from which

$$\frac{e'f'}{ef} = \frac{a}{57.29 \sin a}$$

This ratio tends to unity as a diminishes and becomes, for $a = 30^\circ$,

$$= \frac{60}{57.29}$$

which shows that the degrees upon the projection are a very little smaller than those which correspond to them upon the sphere. From the formula $\rho = ra$ it is clear that the central distances are conserved, and from what we have just seen it is clear that, if the projection be not extended more than 30° from the center, the distances perpendicular to the radii ρ will also be very nearly conserved. With the pole as center this projection has been much used by the French *Bureau des Longitudes* for the charts of eclipses published every year in the *Connaissance des Temps*. This equidistant zenithal projection may be considered as the final one of a series of projections of which the first two terms are the gnomonic and the stereographic projections respectively. For the gnomonic we have

$$\rho = r \tan a$$

for the stereographic

$$\rho = 2r \tan \frac{a}{2}$$

Now write

$$\rho = nr \tan \frac{a}{n}$$

for $n=1, 2$, respectively we have the last two equations. This can be written, however, in the form

$$n \tan \frac{a}{n} = \frac{a}{\cos \frac{a}{n}} \cdot \frac{\sin \frac{a}{n}}{\frac{a}{n}}$$

let $n=\infty$

$$\frac{a}{\cos \frac{a}{n}} = a \qquad \frac{\sin \frac{a}{n}}{\frac{a}{n}} = 1$$

and thus

$$\rho = r\infty \tan \frac{a}{\infty} = ra$$

PROJECTION BY BALANCE OF ERRORS.

We will now take up the projection invented by Sir George B. Airy and give the full account of it that is given by the illustrious author himself in the first part of his paper published in the Philosophical Magazine for December, 1861, having the title "Explanation of a Projection by Balance of Errors for Maps applying to a very large extent of the Earth's Surface; and Comparison of this projection with other projections. By G. B. Airy, Esq., Astronomer Royal." As a slight mistake was made in this paper it will not be given in full exactly as it appears in the Philosophical Magazine, but Captain Clarke's correction of Sir George Airy's error will be interpolated in its proper place. Captain Clarke's account of the correction to be made and the consequent necessary changes in some of Airy's conclusions, is to be found in the Philosophical Magazine for April, 1862, "On Projections for Maps applying to a very large extent of the Earth's Surface. By Col. Sir Henry James, R. E., Director of the Ordnance Survey; and Capt. Alexander R. Clarke, R. E."

In this projection, as in all zenithal projections, any point of the earth's surface may be adopted as the center of reference to be represented by the central point of the map. The projection is subjected to the conditions: (1) that the azimuth of any other point on the earth, as viewed from the center of reference, shall be the same as the azimuth of the corresponding point of the map as viewed from the central point of the map; (2) that equal great-circle distances of other points on the earth from the center of reference, in all directions, shall be represented by equal radial distances from the central point of the map. These conditions include the stereographic projection, Sir Henry James's projection, and others; but they exclude the Mercator projection, and the projections proposed by Sir John Herschel. The two errors, to one or both of which all projections are liable, are, change of area, and distortion, as applying to small portions of the earth's surface. On the one hand, a projection may be invented (called by Airy, "Projection with Unchanged Areas") in which there is no change of area, but excessive distortion for parts far from the center; on the other hand, the stereographic projection has no distortion but has great change of area for distant parts. Between these lie the projections usually adopted by geographers with the tacit purpose of greatly reducing the error of one kind by the admission of a small error of the other kind. The object of the projection invented by Airy "is to exhibit a distinct mathematical process for determining the magnitudes of these errors, so that the result of their combination shall be the most advantageous." The theory is founded upon the following assumptions and inferences:

First. The change of area being represented by

$$\frac{\text{projected area}}{\text{original area}} - 1$$

and the distortion being represented by

$$\frac{\text{ratio of projected sides}}{\text{ratio of original sides}}-1=\frac{\text{projected length}\times\text{original breadth}}{\text{projected breadth}\times\text{original length}}-1$$

(where the length of the rectangle is in the direction of the great circle connecting the rectangle's center with the center of reference, and the breadth is transverse to that great circle), these two errors, when of equal magnitude, may be considered as equal evils.

Second. As the annoyance caused by a negative value of either of these formulas is as great as that caused by a positive value, we must use some even power of the formulas to represent the evil of each. The squares will be used.

Third. The total evil in the projection of any small part may be represented by the sum of these squares.

Fourth. The total evil on the entire map may therefore be represented by the summation through the whole map (respect being had to the magnitude of every small area) of the sum of these squares for every small area.

Fifth. The process for determining the most advantageous projection will therefore consist in determining the laws expressing the radii of map circles in terms of the great-circle radii on the earth (*i. e.*, to determine $\rho=F\alpha$), which will make the total evil, represented as has just been stated, as small as possible.

Let a and b denote the length and breadth of a small rectangle on the earth's surface, and $a+\delta a$, $b+\delta b$ the length and breadth of the representation of this rectangle upon the map, neglect powers of δa and δb above the first. Then the change of area

$$=\frac{\text{projected area}}{\text{original area}}-1=\frac{(a+\delta a)(b+\delta b)}{ab}-1=\frac{\delta a}{a}+\frac{\delta b}{b}$$

And the distortion

$$=\frac{\text{projected length}}{\text{projected breadth}}\times\frac{\text{original breadth}}{\text{original length}}-1=\frac{a+\delta a}{b+\delta b}\cdot\frac{b}{a}-1=\frac{\delta a}{a}-\frac{\delta b}{b}$$

The sum of their squares, or

$$\left(\frac{\delta a}{a}+\frac{\delta b}{b}\right)^2+\left(\frac{\delta a}{a}-\frac{\delta b}{b}\right)^2$$

is

$$=2\left\{\left(\frac{\delta a}{a}\right)^2+\left(\frac{\delta b}{b}\right)^2\right\}$$

and we may therefore use

$$\left(\frac{\delta a}{a}\right)^2+\left(\frac{\delta b}{b}\right)^2$$

as the measure of the evil for each small rectangle.

Let α denote the length, expressed in terms of radius of the arc of a great circle on the earth connecting the center of the small rectangle with the center of reference; ρ the corresponding distance upon the chart, expressed in terms of the same radius, of the projection of the center of the small rectangle from the center of the chart; to find ρ in terms of α. Let the length of a small rectangle on the earth be $\delta\alpha$, the corresponding length on the map $\delta\rho$. Also let β be the infinitesimal azimuthal angle under which, in both cases, the breadth of the rectangle is seen from the center of reference or the center of the map. Then we have

$$a=\delta\alpha \qquad a+\delta a=\delta\rho \qquad \delta a=\delta\rho-\delta\alpha$$

$$b=\beta\sin\alpha \qquad b+\delta b=\beta\rho \qquad \delta b=\beta(\rho-\sin\alpha)$$

$$\left(\frac{\delta a}{a}\right)^2+\left(\frac{\delta b}{b}\right)^2=\left(\frac{\delta\rho}{\delta\alpha}-1\right)^2+\left(\frac{\rho}{\sin\alpha}-1\right)^2$$

This quantity expresses the evil on each small rectangle. The product of the evil by the extent of surface which it affects, omitting the general multiplier β, is

$$\left\{\left(\frac{d\rho}{d\alpha}-1\right)^2+\left(\frac{\rho}{\sin\alpha}-1\right)^2\right\}\sin\alpha\, d\alpha$$

Consequently the summation of the partial evils for the whole map is given by

$$\int d\alpha\left\{\left(\frac{d\rho}{d\alpha}-1\right)^2+\left(\frac{\rho}{\sin\alpha}-1\right)^2\right\}\sin\alpha$$

Or if $\rho-\alpha=y$ and if we put p for $\frac{dy}{d\alpha}$, the expression is

$$\int\left\{p^2\sin\alpha+\frac{(y+\alpha-\sin\alpha)^2}{\sin\alpha}\right\}d\alpha$$

and this integral over the surface to which the map applies is to be a minimum. Just here it will be of interest to give Captain (now Colonel) Clarke's elegant method of obtaining this fundamental equation.

Let P be the point on the sphere which is to be the center of the map, and Q any other point on the sphere such that the arc PQ$=\alpha$; if Q′ be the representation of Q on the development, PQ′$=\rho$. Suppose a very small circle, radius ω, described on the sphere having its center at Q; then the representation of this small circle on the map will be an ellipse having its minor axis in the line PQ′ and its center at Q′; the lengths of the semi-axes will be

$$\omega\frac{d\rho}{d\alpha}\qquad\qquad \rho\frac{\omega}{\sin\alpha}$$

the differences between these quantities and that which they represent, that is, ω, are

$$\omega\left(\frac{d\rho}{d\alpha}-1\right)\qquad\qquad \omega\left(\frac{\rho}{\sin\alpha}-1\right)$$

and the sum of the squares of these errors is the measure of the *misrepresentation* at Q′. The sum for the whole surface from $\alpha=0$ to $\alpha=\tau$ is proportional to

$$\int_0^\tau\left\{\left(\frac{d\rho}{d\alpha}-1\right)^2+\left(\frac{\rho}{\sin\alpha}-1\right)^2\right\}\sin\alpha\, d\alpha$$

which is to be a minimum.

Resuming now the expression

$$\int_0^\tau\left\{p^2\sin\alpha+\frac{(y+\alpha-\sin\alpha)^2}{\sin\alpha}\right\}d\alpha$$

write it as

$$\int_0^\tau V\,d\alpha,$$

Make

$$M=\frac{dV}{d\alpha}\qquad\qquad N=\frac{dV}{dy}\qquad\qquad P=\frac{dV}{dp}=2p\sin\alpha$$

and giving to y only a variation subject to the condition that $\delta y=0$ when $\alpha=0$, the equations of solution are

$$N-\frac{dP}{d\alpha}=0\qquad\qquad P_\tau=0$$

where P_τ is the value of $2p\sin\alpha$ when $\alpha=\tau$. Now

$$N=\frac{2(y+\alpha-\sin\alpha)}{\sin\alpha}\qquad\qquad \frac{dP}{d\alpha}=2\frac{d^2y}{d\alpha^2}\sin\alpha+2\frac{dy}{d\alpha}\cos\alpha$$

and the equation

$$N-\frac{dP}{d\alpha}=0$$

becomes

$$\frac{y+\alpha-\sin\alpha}{\sin\alpha}-\sin\alpha\frac{d^2y}{d\alpha^2}-\cos\alpha\frac{dy}{d\alpha}=$$

or

$$\sin^2\alpha\frac{d^2y}{d\alpha^2}+\sin\alpha\cos\alpha\frac{dy}{d\alpha}-y=\alpha-\sin\alpha$$

For $\alpha-\sin\alpha$ use the symbol A and assume

$$z=\sin\alpha\frac{dy}{d\alpha}+y$$

Then, by actual differentiation and substitution,

$$\sin\alpha\frac{dz}{d\alpha}-z=\sin^2\alpha\frac{d^2y}{d\alpha^2}+\sin\alpha\cos\alpha\frac{dy}{d\alpha}-y$$

or

$$\sin\alpha\frac{dz}{d\alpha}-z=A$$

This equation is integrable when multiplied through by $\frac{1}{\sin^2\frac{\alpha}{2}}$; the solution gives

$$z=\tfrac{1}{2}\tan\frac{\alpha}{2}\int\frac{A\,d\alpha}{\sin^2\frac{\alpha}{2}}$$

therefore

$$\sin\alpha\frac{dy}{d\alpha}+y=\tfrac{1}{2}\tan\frac{\alpha}{2}\int\frac{A\,d\alpha}{\sin^2\frac{\alpha}{2}}$$

This last equation is integrable when multiplied by $\frac{1}{\cos^2\frac{\alpha}{2}}$; the solution gives

$$y=\tfrac{1}{2}\cot\frac{\alpha}{2}\int\frac{\sin\frac{\alpha}{2}\,d\alpha}{\cos^3\frac{\alpha}{2}}\int\frac{A\,d\alpha}{\sin^2\frac{\alpha}{2}}$$

If $\frac{\alpha}{2}=\phi$, the solution may be put in the form

$$y=\cot\phi\int\frac{\sin\phi\,d2\phi}{\cos^3\phi}\int\frac{A\,d\phi}{\sin^2\phi}$$

or

$$y=\frac{1}{2\sin\phi\cos\phi}\int\frac{A\,d\phi}{\sin^2\phi}-\tfrac{1}{2}\cot\phi\int\frac{A\,d\phi}{\sin^2\phi\cos^2\phi}$$

or

$$y=\tfrac{1}{2}\tan\phi\int\frac{A\,d\phi}{\sin^2\phi}-\tfrac{1}{2}\cot\phi\int\frac{A\,d\phi}{\cos^2\phi}$$

Replacing A by its value $\alpha-\sin\alpha$, we have, finally,

$$y=-\alpha-2\cot\frac{\alpha}{2}\log\cos\frac{\alpha}{2}+C\tan\frac{\alpha}{2}+C'\cot\frac{\alpha}{2}$$

In this, as $y=0$, for $\alpha=0$ we must have $C'=0$.

The second condition

$$P_r = \left(\frac{dy}{du}\right) = 0$$

is easily seen to become

$$\operatorname{cosec}^2\frac{\tau}{2}\log\cos\frac{\tau}{2} + \tfrac{1}{2}C\sec^2\frac{\tau}{2} = 0$$

which gives for C

$$C = -2\cot^2\frac{\tau}{2}\log\cos\frac{\tau}{2} = \cot^2\frac{\tau}{2}\,\log\sec^2\frac{\tau}{2}$$

For the center of the chart, where $u=0$, we have, on substituting for y its value,

$$\left(\frac{d\rho}{du}\right)_0 = \frac{1+c}{z}$$

And finally

$$\rho = -2\cot\frac{u}{2}\log\cos\frac{u}{2} + \tan\frac{u}{2}\cot^2\frac{\tau}{2}\log\sec^2\frac{\tau}{2}$$

The logarithms a[illegible]e Naperian, and to transform to common logarithms it will be necessary to divide by M=0.43[illegible]448.

The limiting radius of the map is $R = 2C\tan\frac{\tau}{2}$. This quantity does not increase indefinitely, but is a maximum for $\tau = 120^\circ\ 24'\ 53''$; and, for quarter values of τ, R diminishes. As in all that precedes we have tacitly assumed the radius of the sphere as equal to unity, we must, for any radius r, multiply the found values of ρ and R by r.

Sir Henry James has invented a perspective projection which is nearly enough allied to the present subject to be included under the head of zenithal projections. In the attempt to make the misrepresentation on the map as little as possible, he has been led to choose a position for the point of sight given by the formula $c=1.50r$. Colonel James, wishing to apply this projection to the representation of a portion of the earth's surface greater than a hemisphere, takes for the plane of projection no longer that of a great circle but a plane parallel to the ecliptic and passing through either the tropics, *i. e.*, at a distance from the center of the sphere $= r\sin 23^\circ\ 30'$. The radius of the bounding circle of the chart is $= r\cos 23\tfrac{1}{2}^\circ = 0.91706\ r$ and the equal lengths upon the spherical surface are at the boundaries of the chart only one-sixth greater than towards the middle. Colonel James has chosen this particular position for the plane of projection because, with the central point assumed upon the tropic of Cancer, in this circle of projection can be represented Europe, Asia, Africa, and America. The name chosen by Colonel James, "a projection of two-thirds of the sphere," is not exactly correct, inasmuch as in reality seven-tenths of the surface of the earth is represented.

For the projection of a hemisphere the above system is the best possible, for in it the misrepresentation is a minimum; but for extending the projection from 90° to 113° 30′ Captain Clarke has shown that the true position for the point of sight is at a distance from the surface given by the formula $c=1$[illegible]r instead of $c=1$[illegible]r. Taking the radius of the sphere as unity, and making h the distance of the eye from the center, and k the distance of the plane of projection from the same point, we have

$$\rho = \frac{k\sin u}{h+\cos u}$$

for the radius of any almucantar. This involves two arbitrary constants k and h, and these may be so determined as to render the integral

$$\int_0 U\sin u\,du$$

a minimum, where

$$U = \left(\frac{d\rho}{du} - 1\right)^2 + \left(\frac{\rho}{\sin u} - 1\right)^2$$

Replacing $\frac{d\rho}{d\alpha}$ by its value, $k\frac{(h\cos\alpha+1)}{(h+\cos\alpha)^2}$, and ρ by its value, we have

$$Q=\int_0^\beta\left\{\left(\frac{k}{h+\cos\alpha}-1\right)^2+\left(\frac{k(h\cos\alpha+1)}{(h+\cos\alpha)^2}-1\right)^2\right\}\sin\alpha d\alpha$$

This must be a minimum with respect to k and h. Effecting the integration, we get

$$Q=k^2H_1+2kH_2+4\sin^2\frac{\beta}{2}$$

where the symbols H_1 and H_2 are

$$H_1=\frac{1+h^2}{G}-\frac{h(h^2-1)}{G^2}+\frac{(h^2-1)^2}{3G^3}-\frac{1}{3}\frac{(1-h)^3}{1+h}-1 \qquad H_2=(1+h)\log\frac{G}{1+h}+\frac{h^2-1}{G}-(h-1)$$

and

$$G=h+\cos\beta$$

Now, the conditions for a minimum are

$$\frac{dQ}{dk}=0 \qquad \frac{dQ}{dh}=0$$

or

$$kH_1+H_2=0 \qquad k\frac{dH_1}{dh}+2\frac{dH_2}{dh}=0$$

from which

$$k=-\frac{H_2}{H_1} \qquad H_2\frac{dH_1}{dh}-2H_1\frac{dH_2}{dh}=0$$

hence

$$Q=4\sin^2\frac{\beta}{2}-\frac{H_2^2}{H_1}$$

Now h must be so determined that $\frac{H_2^2}{H_1}$ shall be a maximum. This is most easily done by calculating the values corresponding to assumed values of h. We have the following:

h	$\log H_2^2-\log H_1$
1.35	0.420732
1.36	0.420756
1.37	0.420762
1.38	0.420747
1.39	0.420665

By interpolation the maximum is found to be

h	$\log H_2^2-\log H_1$
1.36763	0.4207623

therefore

$$\frac{H_2^2}{H_1}=2.634889$$

and consequently

$$Q=0.16261$$

The point of sight is here at the distance of about 3/8 of the radius from the surface of the sphere, instead of 1/2 as in the previously-described projection by Colonel James. We have also

$$\rho=\frac{1.66261\,r\sin\alpha}{1.36763+\cos\alpha}$$

For $\omega=113^\circ\ 30'$, the radius of the limiting circle of the chart is

$$R=1.5737$$

and by Sir George Airy's "Balance of Errors"

$$R=1.5760$$

The values of ρ from the above formula are found in Table XXVI. This species of perspective projection is very useful for representing large portions of the earth's surface, and for the construction of physical or geological charts; for the large star maps it is preferable to the stereographic projection, because it is capable of representing with but little error at the limits a very large portion of the sky, but, not possessing the important attribute of conserving the angles, does not show the constellations in their true forms.

GLOBULAR PROJECTION.*

This projection, which is often wrongly confounded with Postel's equidistant projection, is very simple of construction, all of the lines drawn to represent meridians or parallels being arcs of circles.

A circle is drawn of arbitrary radius, and two rectangular diameters are also drawn, representing the first meridian and the equator respectively. The parallels are drawn passing through points of equal division of the first meridian, and the meridians are made to pass through the two poles and the points of equal division of the equator. If the centers are too far removed to construct the circles by that means, they can be constructed by means of points. The formulas for the radii of meridians and parallels are readily found to be

$$\rho_m=\frac{r^2}{2\delta}-\frac{\delta}{2}$$

where δ is the distance from the point of intersection of the meridian with the equator to the center of the meridian which limits the projection, and r is the radius of the sphere, and

$$\rho_p=\frac{r^2+\delta_1^2-2h\delta_1}{2(h-\delta_1)}$$

where δ_1=the distance from the point of intersection of the parallel with the central meridian to the center of the hemisphere, and $h=r\sin\omega$. If n denote the number of points of subdivision of the equator and central meridian, then

$$\delta=\frac{r}{n} \qquad \delta'=\frac{r}{n}$$

and, in consequence,

$$\rho_m=\frac{r}{2n}(n^2+1) \qquad \rho_p=r\frac{(n^2-2n\sin\omega+1)}{2(n\sin\omega-1)}$$

Before leaving this subject we will study briefly the alterations in angles, lengths, and areas caused by projection, a general investigation of the theory of alteration being reserved for the purely theoretical part of this work.

Denote by θ the angle made by an arbitrary line upon the sphere with a meridian, and by θ' the corresponding line upon the chart, and make, for brevity,

$$\phi=\theta'-\theta$$

We have now β, as before, denoting the azimuth,

$$\tan\theta=\frac{d\beta}{d\alpha}\sin\alpha \qquad \tan\theta'=\rho\frac{d\beta}{d\rho}$$

* Improperly called "Arrowsmith's projection;" it was invented by J. B. Nicolosi, of Paterno, Sicily, in 1660, while Arrowsmith did not use it until 1794.

These two equations give by elimination of $d\beta$ (and consequently of β)

$$\tan \Psi = J \tan \theta$$

where

$$J = \frac{\rho\, d\alpha}{d\rho \sin \alpha} = \frac{d \log \tan \frac{\alpha}{2}}{d \log \rho}$$

and also

$$\tan \phi = \frac{(J-1) \tan \theta}{1 + J \tan^2 \theta}$$

It is clear from this that ϕ, which is the angular alteration, is independent of the azimuth β. We can now calculate ϕ in all cases where the form of the function F is known, which defines ρ by means of the relation $\rho = F(\alpha)$. Since $\tan \phi = 0$, for $\theta = 0, \frac{\pi}{2}$, we have for these two cases $\theta = \Psi$, which we already knew, this only expressing the fact the projections of the almucantars and azimuthal circles are orthogonal as are the lines themselves upon the sphere. We will now determine the maximum of ϕ regarding θ as the independent variable; denote the resulting values of θ, Ψ, and ϕ by the same symbols with the suffix $(_0)$. Forming the first differential coefficient of $\tan \phi$, we are conducted at once to the relation

$$1 - J \tan^2 \theta = 0$$

from which

$$\tan \theta_0 = \pm J^{-\frac{1}{2}} = \pm \sqrt{\frac{d \log \rho}{d \log \tan \frac{\alpha}{2}}}$$

There are thus two directions upon the sphere symmetrically situated with respect to the meridian, the angle between which is most altered. The same is, of course, true upon the chart, and we have, in fact, for $\tan \Psi$,

$$\tan \Psi_0 = \pm J^{\frac{1}{2}} = \sqrt{\frac{d \log \tan \frac{\alpha}{2}}{d \log \rho}}$$

Combining these two, we find

$$\tan \theta_0 \tan \Psi_0 = 1$$

or

$$\theta_0 + \Psi_0 = \frac{\pi}{2}$$

excluding negative arcs and arcs $> \frac{\pi}{2}$. We are able now to find the maximum of $\tan \phi$, *i. e.*,

$$\tan \phi_0 = \frac{J-1}{2\sqrt{J}}$$

Substituting for J its value

$$\tan \phi_0 = \frac{-d \log \rho + d \log \tan \frac{\alpha}{2}}{2\sqrt{d \log \rho \; d \log \tan \frac{\alpha}{2}}}$$

This gives $\phi_0 = 0$ for $\alpha = \rho = 0$, or there is never any alteration at the center. For $J = 1$, that is, for a projection in which the angular alteration is everywhere zero, we find

$$d \log \rho = d \log \tan \frac{\alpha}{2}$$

and consequently (there being no constant)

$$\rho = \tan \frac{\alpha}{2}$$

which is the law of the stereographic projection.

ALTERATION OF LENGTHS.

Denoting upon the sphere the co-ordinates of two infinitely near points by α, β and $\alpha+d\alpha$, $\beta+d\beta$, we have for the distance between them

$$\delta_1=\sqrt{\sin^2\alpha d\beta^2+d\alpha^2}$$

and for the corresponding distance upon the chart

$$\delta_2=\sqrt{d\rho^2+\rho^2 d\beta^2}$$

Make

$$\frac{\delta_2}{\delta_1}=m$$

thence

$$d\rho^2+\rho^2d\beta^2=m^2(\sin^2\alpha d\beta^2+d\alpha^2)$$

The ratio m is, of course, different in different points of the chart and also, in general, differs with the direction of the element under consideration. This last is equivalent to saying that m is a function of θ and of course varying with

$$\tan\theta=\frac{d\beta}{d\alpha}\sin\alpha_1$$

To find, then, the value of θ which makes m a maximum or minimum. The above expression for m^2 gives

$$m^2=\frac{d\rho^2+\rho^2d\beta^2}{\sin^2\alpha d\beta^2+d\alpha^2}$$

Dividing numerator and denominator by $d\alpha^2$ and substituting $\tan\theta$ for its value,

$$m^2=\frac{d\rho^2}{d\alpha^2}\cos^2\theta+\rho^2\frac{\sin^2\theta}{\sin^2\alpha}$$

Equating to zero the derivative of this expression with respect to θ, there results

$$\sin\theta\cos\theta\left(\frac{\rho^2}{\sin^2\alpha}-\frac{d\rho^2}{d\alpha^2}\right)=0$$

This shows that m is a maximum or minimum for $\theta=0$ or $\frac{\pi}{2}$; *i. e.*, either for an almucantar or an azimuth (the diameter of the almucantar). Taking the second derivative, we are conducted to

$$\cos 2\theta\left(\frac{\rho^2}{\sin^2\alpha}-\frac{d\rho^2}{d\alpha^2}\right)=0$$

We must examine the sign of this under the two suppositions of $\theta=0$, or $\frac{\pi}{2}$. Suppose first that the factor in parenthesis is positive; *i. e.*,

$$\frac{d\rho^2}{\rho^2}-\frac{d\alpha^2}{\sin^2\alpha}<0$$

there results at once

$$\rho<\tan\frac{\alpha}{2}$$

This condition being supposed satisfied, by any means whatever, we will have the second derivative positive for $\theta=0$ and negative for $\theta=\frac{\pi}{2}$; consequently m will be a minimum for $\theta=0$, and a maximum for $\theta=\frac{\pi}{2}$. If the same factor is negative, we must have

$$\rho>\tan\frac{\alpha}{2}$$

and the converse of the above will take place, m being a maximum for $\theta=0$, that is, along the projection of the azimuthal circle, and a minimum for $\theta=\frac{\pi}{2}$, *i. e.*, along the projection of the almucantar. From

$$\frac{\rho^2}{\sin^2 \alpha}-\frac{d\rho^2}{d\alpha^2}=0$$

there follows

$$\rho=\tan\frac{\alpha}{2}$$

the stereographic projection. This is, then, the only system in which the ratio of corresponding elements upon the sphere and upon the chart is independent of θ. To find the actual maximum and minimum values of m, substitute $\theta=0$ and $\frac{\pi}{2}$ in the formula

$$m^2=\frac{d\rho^2}{d\alpha^2}\cos^2\theta+\frac{\rho^2}{\sin^2\alpha}\sin^2\theta$$

For brevity, write $m_0=m$ for $\theta=0$, and $m_1=m$ for $\theta=\frac{\pi}{2}$; then

$$m_0=\frac{d\rho}{d\alpha} \text{ upon the radius} \qquad m_1=\frac{\rho}{\sin\alpha} \text{ upon the almucantar}$$

The alteration of areas will obviously be expressed by the product $m_0 m_1=\mu^2$; then

$$\mu^2=\frac{\rho d\rho}{\sin\alpha d\alpha}$$

or

$$\mu^2=-\frac{1}{2}\frac{d\rho^2}{d\cos\alpha}$$

The application of the preceding results to the different cases of zenithal projection that we have studied is very simple. Beginning first with the alteration of angles, we have for the radius of the almucantars (or parallels) in perspective projection

$$\rho=\frac{c\sin\alpha}{c+\cos\alpha}$$

and consequently

$$J=\frac{d\log\tan\frac{\alpha}{2}}{d\log\rho}=\frac{c+\cos\alpha}{1+c\cos\alpha}$$

Now

$$\tan\theta_0=\pm\sqrt{\frac{c\cos\alpha+1}{c+\cos\alpha}}$$

and

$$\tan\phi_0=\frac{(c-1)(1-\cos\alpha)}{2\sqrt{(c+\cos\alpha)(c\cos\alpha+1)}}$$

In Colonel James's projection,

$$c=1.50 \qquad J=\frac{1.50+\cos\alpha}{1.50\cos\alpha+1} \qquad \tan\theta_0=\sqrt{\frac{1.50\cos\alpha+1}{1.50+\cos\alpha}}$$

and

$$\tan\phi_0=\frac{0.25(1-\cos\alpha)}{\sqrt{(1.50+\cos\alpha)(1.50\cos\alpha+1)}}$$

For $\alpha=90^\circ$,

$$\tan\theta_0=\sqrt{\frac{1}{1.50}}=\sqrt{.666+} \qquad \theta_0=39^\circ\ 14' \text{ nearly,}$$

that is, at a distance of 90° from the center the angle most altered is about $=39^\circ\ 14'$; this is represented upon the map by 50° 40′, the alteration being 11° 32′. In Captain Clarke's modification of this projection,

$$c=1.36763 \qquad \tan\phi_0=\frac{0.36763\,(1-\cos\alpha)}{2\sqrt{(1.36763+\cos\alpha)(1.36763\cos\alpha+1)}}$$

In this case, making $\alpha=90°$, we find for the maximum alteration at that distance from the center

$$\phi_0=8°\ 56'$$

In the equidistant zenithal projection

$$\rho=\alpha \qquad J=\frac{\alpha}{\sin\alpha} \qquad \tan\theta_0=\pm\sqrt{\frac{\sin\alpha}{\alpha}} \qquad \tan\phi_0=\frac{\alpha-\sin\alpha}{2\sqrt{\alpha\sin\alpha}}$$

At the center, $\alpha=0$, J is equal to unity, and in consequence

$$\theta_0=\pm 45° \qquad \varpi_0=\pm 45° \qquad \phi_0=0$$

θ_0 now decreases until $\alpha=\frac{\pi}{2}$; then $\tan\theta_0=\pm\sqrt{\frac{2}{\pi}}$, or $\theta_0=38°\ 35'\ 10''$, and also $\varpi_0=51°\ 24'\ 30''$; the maximum deviation being $\phi_0=12°\ 49'\ 40''$. In the projection by Balance of Errors we must calculate θ_0 and ϖ_0 by means of the rather complicated formula

$$\rho=-2\cot\frac{\alpha}{2}\log\cos\frac{\alpha}{2}+\tan\frac{\alpha}{2}\left(\cot^2\frac{\gamma}{2}\log\sec^2\frac{\gamma}{2}\right)$$

The angular alterations in the gnomonic and orthographic projection are readily found to be equal with contrary signs.

For perspective projections we have more generally

$$\rho=\frac{c'\sin\alpha}{c+\cos\alpha}$$

from which we can find

$$m_0=\frac{d\rho}{d\alpha}=\frac{c'\,(c\cos\alpha+1)}{(c+\cos\alpha)^2} \qquad m_1=\frac{\rho}{\sin\alpha}=\frac{c'}{c+\cos\alpha} \qquad \mu^2=\frac{c'^2\,(c\cos\alpha+1)}{(c+\cos\alpha)^3}$$

For stereographic projection,

$$c'=2 \qquad m_0=\frac{1}{\cos^2\frac{\alpha}{2}} \qquad m_1=\frac{1}{\cos^2\frac{\alpha}{2}} \qquad \mu^2=\frac{1}{\cos^4\frac{\alpha}{2}}$$

The ratios of alteration thus depend only upon α, beginning at the center, or $\alpha=0$, with the values of unity, at the circumference or $\alpha=90°$,

$$m_0=m_1=2 \qquad \mu^2=4$$

In Colonel James's original projection the plane of projection coincides with one of the tropics, and $c'=1.1012$. Consequently for the center

$$m_0=m_1=\frac{1}{2.270}$$

At 90° from the center

$$m_0=\frac{1}{2.043} \qquad m_1=\frac{1}{1.362} \qquad \mu^2=\frac{1}{2.7825}$$

At the limit of 113° 30′

$$m_0=\frac{1}{1.201} \qquad m_1=2.365 \qquad \mu^2=1.88$$

In Captain Clarke's modification of this projection we have, for the same limits,

$$m_0=1.149 \qquad m_1=2.436 \qquad \mu^2=2.799$$

For the equidistant zenithal projection

$$m_2=1 \qquad m_1=\frac{s}{\sin s} \qquad \mu'=\frac{s}{\sin s}$$

At the center, $s=0$, the ratios m_1 and μ' are $=1$; from this point they increase until $s=90°$, when

$$m_1=\mu'=\frac{\pi}{2}=1.5708$$

s continuing to increase, m_1 and μ' also increasing until $s=180°$, when

$$m_1=\mu'=\infty$$

---

§ VI.

EQUIVALENT PROJECTIONS.

The only condition which is to be fulfilled in this class of projections is the equivalence of an elementary quadrilateral upon the spheroid with the corresponding quadrilateral upon the map. The quadrilateral upon the surface to be projected can be formed by two meridians and two parallels, each indefinitely near the other; the corresponding quadrilateral upon the map will also be very approximately a rectangle. The general mathematical investigation of this kind of projection will be given in another place, and for the present we confine ourselves merely to the equivalent projection of the sphere. Denote, as usual, latitude and longitude by θ and ω respectively; call ρ the radius of the parallel of latitude θ, and s the meridional distance of a point from the pole; then for the area of the small quadrilateral included between two infinitely near meridians and two infinitely near parallels, we have

$$r\rho ds d\omega$$

Writing here, since ρ and s are functions of θ,

$$r\rho ds=\Theta d\theta$$

where Θ is a function of θ, the element of areas is

$$=\Theta d\theta d\omega$$

For the earth we have, without any difficulty,

$$\Theta=\frac{r^2(1-\epsilon^2)\cos\theta}{(1-\epsilon^2\sin^2\theta)^2}$$

The co-ordinates of the corners of the infinitely small quadrilateral on the sphere are

$$\theta,\ \omega \qquad \theta,\ \omega+d\omega \qquad \theta+d\theta,\ \omega \qquad \theta+d\theta,\ \omega+d\omega$$

Taking ξ, η as the projection of θ, ω, Taylor's Theorem gives for the remaining co-ordinates

$$\xi+\frac{d\xi}{d\omega}d\omega \qquad \eta+\frac{d\eta}{d\omega}d\omega$$

$$\xi+\frac{d\xi}{d\theta}d\theta \qquad \eta+\frac{d\eta}{d\theta}d\theta$$

$$\xi+\frac{d\xi}{d\theta}d\theta+\frac{d\xi}{d\omega}d\omega \qquad \eta+\frac{d\eta}{d\theta}d\theta+\frac{d\eta}{d\omega}d\omega$$

and, consequently, for the area of this parallelogram

$$\begin{vmatrix} \frac{d\xi}{d\omega}, & \frac{d\xi}{d\theta} \\ \frac{d\eta}{d\omega}, & \frac{d\eta}{d\theta} \end{vmatrix} d\omega\, d\theta$$

The condition of projection thus requires

$$\begin{vmatrix} \frac{d\xi}{d\omega}, & \frac{d\xi}{d\theta} \\ \frac{d\eta}{d\omega}, & \frac{d\eta}{d\theta} \end{vmatrix} = \theta$$

It will clearly be necessary to choose some value for either ξ or η which shall be a function of θ and ω; suppose we take

$$\xi = f(\theta, \omega)$$

then the differential coefficients $\frac{d\xi}{d\omega}$ and $\frac{d\xi}{d\theta}$ will also be known functions of θ and ω; writing

$$\mathrm{P} = \frac{d\xi}{d\omega} \qquad \mathrm{Q} = \frac{d\xi}{d\theta}$$

the above equation of condition can be written in the form

$$\mathrm{P}\frac{d\eta}{d\theta} - \mathrm{Q}\frac{d\eta}{d\omega} = \theta$$

To integrate a partial differential equation of this form we know (*vide* Boole's Differential Equations, page 324) that it is necessary to form the system of simultaneous differential equations

$$\frac{d\theta}{\mathrm{P}} = -\frac{d\omega}{\mathrm{Q}} = \frac{d\eta}{\theta}$$

deduce their general integrals in the form $u = a$, $v = b$, and construct the equation $\mathrm{F}(u, v) = 0$ which will be the general solution sought. We can proceed directly as follows: From the first and second differential expressions given above we have

$$\mathrm{P}d\omega + \mathrm{Q}d\theta = 0$$

which leads at once to the integral

$$f(\theta, \omega) = c$$

when c is an arbitrary constant. Solve this equation for ω and we obtain

$$\omega = f_1(\theta, c)$$

which being substituted in P will make that quantity a function of θ only; designate this form of P by the symbol P_θ. Of course, P_θ contains c also, but that will not affect the integrations, so we need not indicate its presence. Now we have again, from the above simultaneous equations,

$$d\eta = \frac{\theta d\theta}{\mathrm{P}}$$

This becomes, on substituting the new value of P,

$$d\eta = \frac{\theta d\theta}{\mathrm{P}_\theta}$$

from which

$$\eta = c' + \int \frac{\theta d\theta}{\mathrm{P}_\theta}$$

Now, in general, since $f(\theta, \omega) = d\xi$, we have

$$\mathrm{P}d\omega + \mathrm{Q}d\theta = d\xi$$

writing then

$$\xi = c \qquad \eta - \int \frac{\theta d\theta}{\mathrm{P}_\theta} = c'$$

and forming the arbitrary relation $c' = \mathrm{F}(c)$, we obtain

$$\eta - \int \frac{\theta d\theta}{\mathrm{P}_\theta} = \mathrm{F}(\xi)$$

and for the equations of projections without alterations of areas

$$\xi=f(\theta,\omega) \qquad \eta=\mathrm{F}(\xi)+\int\frac{\theta d\theta}{\mathrm{P}_\theta}$$

or, finally, regarding ξ as constant in the expression for $\frac{d\xi}{d\omega}$,

$$\text{(I)} \qquad \xi=f(\theta,\omega) \qquad \eta=\mathrm{F}(\xi)+\int\frac{\theta d\theta}{\left(\frac{d\xi}{d\omega}\right)_\theta}$$

Designating the integral $\int\frac{\theta d\theta}{\left(\frac{d\xi}{d\omega}\right)_\theta}$ by $\Psi(\theta,\xi)$ and forming the partial derivatives of η, we have

$$\frac{d\eta}{d\omega}=\frac{d\xi}{d\omega}\mathrm{F}'(\xi)+\frac{d\xi}{d\omega}\frac{d\Psi}{d\xi} \qquad \frac{d\eta}{d\theta}=\frac{d\xi}{d\theta}\mathrm{F}'(\xi)+\frac{d\xi}{d\theta}\frac{d\Psi}{d\xi}+\frac{d\Psi}{d\theta}$$

Eliminating $\mathrm{F}'(\xi)$ and $\frac{d\Psi}{d\xi}$, we have

$$\begin{vmatrix} \frac{d\xi}{d\omega}, & \frac{d\xi}{d\theta} \\ \frac{d\eta}{d\omega}, & \frac{d\eta}{d\theta} \end{vmatrix}=\frac{d\Psi}{d\theta}\frac{d\xi}{d\omega}=\theta$$

which is the equation implying the principle of the conservation of areas. Therefore, if equations (I) are satisfied, the areas of corresponding parts of the surface and the plane of projection will be the same.

From the usual considerations we know that the cosine of the angle between the projections of a meridian and a parallel is proportional to

$$\frac{d\xi}{d\omega}\frac{d\xi}{d\theta}+\frac{d\eta}{d\omega}\frac{d\eta}{d\theta}$$

and if this angle is right, we have the condition

$$\frac{d\xi}{d\omega}\frac{d\xi}{d\theta}+\frac{d\eta}{d\omega}\frac{d\eta}{d\theta}=0$$

which, combined with

$$\frac{d\xi}{d\omega}\frac{d\eta}{d\theta}-\frac{d\xi}{d\theta}\frac{d\eta}{d\omega}=\theta$$

removes the indeterminate nature of the functions F and f. If the angle is not right, we have still the relation

$$\tan\omega=\frac{\frac{d\xi}{d\omega}\frac{d\eta}{d\theta}-\frac{d\xi}{d\theta}\frac{d\eta}{d\omega}}{\frac{d\xi}{d\omega}\frac{d\xi}{d\theta}+\frac{d\eta}{d\omega}\frac{d\eta}{d\theta}}$$

The arbitrary functions of the problem might also be determined by the condition that the meridians and parallels should be projected into curves of a given kind. The general solution of the problem in this case would be extremely difficult, if, indeed, not quite impossible, though particular cases are readily solved. Suppose, for example, that we wish the parallels to be projected in right lines. Take the equator for axis of ξ, then η must be independent of the longitude, and thus

$$\eta=\varphi(\theta)$$

Therefore in the general formulas it is necessary to place $\mathrm{F}(\xi)=0$, and to have the differential coefficient $\frac{d\xi}{d\omega}$ independent of ω; that is, ξ must have the form

$$\xi=\omega f(\theta)+\Psi(\theta)$$

or, if ξ is to vanish with ω,

$$\xi=\omega f(\theta)$$

then

$$\eta=\int\frac{\theta d\theta}{f(\theta)}=\varphi(\theta)$$

from which

$$f(\theta)=\frac{\theta}{\varphi'(\theta)}$$

and consequently

$$\xi=\frac{\omega\theta}{\varphi'(\theta)} \qquad \eta=\varphi(\theta)$$

are the equations of the problem. These become for a spherical earth

$$\eta=\varphi(\theta) \qquad \xi=\frac{r^2\omega\cos\theta}{\varphi'(\theta)}$$

For a spheroidal earth they are

$$\eta=\varphi(\theta) \qquad \xi=\frac{r^2\omega\cos\theta\,(1-e^2)}{(1-e^2\sin^2\theta)^2\varphi'(\theta)}$$

If n is any arbitrary constant, we can write for the simplest value of η

$$\eta=nr\theta=\varphi(\theta)$$

from which $\varphi'(\theta)=nr$, and consequently

$$\xi=\frac{r\omega}{n}\cdot\frac{\cos\theta\,(1-e^2)}{(1-e^2\sin^2\theta)^2}$$

or, for a spherical earth,

$$\xi=\frac{r\omega\cos\theta}{n} \qquad \eta=nr\theta$$

If we place $n=1$ we have a projection in which the lengths of the degrees of the parallels are conserved upon the projection. This is Sanson's projection, ordinarily and improperly called *Flamsteed's*. It has been often employed to represent the entire surface of the globe. The general appearance of the projection is that of a species of oval, with the major axis twice the length of the shorter. The angle of intersection of the representations of meridians and parallels is seen to be right only upon the central meridian and the equator,

$$\frac{d\xi}{d\omega}\frac{d\xi}{d\theta}+\frac{d\eta}{d\omega}\frac{d\eta}{d\theta}=-r^2\omega\cos\theta\sin\theta$$

the equations for the projection being

$$\xi=r\theta \qquad \eta=r\omega\cos\theta$$

and this is zero only for $\omega=0$ or $\theta=0$. The value of the right-hand member, or $\frac{r^2\omega}{2}\sin 2\theta$ increases very rapidly with θ, and the angle of intersection consequently changes very rapidly from its value of 90°. For the equation of the projection of the meridians we have

$$\eta=r\omega\cos\frac{\xi}{r}$$

a transcendental curve.

The co-ordinates of this projection are now, for a sphere,

$$\xi=\frac{\pi}{180}\omega r\cos\theta \qquad \eta=\frac{\pi}{180}r\theta$$

for a spheroid they are

$$\xi=\frac{\pi}{180}\frac{r\omega\cos\theta}{(1-e^2\sin^2\theta)^{\frac{1}{2}}} \qquad \eta=\frac{\pi}{180}\frac{r\theta(1-e^2)}{(1-e^2\sin^2\theta)^{\frac{3}{2}}}$$

If ϵ is the complement of the angle between a meridian and parallel

$$\tan\epsilon=\frac{\pi}{180}\omega\sin\theta$$

If B is the angle on the sphere formed by a meridian and any other curve, and β the projection of this angle, we have

$$\tan(\beta-\alpha)=\tan B-\tan\alpha=\frac{\sin(B-\alpha)}{\cos B\cos\alpha}$$

For the element $d\sigma$ of a meridian on the projection we have

$$d\sigma=\frac{ds}{\cos\alpha}$$

ds being the corresponding length on the sphere; this is, for a sphere,

$$d\sigma=r\,d\theta\sqrt{1+\omega^2\sin^2\theta}$$

and consequently

$$\sigma=r\int(1+\omega^2\sin^2\theta)^{\frac{1}{2}}\,d\theta$$

This is an elliptic integral, and depends for its solution on the rectification of an arc of an ellipse with semi-axes $=r\sqrt{1+\omega^2}$ and r respectively.

If the meridians are to be projected into right lines perpendicular to the equator, we must make ξ a function of ω only, thus:

$$\xi=\frac{\omega}{n}$$

and

$$\varphi'(\theta)=n\cos\theta$$

then

$$\eta=\varphi(\theta)=n\sin\theta$$

This projection is a projection of Lambert's, called by Germain "Lambert's isocylindric projection." For $n=1$ and a radius of r these formulas are

$$\xi=r\omega\qquad\qquad \eta=r\sin\theta$$

We have thus a projection consisting of a series of equidistant parallels, straight lines at right angles to the equator, representing the meridians, and another series of parallels at right angles to the first, whose distances from the equator vary as $\sin\theta$, representing the parallels of latitude. The value of Θ is, for the sphere, $\Theta=r^2\cos\theta$, and for the spheroid

$$\Theta=\frac{r^2(1-e^2)\cos\theta}{(1-e^2\sin^2\theta)^2}$$

making $\sin\theta=x$

$$\eta=\frac{1}{r}\int\Theta\,d\theta=r(1-e^2)\int\frac{dx}{(1-e^2x^2)^2}=r(1-e^2)\left[\frac{x}{2}\cdot\frac{1}{1-e^2x^2}+\frac{1}{4e}\log\frac{1+ex}{1-ex}+c\right]$$

or

$$\eta=r(1-e^2)\left\{\frac{\sin\theta}{2(1-e^2\sin^2\theta)}+\frac{1}{4e}\log\frac{1+e\sin\theta}{1-e\sin\theta}+c\right\}$$

If η is to vanish with θ, c must $=0$. Developing the value of η,

$$\frac{1}{1-e^2\sin^2\theta}=1+e^2\sin^2\theta+e^4\sin^4\theta+\ .\ .\ .\ .\ .\ e^{2n}\sin^{2n}\theta+\ .\ .\ .\ .\ .$$

$$\frac{1}{2}\log\frac{1+e\sin\theta}{1-e\sin\theta}=e\sin\theta+\frac{1}{3}e^3\sin^3\theta+\frac{1}{5}e^5\sin^5\theta+\ .\ .\ .\ .\ .\ \frac{1}{2i+1}e^{2i+1}\sin^{2i+1}\theta+\ .\ .\ .\ .\ .$$

Neglecting powers of e higher than the third

$$\eta=r(1-e^2)(\sin\theta+\tfrac{2}{3}e^2\sin^3\theta)$$

or

$$\eta=r\sin\theta-re^2\sin\theta\,(1-\tfrac{2}{3}\sin^2\theta)$$

Another simple example of the use of the general formulas is to write

$$\xi=\rho\omega=\frac{r\omega\cos\theta}{\sqrt{1-e^2\sin^2\theta}}$$

Then

$$\left(\frac{d\xi}{d\omega}\right)_\theta = \frac{r\cos\theta}{\sqrt{1-\epsilon^2\sin^2\theta}}$$

and consequently

$$\eta = r(1-\epsilon^2)\int\frac{d\theta}{(1-\epsilon^2\sin^2\theta)^{\frac{3}{2}}}$$

This is, of course, an elliptic integral, but as it is quite simple we may reduce it a little further. Passing at once to the usual notation employed in elliptic functions,

$$\eta = r(1-\epsilon^2)\int\frac{d\theta}{[\Delta(\theta)]^3}$$

As ϵ is the modulus, $\sqrt{1-\epsilon^2}$ is the complementary modulus, and as usual write it ϵ'; then

$$\eta = r\epsilon'^2\int\frac{d\theta}{[\Delta(\theta)]^3}$$

Now

$$\frac{d}{d\theta}\,\frac{\sin\theta\cos\theta}{\Delta(\theta)} = \frac{1-2\sin^2\theta+\epsilon^2\sin^4\theta}{[\Delta(\theta)]^3}$$

Writing, for more convenience, Δ instead of $\Delta(\theta)$,

$$\epsilon^2\frac{d}{d\theta}\,\frac{\sin\theta\cos\theta}{\Delta} = \frac{\Delta^4-\epsilon'^2}{\Delta^3} = \Delta - \frac{\epsilon'^2}{\Delta^3}$$

and thence by integration

$$\int\frac{d\theta}{\Delta^3} = \frac{1}{\epsilon'^2}\int\Delta\,d\theta - \frac{\epsilon^2\sin\theta\cos\theta}{\epsilon'^2\Delta}$$

Denoting as usual the elliptic integral of the second kind by $E(\theta)$, the modulus being understood to be ϵ, we obtain for η

$$\eta = rE(\theta) - \frac{\epsilon^2\sin\theta\cos\theta}{\Delta}$$

The further discussion of this general value of η would only be interesting from a purely mathematical point of view; so we shall not dwell any longer upon it, but will expand Δ in order to get the approximate value of η. Expanding Δ and neglecting higher powers of ϵ than the third,

$$\int\sqrt{1-\epsilon^2\sin^2\theta}\,d\theta = \int d\theta[1-\tfrac{1}{2}\epsilon^2\sin^2\theta] = \int d\theta[1-\tfrac{1}{4}\epsilon^2+\tfrac{1}{4}\epsilon^2\cos 2\theta] = (1-\tfrac{1}{4}\epsilon^2)\,\theta+\tfrac{1}{8}\epsilon^2\sin 2\theta$$

also, under the same conditions,

$$\frac{r\epsilon^2\sin\theta\cos\theta}{\sqrt{1-\epsilon^2\sin^2\theta}} = \tfrac{1}{2}r\epsilon^2\sin 2\theta$$

Thus we have for η, if the constant of integration be assumed $=0$,

$$\eta = r(1-\tfrac{1}{4}\epsilon^2)\,\theta - \tfrac{3}{8}r\epsilon^2\sin 2\theta$$

which, for the sphere, becomes

$$\eta = r\theta$$

and also

$$\xi = r\omega\cos\theta$$

the Sanson Projection. We have seen that the isocylindric projection of Lambert had the equator divided into equal subdivisions, and the central meridian into divisions the upper (or lower, if below the equator) points of which were at distances from the equator proportional to the sine of the latitude. This would obviously not be a very advantageous projection for countries which, like America, have their greatest extent in a north-and-south direction. For such a case, Lambert proposed, so to speak, to turn the preceding projection through a right angle, dividing the central meridian into equal parts and the equator into parts depending upon the sine of the longitude. This projection is Lambert's transverse isocylindric projection. The conditions of the problem are that for $\omega=0$, $\xi=0$ and $\eta=\theta$; and that for $\theta=0$, $\xi=\sin\omega$ and $\eta=0$. These require for ξ the value

$$\xi=r\sin\omega\cos\theta$$

This gives

$$\frac{d\xi}{d\omega}=r\cos\omega\cos\theta \qquad \left(\frac{d\xi}{d\omega}\right)_\theta=\sqrt{r\cos^2\theta-\xi^2}$$

As η is to be $=0$, for $\theta=0$ we must have $F(\xi)=0$; then

$$\eta=\int\frac{\cos\theta d\theta}{\sqrt{r\cos^2\theta-\xi^2}}=\sin^{-1}\frac{\sin\theta}{\sqrt{1-\xi^2}}$$

The equations for the transverse isocylindric projection are then

$$\xi=r\sin\omega\cos\theta \qquad \sin\eta=\frac{\sin\theta}{\sqrt{1-r^2\sin^2\omega\cos^2\theta}}$$

or

$$\tan\eta=\frac{\tan\theta}{\sqrt{1-r^2\sin^2\omega}}$$

Eliminating θ we have for the meridians the equation

$$\xi^2(1+\tan^2\eta\cos^2\omega)=r^2\sin^2\omega$$

and similarly for the parallels

$$\tan^2\eta(r^2\cos^2\theta-\xi^2)=r^2\sin^2\theta$$

These are transcendental curves, and in order to construct them a series of points must be found, and the curves drawn through them. This projection is symmetrical with respect to the equator and the central meridian, and this allows us to obtain at once the four points which have for geographical co-ordinates the same northern or southern latitude, or the east or west longitude. The formula

$$\xi=r\sin\omega\cos\theta$$

shows further that we obtain the same value of ξ for the points

$$\theta=50 \qquad \omega=60^\circ$$

and

$$\theta'=90^\circ-\omega=30^\circ \qquad \omega'=90^\circ-\theta=40^\circ$$

From the equation of the meridians we obtain for the tangent of the angle made with the axis of ξ by the tangent to any meridian

$$\frac{d\eta}{d\xi}=-\frac{(1+\tan^2\theta)\cos\omega}{(\cos^2\omega+\tan^2\theta)\sin\omega\sin\theta}$$

For $\theta=0$, $\frac{d\eta}{d\xi}=\infty$, which shows that the meridians cut the equator at right angles. For $\theta=90^\circ$, *i. e.*, at the pole, the tangent of the angle made by a meridian with the first meridian is $\frac{d\xi}{d\eta}$ and is equal to

$$-\left\{\frac{(\cos^2\omega\ \tan^2\theta)\sin\omega}{(1+\tan^2\theta)\cos\omega}\right\}_{\theta=90}=\left\{\frac{\left(\frac{\cos\omega}{\tan^2\theta}+1\right)\sin\omega}{\left(\frac{1}{\tan^2\theta}+1\right)\cos\omega}\right\}_{\theta=90}=\tan\omega$$

or, at the poles, the meridians make their true angles with the principal meridian. In like manner we find for the parallels

$$\frac{d\eta}{d\xi}=\frac{\sin\theta\tan\omega}{\cos^2\theta(\cos^2\omega+\tan^2\theta)}.$$

For $\omega=0$, $\frac{d\xi}{d\eta}=0$, or the parallels are perpendicular to the central meridian; and for $\omega=90$, $\frac{d\eta}{d\xi}=\infty$, which shows that the parallels are perpendicular to the meridian of 90°, or that this meridian is parallel to the equator.

Considering the earth as spherical, let θ denote the angle made with a given meridian by any other curve on the sphere, and let τ denote the corresponding angle on the plane of projection. Of course this θ has no reference or connection with the function θ already used in this chapter. For the determination of τ we have the formula

$$\tan\tau=\frac{\tan\theta}{\cos^2\phi}$$

As we already know, $\tau=\theta$ for $\theta=0$, or $\theta=\frac{\pi}{2}$. The maximum alteration at each point corresponds to

$$\tan\theta_0=\pm\cos\phi$$

or to

$$\tan\tau_0=\pm\frac{1}{\cos\phi}$$

The alteration ϕ_0 then, or $\tau_0-\theta_0$, is given by

$$\tan\phi_0=\frac{\sin^2\phi}{2\cos\phi}$$

This is $=0$ for $\phi=0$, and increases rapidly, with ϕ becoming $=\infty$ for $\phi=\frac{\pi}{2}$.

Resuming the formulas by means of which the parallels are projected into straight lines

$$\xi=\frac{r^2\,\omega\cos\phi}{\theta'(\phi)}\qquad\qquad \eta=\theta(\phi)$$

let us find the equations of the system which permit the parallels to be projected into right lines and the meridians into a group of right lines, passing through the projection of the pole.

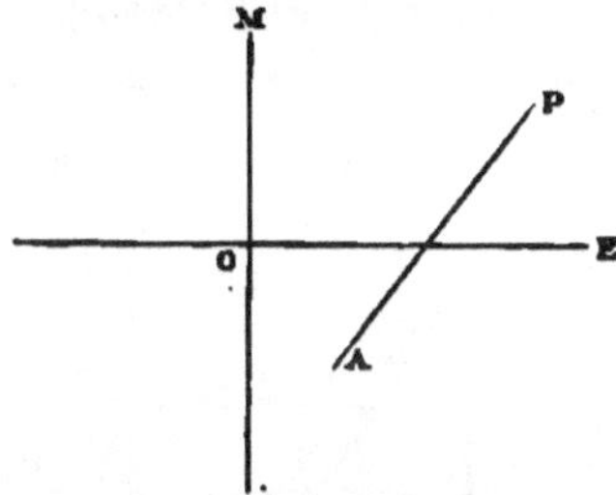

Fig. 28.

Take O, Fig. 28, the origin, at the intersection of the equator and first meridian, and let PA denote the projection of any meridian passing through P, the projection of the pole. Then the equation of PA is

$$\xi=(b-\eta)f(\omega)$$

b denoting the ordinate of P. Now, since $\eta=\theta(\phi)$,

$$\theta'(\phi)=\frac{d\eta}{d\phi}$$

therefore

$$(b-\eta)f(\omega)=\frac{r^2\,\omega\cos\phi}{\frac{d\eta}{d\phi}}$$

The right-hand side of this equation is linear in ω, $\frac{d\eta}{d\phi}$ being a function of ϕ only; the left-hand side must therefore be also of the first degree, or $f(\theta)=c\theta$, c being a constant. Clearing of fractions, we obtain the differential equation

$$c(b-\eta)\,d\eta=r^2\cos\theta\,d\theta$$

and by integration

$$c\left(b\eta-\frac{\eta^2}{2}\right)=c'+r^2\sin\theta$$

Now η vanishes with θ, therefore $c'=0$; and since $\eta=b$ for $\theta=90^\circ$, we must also have

$$\frac{cb^2}{2}=r^2 \qquad c=\frac{2r^2}{b^2}$$

The equations of this system are thus

$$2\left(b\eta-\frac{\eta^2}{2}\right)=b^2\sin\theta \qquad \xi=2(b-\eta)\frac{r^2}{b^2}\omega$$

M. Collignon determined the constant b as a function of r by making the projection fulfill the condition of being in the form of a square; that is, by making the limiting meridians compose the sides of a square. This is accomplished by assuming that for $\theta=0$ and $\omega=\pm\frac{\pi}{2}$, we must have $\xi=\pm b$. On the substitution of these values we come immediately to the relation $b=r\sqrt{\pi}$, and thus the equations of projection are seen to become

$$\xi=\frac{2\omega}{\pi}(r\sqrt{\pi}-\eta) \qquad \eta^2-2r\sqrt{\pi}\eta+\pi r^2\sin\theta=0$$

Since η is always $<r\sqrt{\pi}$, we need only to use the root

$$\eta=r\sqrt{\pi}\,(1-\sqrt{1-\sin\theta})$$

or

$$\eta=r\sqrt{\pi}\,[1-\sqrt{2}\sin\tfrac{1}{2}(90^\circ-\theta)]$$

The figure being symmetrical with respect to the equator, its construction is very simple, and we need not go into the details here.

The next case that we shall take up is that of equivalent projection, when the parallels are projected into concentric circles.

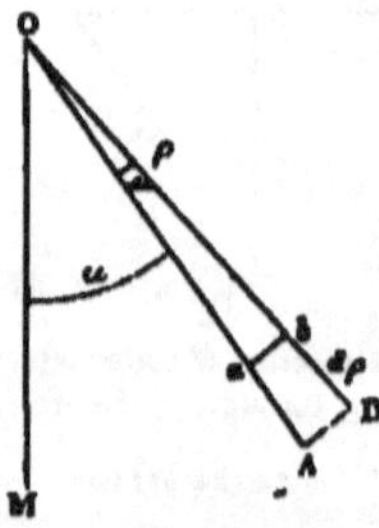

FIG. 29.

Take O, Fig. 29, as the center of the projections of all the parallels, OM as the initial line of a system of polar co-ordinates (ρ, α), and, for final simplicity, choose OM as the projection of the first meridian; then of course $\alpha=0$, whence $\omega=0$. The area of the small quadrilateral, as AB, included between two infinitely near concentric parallels and two meridians, making an infinitely small angle, $d\alpha$, with each other, is $\rho d\rho d\alpha$. Equating this to the corresponding element on the surface, we have

$$\rho d\rho\, d\alpha=\Theta d\omega\, d\theta$$

from which is obtained

$$\frac{d\alpha}{d\omega}=\frac{\theta}{\rho\frac{d\rho}{d\theta}}$$

For the integration of this it is only necessary to note that ρ is a function of θ alone, and as, for the surfaces which we are considering, the sphere and spheroid, θ is also only a function of θ, the right-hand side of this equation is independent of ω, and consequently we have at once, by integration,

$$\alpha=\frac{\theta\omega}{\rho\frac{d\rho}{d\theta}}+F(\theta)$$

the arbitrary function $F(\theta)$ being added instead of an absolute constant; but we assumed $\alpha=0$ when $\omega=0$, therefore $F(\theta)$, and we have, for the equations defining this projection,

$$\alpha=\frac{\omega\theta}{\rho\frac{d\rho}{d\theta}} \qquad \rho=f(\theta)$$

The simplest case that we can assume is when ρ is a linear function of θ, or $\rho=r(a'+a\theta)$, or simply $\rho=m+n\theta$. Then for the spheroid

$$\rho\alpha=\frac{r^2(1-e^2)\omega\cos\theta}{n(1-e^2\sin^2\theta)^2}$$

and for the sphere

$$\rho\alpha=\frac{r^2\omega\cos\theta}{n}$$

The constants m, n, can be determined by subjecting the projection to further conditions. First, assume the projection of the pole for the centre of the circles, then, since $\theta=\frac{\pi}{2}$ gives $\rho=0$, we have

$$m=n\frac{\pi}{2}$$

and consequently

$$\rho=n\left(\theta-\frac{\pi}{2}\right)$$

or, since the absolute value of ρ is all that we are concerned with,

$$\rho=n\left(\frac{\pi}{2}-\theta\right)$$

The simplest supposition that we can make with reference to the constant n is $n=r$; then the equations of projection become

$$\rho=r\left(\frac{\pi}{2}-\theta\right) \qquad \alpha=\frac{\omega\cos\theta}{\frac{\pi}{2}-\theta}$$

The radii of the projections of the parallels are in this case equal to the complements of the arcs of the meridians upon the sphere which measure the latitude; from the second of these equations we also have

$$\rho\alpha=r\omega\cos\theta$$

or the degrees of longitude are projected in their true lengths. This projection was invented by Johann Werner, of Nürnberg, in 1514; it is obviously a desirable way of representing polar regions. Still another method of determining the arbitrary constants is to assume that for $\theta=\theta_1$ we have $\rho=\rho_1$. This gives

$$\rho_1=m+n\theta$$

Subtracting this from $\rho=m+n\theta$,

$$\rho=\rho_1+n(\theta-\theta_1)$$

If, in the case of a sphere, we assume for ρ_1 the value $\rho_1 = r \cot \theta_1$ and also $n = -r$, we obtain

$$\rho = r \cot \theta_1 + r(\theta - \theta_1)$$

This gives us Bonne's projection which has been treated at length in another place.

Resume now the formulæ which determine the projections of the parallels as concentric circles; these are, for the case of the sphere,

$$\rho = f(\theta) \qquad \alpha = \frac{-\omega r^2 \cos \theta}{\rho \frac{d\rho}{d\theta}}$$

The function f may be determined by introducing the condition that the meridians and parallels shall cut at right angles. Since the parallels are concentric circles, the meridians must clearly be diameters of these circles making equal angles with each other. If we assume that the angle between the projection of any two meridians is to the angle between the meridians themselves as 1 is to n, we shall have

$$\alpha = \frac{\omega}{n}$$

and, as an easy consequence,

$$\rho d\rho = n r^2 \cos \theta d\theta$$

from which, by integrating,

$$\frac{\rho^2}{2} = c - n r^2 \sin \theta$$

The minus sign is necessary, for as ρ increases, θ diminishes, and thus $d\rho$ and $d\theta$ have opposite signs. Since ρ is $=0$ at the pole, or for $\theta = 90^\circ$, we have for the constant of integration

$$c = n r^2$$

Therefore

$$\frac{\rho^2}{2} = n r^2 (1 - \sin \theta)$$

or, introducing the complementary angle $\varphi = 90^\circ - \theta$,

$$\rho^2 = 4 r^2 n \sin^2 \frac{\varphi}{2}$$

The equations of this projection are now

$$\rho = 2\sqrt{n}\, r \sin \frac{\varphi}{2} \qquad Q = \frac{\omega}{n}$$

the value of ρ is very easy to construct from this formula, in which $2r \sin \frac{\varphi}{2}$ denotes the chord of the arc of the meridian which joins the pole to the parallel of latitude θ.

The coefficient n being arbitrary, we can give it what value we please, or can determine it by subjecting the projection to some other condition. Representing by φ the mean latitude of the region to be projected, required to determine n in such a manner that the degree of mean latitude θ_1 shall preserve its true ratio to the degree of longitude. For this condition it is necessary that

$$\rho \alpha = \rho \frac{\omega}{n} = \rho \omega \sin \varphi_1$$

Now

$$\rho = 2r \sqrt{n} \sin \frac{\varphi_1}{2}$$

then

$$\frac{2r}{\sqrt{n}} \sin \frac{\varphi_1}{2} = r \sin \varphi_1$$

from which

$$n = \frac{1}{\cos^2 \frac{\varphi_1}{2}} = \frac{2}{1 + \cos \varphi_1}$$

If we take $\varphi_1=45°$, we find $n=1.171$, and the entire angle at the pole of the hemisphere equals 307° 25′. This projection has been called by Germain "Lambert's isospherical stenoteric projection."* For $n>1$ it is an actual conic projection, *i. e.*, one obtained by the development of a cone either tangent or secant to the sphere.

Before taking up the important case where $n=1$ it will be well to speak of Alber's projection,† obtained by developing a secant cone which passes through two parallels of latitude whose lengths it is required to preserve. Call these two parallels θ and θ_1; then if A be the area of the zone included between them

$$A=2\pi r^2(\sin\theta_2-\sin\theta_1)=2\pi r^2\cos\frac{\theta_2+\theta_1}{2}\sin\frac{\theta_2-\theta_1}{2}$$

The area of the corresponding portion of the cone is

$$A'=\pi r\delta(\cos\theta_2+\cos\theta_1)=2\pi r\delta\cos\frac{\theta_2+\theta_1}{2}\cos\frac{\theta_2-\theta_1}{2}$$

when $\delta=\rho_1-\rho_2$ the difference of the radii of these two parallels upon the chart. From the condition of equivalence $A=A'$ we derive

$$\delta=2r\tan\frac{\theta_2-\theta_1}{2}$$

The radius ρ_2 is readily found to be

$$\rho^2=\frac{\delta\cos\theta_2}{\cos\theta_1-\cos\theta_2}=\frac{r\cos\theta_2}{\sin\frac{\theta_2+\theta_1}{2}\cos\frac{\theta_2-\theta_1}{2}}$$

and, since $\rho_1=\rho_2+\delta$,

$$\rho_1=\frac{r\cos\theta_1}{\sin\frac{\theta_2+\theta_1}{2}\cos\frac{\theta_2-\theta_1}{2}}$$

These may be written simply

$$\rho_1=\frac{r\cos\theta_1}{k}\qquad\qquad\rho_2=\frac{r\cos\theta_2}{k}$$

in which

$$k=\frac{1}{\sin\frac{\theta_2+\theta_1}{2}\cos\frac{\theta_2-\theta_1}{2}}$$

We have, obviously, by equating the length of the arc (θ_1) on the projection to the corresponding arc on the sphere,

$$\rho_1\alpha=\omega r\cos\theta_1$$

or

$$\alpha=\frac{\omega}{k}$$

that is, the angles between the meridians are altered in the ratio $1:k$.

The area of the infinitesimal element of the conical surface comprised between two consecutive parallels, and two consecutive meridians is $=\alpha\rho d\rho$; the same element upon the sphere is $\omega r^2\cos\theta d\theta$; then, as $d\rho$ and $d\theta$ are of opposite signs, we have

$$\alpha\rho d\rho=-\omega r^2\cos\theta d\theta$$

Substituting here for α its value $\frac{\omega}{k}$, we obtain immediately, on integrating,

$$\rho^2=2r^2k(\sin\theta_1-\sin\theta)+\rho_1^2$$

* Projection isosphérique sténotère de Lambert.

† Beschreibung einer neuen Kegelprojection, von H. C. Albers, Zach's Monatliche Correspondenz, 1805.

From this we can derive the radius of the projection of any parallel; for example, make $\theta=0$, *i. e.*, the equator, then for the radius of its projection there results

$$\rho_0^2=2r^2k \sin \theta_1+\rho_1^2$$

Again, the pole is projected into an arc of a circle; for make $\theta=\frac{\pi}{2}$ and then

$$\rho^2=\rho_1^2-2r^2k(1-\sin \theta_1)$$

When the difference of latitude of the two parallels whose length is to be preserved is very small the alteration in this system is very slight. The distances in the central zone are increased from north to south, and diminished from east to west, and the greatest error is upon the central parallels.

CENTRAL EQUIVALENT PROJECTION.

The projection that we designate by this title is spoken of by Germain as "Zenithal equivalent," but in adopting the above title the author has preferred to choose a term as nearly as possible like that adopted by Collignon when he described the projection. This was "Système central d'égale superficie." * This system is founded upon the principle of elementary geometry that the area of a zone equals the product of the circumference of a great circle by the height of the zone. The same law of area holding for a spherical segment or zone of one base, we have, calling h the altitude of the zone, (area of zone or segment) $=\pi 2rh$. But $2rh=$(chord of half the arc)^2; therefore the area of the zone is equal to the area of the circle whose radius is equal to the rectilinear distance from the pole of the zone to the circumference which serves as a base. If from the pole of the zone we draw two arcs of great circles including a certain definite angle, and from the center of the equivalent circle two radii including the same angle, the portion of the zone bounded by its base and these two arcs will be equal to the sector of the circle cut out by the two corresponding radii. This gives us, then, an obvious manner of representing any portion of a given spherical surface without alteration of area; any point can be assumed upon the sphere as center, so for simplicity the pole of the equator is chosen; the parallels are seen to be transformed into concentric circles, and the meridians into straight lines passing through the common center.

Taking now the projection of the principal meridian as the axis of ξ, and as usual writing $\varphi=90^\circ-\theta$, we have, for the equation of the meridians,

$$\eta=\xi \tan \omega$$

and for the parallels

$$\xi^2+\eta^2=4r^2 \sin^2 \frac{\varphi}{2}$$

from which

$$\xi=2r \sin \frac{\varphi}{2} \cos \omega \qquad \eta=2r \sin \frac{\varphi}{2} \sin \omega$$

and consequently

$$\frac{d\xi}{d\omega}=-2r \sin \frac{\varphi}{2} \sin \omega \qquad \frac{d\xi}{d\theta}=-r \cos \frac{\varphi}{2} \cos \omega$$

$$\frac{d\eta}{d\omega}=2r \sin \frac{\varphi}{2} \cos \omega \qquad \frac{d\eta}{d\theta}=r \cos \frac{\varphi}{2} \sin \omega$$

Substituting these in our general differential equation

$$\begin{vmatrix} \frac{d\xi}{d\omega}, & \frac{d\xi}{d\theta} \\ \frac{d\eta}{d\omega}, & \frac{d\eta}{d\theta} \end{vmatrix}=\theta$$

and we find

$$\frac{d\xi}{d\omega}\frac{d\eta}{d\theta}-\frac{d\xi}{d\theta}\frac{d\eta}{d\omega}=2r^2 \sin \frac{\varphi}{2} \cos \frac{\varphi}{2} \sin^2 \omega+2r^2 \sin \frac{\varphi}{2} \cos \frac{\varphi}{2} \cos^2 \omega=r^2 \sin \varphi=r^2 \cos \theta$$

* Journal de l'École Polytechnique, cahier 41; "representation de la surface du globe terrestre"; E. Collignon

which verifies our supposition of equal areas. It is also easy to see that

$$\frac{d\xi}{d\omega}\frac{d\xi}{d\theta}+\frac{d\eta}{d\omega}\frac{d\eta}{d\theta}=0$$

or the meridians and parallels cut at right angles on the chart as on the sphere.

ALTERATION OF ANGLES.

The alteration of angles is zero at the center of the chart. At any point whatever of the chart, M, Fig. 30, draw a line MM′ such that the corresponding direction upon the sphere shall make an angle θ with the meridian; we wish to find the angle Ψ upon the chart made by this line

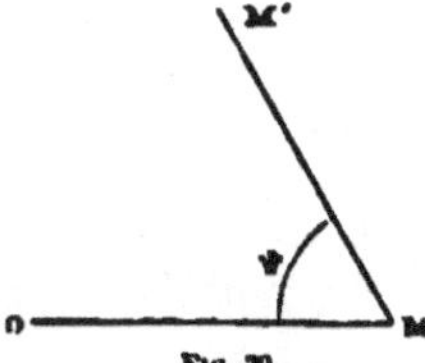

FIG. 30.

with the projection of the meridian, *i. e.*, with the line drawn from M to the center O. Let φ and ω represent the geographical co-ordinates of M, and $\varphi+d\varphi$, $\omega+d\omega$ the geographical co-ordinates of M′ infinitely near to M; then

$$\tan\theta=\cos\varphi\frac{d\omega}{d\varphi}=\sin\varphi\frac{d\omega}{d\varphi}\qquad\qquad \tan\Psi=2\tan\frac{\varphi}{2}\frac{d\omega}{d\varphi}$$

from which

$$\tan\Psi=\frac{\tan\theta}{\cos^2\frac{\varphi}{2}}$$

Since

$$\tan\phi=\frac{\tan\Psi-\tan\theta}{1+\tan\Psi\tan\theta}=\frac{\tan\theta\left(1-\cos^2\frac{\varphi}{2}\right)}{\cos^2\frac{\varphi}{2}+\tan^2\theta}$$

the maximum of alteration $\Psi-\theta$, or ϕ, corresponds to the direction for which

$$\Psi+\theta=\frac{\pi}{2};$$

and in seeking for the maximum of this, since $1-\cos^2\frac{\varphi}{2}$ is constant, we need only consider the factor

$$\frac{\tan\theta}{\cos^2\frac{\varphi}{2}+\tan^2\theta}.$$

Equating to zero the derivative of this with respect to θ, there results simply

$$\tan\theta=\pm\cos\frac{\varphi}{2}$$

Consequently

$$\tan\Psi=\pm\frac{1}{\cos\frac{\varphi}{2}}$$

the upper signs being taken together, and also the lower ones. From these follows

$$\tan\theta \tan\theta'=1$$

from which, as in a former case, excluding negative arcs and arcs greater than $\frac{\pi}{2}$, there results,

$$\theta+\theta'=\frac{\pi}{2}$$

We can deduce from this that the maximum deviation for the direction OM is given by

$$\tan(\theta'-\theta)=\tfrac{1}{2}(\tan\theta'-\tan\theta)=\tfrac{1}{2}\tan\frac{\varphi}{2}\sin\frac{\varphi}{2}$$

The angle θ, upon the sphere, of maximum deviation is $=45^\circ$ for $\varphi=0$, *i. e.*, at the center of the chart; θ then decreases while θ' (and consequently φ) increases. When $\varphi=\frac{\pi}{2}$

$$\tan=\theta\ \frac{1}{\sqrt{2}} \qquad \tan\theta'=\sqrt{2}$$

The angular alteration is thus seen to increase continuously from the center to that point of the sphere which is diametrically opposite the assumed center. It is evidently useless to prolong the chart so far as that, and, indeed, the custom is, in this projection, to represent the map in two parts, one for each hemisphere.

ALTERATION OF LENGTHS.

In the direction OM the projection substitutes for the arc on the sphere the chord of the same arc. Let φ, as usual, represent the angular distance OM; then the length of this line upon the sphere is $=r\varphi$ and its length upon the chart, *i. e.*, the length of the chord of the arc OM, is $=2r\sin\frac{\varphi}{2}$. Differentiation of each of these gives us the lengths of the element of the meridian upon the sphere and upon the chart; these are $rd\varphi$ and $r\cos\frac{\varphi}{2}d\varphi$. Thus the meridional elements are reduced upon the chart in the ratio $\cos\frac{\varphi}{2}:1$. The converse is true concerning the elements of the parallels; they are augmented in the ratio $1:\cos\frac{\varphi}{2}$; this is obvious on account of the necessity for conserving the areas.

Suppose now that upon the sphere we take any element ds making the angle θ with the meridian OM; its projection upon MO will $=ds\cos\theta$, and perpendicular to MO will be $=ds\sin\theta$; similarly, if $d\sigma$ correspond upon the chart to ds upon the sphere, $d\sigma\cos\theta'$ will be the projection of $d\sigma$ upon the radius OM, and $d\sigma\sin\theta'$ will be the projection of the same element in the direction perpendicular to OM. Now, since the projection does not alter the right angle at which $ds\cos\theta$ and $ds\sin\theta$ cut each other, we will have

$$ds\cos\theta\cos\frac{\varphi}{2}=d\sigma\cos\theta' \qquad ds\sin\theta\frac{1}{\cos\frac{\varphi}{2}}=d\sigma\sin\theta'$$

from which, by squaring and adding,

$$d\sigma^2=ds^2\left(\cos^2\theta\cos^2\frac{\varphi}{2}+\sin^2\theta\frac{1}{\cos^2\frac{\varphi}{2}}\right)$$

Now the expression in parenthesis reduces to unity when upon the sphere,

$$\tan\theta=\cos\frac{\varphi}{2}$$

or, when upon the chart,

$$\tan\theta'=\frac{1}{\cos\frac{\varphi}{2}}.$$

that is, for the direction of maximum deviation. This direction then possesses the remarkable property of conserving the lengths.

Now, through any given point upon the sphere, and upon the chart, as M, we can draw two curves which shall cut all the meridians MO of the sphere, and the radii MO of the chart under the angles θ and θ' in such a way that the distances on these two curves between any two corresponding points shall be the same. The curves so constructed are called by Collignon "isoperimetric curves." The curve upon the sphere passes through O′, the antipodal point to O, and winding round the sphere becomes indefinitely near to O, a logarithmic spiral which cuts the meridians at an angle of 45°. Upon the chart, the isoperimetric curve for small values of φ, that is for points near the center, is very nearly the logarithmic spiral which cuts the radii under the angle of 45°; for increasing values of φ, θ also increases and is $=90°$ for $\varphi=180°$; the curve then touches the circle into which the point O′ has been transformed and is continued beyond this point in a branch symmetrical to the first.

To obtain the polar equation of the isoperimetric curve upon the chart, take $\rho=d\omega$ and ω the angle between ρ and some fixed axis. Now

$$\rho\frac{d\omega}{d\rho}=\tan\theta'=\frac{1}{\cos\frac{\varphi}{2}}$$

but

$$\rho=2r\sin\frac{\varphi}{2}$$

therefore

$$d\omega=\frac{d\rho}{\rho\sqrt{1-\frac{\rho^2}{4r^2}}}$$

the differential equation of the sought curve. For the integration, observe that we have

$$d\rho=r\cos\frac{\varphi}{2}d\varphi$$

which, substituted in the first written equation, gives

$$d\omega=\frac{\frac{d\varphi}{2}}{\sin\frac{\varphi}{2}}$$

and by integration

$$\omega=\log\tan\frac{\varphi}{4}+c$$

This equation joined with $\rho=2r\sin\frac{\varphi}{2}$ gives the means of constructing the curve.

For the element of arc of the isoperimetric curve we have obviously

$$ds=\sqrt{d\rho^2+\rho^2d\omega^2}=d\varphi\sqrt{r^2\cos^2\frac{\varphi}{2}+4r^2\sin^2\frac{\varphi}{2}\frac{1}{4\sin^2\frac{\varphi}{2}}}$$

or

$$ds=rd\varphi\sqrt{1+\cos^2\frac{\varphi}{2}}$$

If we write $\frac{\varphi}{2}=\theta$, this equation becomes very simply

$$ds=\sqrt{2}\,r\sqrt{1-\tfrac{1}{2}\sin^2\theta}\,d\theta$$

or

$$s=\sqrt{2}\,r\int\Delta(k,\theta)\,d\theta \qquad k=\sqrt{\tfrac{1}{2}}$$

an elliptic integral of the second kind, which gives the rectification of the arc of the ellipse, whose eccentricity is $=\sqrt{\frac{1}{2}}$.

The element of the isoperimetric curves is, in polar co-ordinates,

$$\tfrac{1}{2}\rho^2\, d\alpha = r^2 \sin\frac{\varphi}{2}\, d\varphi$$

and the integral of this is

$$= \text{const.} - 2r^2 \cos\frac{\varphi}{2}$$

TRANSFORMATION OF A GREAT CIRCLE.

The angle between the planes of two great circles on the sphere is measured by the arc of a great circle joining their poles. This property affords the means of determining the differential equation of the curve upon the chart which represents the great circle on the sphere.

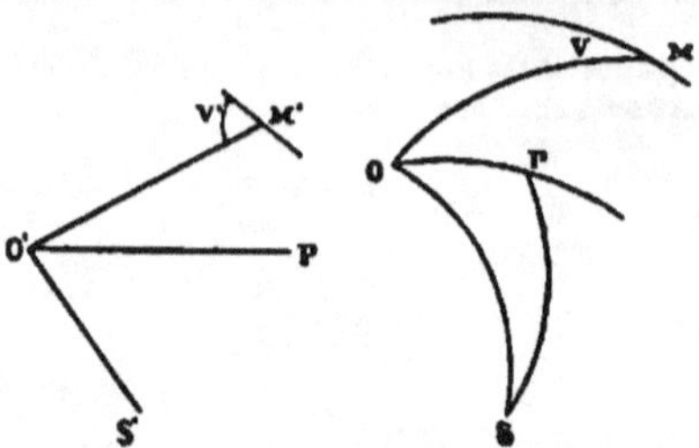

FIG. 31.

Take O, Fig. 31, for the central point, and P for the pole of a great circle which passes through a point, M. The same letters accented denote the corresponding points upon the chart. It is proposed at M′ to draw a tangent to the curve which passes through this point and represents the great circle through M. Join O′ and M′, and call O′M′ $=\rho$ and M′O′P′ $=\alpha$, the line O′P′ being taken as the initial line. Let S, upon the sphere, denote the pole of the great circle OM, which passes through the center, O, and cuts the given circle at M; this point S will be found in the plane of a great circle, OS, perpendicular to that of OM at the point O; the angle V is measured by the arc SP. We have now, in the spherical triangle OSP,

$$\cos SP = \cos OS \cos OP + \sin OS \sin OP \cos POS$$

or, since OS is a quadrant,

$$\cos SP = \sin OP \cos POS = \sin OP \sin \alpha$$

OP is a constant arc that we may call λ, then we have

$$\cos V = \sin \lambda \sin \alpha$$

The angle V on the sphere of course corresponds with V′ upon the chart, and the connecting relation is

$$\tan V = \frac{\tan V'}{\cos^2 \frac{\varphi}{2}}$$

φ being the angular distance OM. But taking the radius of the sphere as unity,

$$\tan V' = \frac{\rho\, d\alpha}{d\rho} \pm \qquad \rho = 2 \sin\frac{\varphi}{2}$$

Eliminating V, V′, and φ between these four equations, we will arrive at the differential equation sought. The first of these equations affords the relation

$$\sin V = \sqrt{1 - \sin^2 \lambda \sin^2 \alpha}$$

and, consequently,

$$\tan V = \frac{\sqrt{1-\sin^2\lambda\sin^2\omega}}{\sin\lambda\sin\omega}$$

and then

$$\frac{\rho d\omega}{d\rho} = \frac{\sqrt{1-\sin^2\lambda\sin^2\omega}}{\sin\lambda\sin\omega} \cdot \frac{1}{\sqrt{1-\frac{\rho^2}{4}}}$$

The constant of integration will be determined by observing that the great circle of which P is the pole passes through the pole of the great circle OP; so, for $\omega=\frac{\pi}{2}$ or $\frac{3\pi}{2}$, we should have $\rho=\sqrt{2}$.

The equation of the projected great circles can be better arrived at in another manner, to the explanation of which we shall now proceed.

Conceive, first, that a stereographic projection has been made—that is, the parallels and meridians have been constructed—with the point of sight at the center, or the antipodal point to the center, of the proposed central equivalent projection.

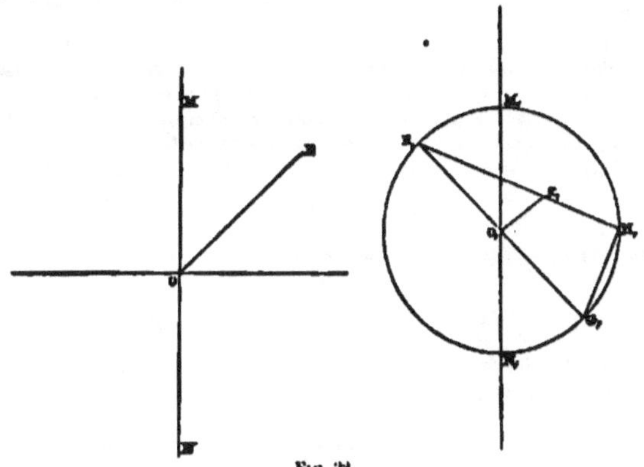

FIG. 32.

Let E_1 (Fig. 32) denote any point of the stereographic projection, and O_1 the center, or point of sight, represented on the central equivalent projection by $O_1M_1N_1$, the meridian through O represented on the other chart by MN; O_1M_1 is equal to the radius of the sphere. Required to find the position of the point on the central equivalent projection represented by E_1 on the stereographic projection. Lay off at O the angle $MOE=M_1O_1E_1$. The point sought is on the line OE, and the distance OE is consequently all that has to be determined. Draw the diameter F_1G_1 perpendicular to O_1E_1; join F_1E_1 and produce it to H_1; join G_1H_1; then G_1H_1 is the distance required.

Another method for constructing the central equivalent from a stereographic projection is as follows:

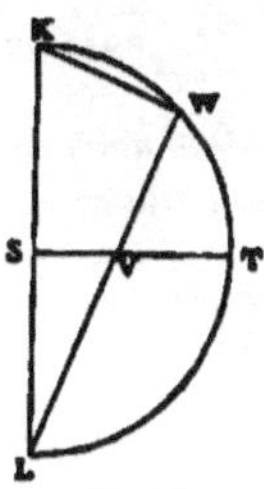

FIG. 33.

In Fig. 33 the length KW is the distance on the central equivalent corresponding to SV on the stereographic projection. The similar triangles KWL and VSL give

$$SV : KW :: LV : KL$$

Dividing through by ST, the radius, and observing that $LV=\sqrt{SV^2+SL^2}$, we find

$$\frac{SV}{ST} : \frac{KW}{ST} = \sqrt{1+\left(\frac{SV}{ST}\right)^2} : 2$$

which gives the ratio $\frac{KW}{ST}$ as a function of $\frac{SV}{ST}$. Calling the former of these ratios C, and the latter S, we have for the formula of transformation from the stereographic, or S, projection to the central equivalent, or C, projection

$$C=\frac{2S}{\sqrt{1+S^2}}$$

If we write $S=\tan\psi$, we have

$$C=2\sin\psi$$

We are now prepared to solve a much more general problem than the one proposed above, viz, to find the equation of the central equivalent projection of any circle of the sphere, whether great or small. Denoting as usual by ξ, η the rectangular co-ordinates on the required projection, let ξ', η' denote rectangular co-ordinates on the auxiliary stereographic projection. The circle of the sphere will be a circle upon the stereographic chart, and if its center is at α', β', its radius ρ' will be given by

$$(\xi'-\alpha')^2+(\eta'-\beta')^2=\rho'^2$$

But according to the proposed plan of transformation

$$\xi'=\xi\sqrt{\frac{r^2}{4r^2-(\xi^2+\eta^2)}} \qquad \eta'=\eta\sqrt{\frac{r^2}{4r^2-(\xi^2+\eta^2)}}$$

Consequently we have for the equation of the curve on the central equivalent projection, which represents a circle on the sphere,

$$\left(\xi\sqrt{\frac{r^2}{4r^2-(\xi^2+\eta^2)}}-\alpha'\right)^2+\left(\eta\sqrt{\frac{r^2}{4r^2-(\xi^2+\eta^2)}}-\beta'\right)^2=\rho'^2$$

or, transforming to polars by means of the formulas $\xi=\rho\cos\theta$, $\eta=\rho\sin\theta$,

$$\left(\frac{r\rho\cos\theta}{\sqrt{4r^2-\rho^2}}-\alpha'\right)^2+\left(\frac{r\rho\sin\theta}{\sqrt{4r^2-\rho^2}}-\beta'\right)^2=\rho'^2$$

A still further simplification is possible by writing

$$k=\sqrt{\alpha'^2+\beta'^2} \qquad \psi=\tan^{-1}\frac{\beta'}{\alpha'}$$

The equation becomes now

$$\left(\frac{r\rho}{\sqrt{4r^2-\rho^2}}\right)^2-\frac{2r\rho k}{\sqrt{4r^2-\rho^2}}\cos(\theta-\psi)+(k^2-\rho'^2)=0$$

This is merely the polar equation of the circle in which the stereographic radius vector ρ, has been replaced by its value $\frac{r\rho}{\sqrt{4r^2-\rho^2}}$ as a function of the radius vector in the central equivalent system. The equation in Cartesian co-ordinates shows that the curve is of the fourth degree. The equation in polar co-ordinates enables us readily to determine the condition that the curve shall represent a great circle of the sphere. Make $\theta=\psi$ then

$$\frac{r\rho}{\sqrt{4r^2-\rho^2}}=k\pm\rho'$$

From which we obtain

$$\rho=\pm\sqrt{\frac{4(k\pm\rho')^2 r^2}{r^2+(k\pm\rho')^2}}$$

This affords four real values for ρ. The signs + and − in the numerator and denominator of this quantity are to be taken in this manner

$$\begin{matrix}+\\+\end{matrix}\ ,\quad \begin{matrix}-\\-\end{matrix}$$

Now, in order that the polar equation shall represent a great circle of the sphere, it is necessary and sufficient that the sums of the squares of the two values of ρ obtained by taking ρ' first with the + and second with the − sign under the radical shall be equal to $4r^2$, or that we shall have

$$\frac{(k+\rho')}{r^2+(k+\rho')^2}+\frac{(k-\rho')^2}{r^2+(k-\rho')^2}=1$$

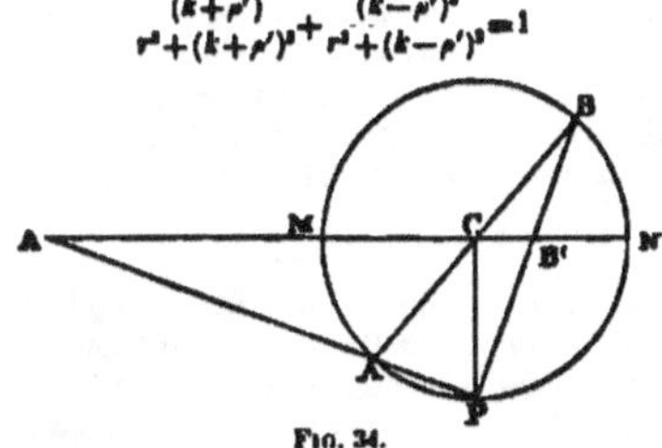

FIG. 34.

That this is a correct formula is easily seen from the following simple geometrical considerations. Let C (Fig. 34) denote the center, and AMBNP the orthographic projection of the sphere; P is the point of sight of the orthographic projection, and the plane MN parallel to the tangent plane at P is the plane of this projection; let AB denote the trace of a plane cutting a great circle from the sphere; and finally let A'B' denote the projection of this great circle; then we have

$$CA'=k+\rho' \qquad CB'=k-\rho'$$

and also, since $PA'^2=CP^2+CA'^2$,

$$\frac{(k+\rho')^2}{r^2+(k+\rho')^2}=\cos^2 PA'C \qquad \frac{(k-\rho')^2}{r^2+(k-\rho')^2}=\cos^2 PB'C$$

But APB = A'PB' is a right angle, and consequently PB'C and PA'C are complementary angles, and the sum of the squares of their cosines is equal to unity. *Q. E. D.*

LOXODROMIC CURVES.

The pole being taken as center, it is very easy to obtain the loxodromic curve. Denote by θ the angle made on the sphere by such a curve with a meridian; then, ψ denoting the corresponding angle on the chart, we have

$$\tan\psi=\tan\theta\,\frac{1}{\cos^2\frac{\varphi}{2}}$$

Now, $\tan\theta$ is constant, and, for $r=1$, $2\sin\frac{\varphi}{2}=\rho$, and also $\tan\psi=\frac{\rho d\alpha}{d\rho}$. The differential equation of the curve is then

$$\frac{\rho d\alpha}{d\rho}=\frac{\tan\theta}{1-\frac{\rho^2}{4}}$$

from which follows

$$d\alpha=\tan\theta\,\frac{d\rho}{\rho\left(1-\frac{\rho^2}{4}\right)}$$

and integrating

$$\alpha = \tan\theta \log \frac{\rho}{\sqrt{4-\rho^2}} + c$$

PROJECTION UPON THE PLANE OF A MERIDIAN.

We will now take up the case of the projection upon the plane of any meridian of the parallels and meridians of the terrestrial sphere; the center will be upon the equator, and the given meridional plane will cut the equator in two points, distant each 90° from the center.

A few definitions will be adopted, both for brevity and clearness of language.

The *central station* is the point of the sphere chosen as center by the map; this we shall designate by O upon the sphere, and by O′ upon the projection.

The *central distance* of a point N of the sphere is the ratio of the length of the arc MO of a great circle to the radius of the sphere; this we shall denote by λ; it is the quantity that, in the case of the pole being taken as the central station, we have heretofore denoted by φ.

The *radius vector* ρ of the point N′ upon the chart is the distance O′N′ of this point from the central station. As usual, r denoting the radius of the sphere, we have

$$\rho = 2r \sin \tfrac{1}{2}\lambda$$

The *azimuthal angle* of the point M upon the sphere is the angle α formed by the arc OM with the meridian through O; upon the chart it is the equal angle formed by the right angle O′ M′ with the meridian through O′, which is also, as we know, a right line.

Now, having given the position of O, we wish to determine the values of ρ and α in terms of the geographical co-ordinates (θ, ω) of any point whatever, as M. We have already resolved the problem for the case when O is assumed as the pole of the sphere, and a very simple transformation of co-ordinates enables us to resolve it for this more general case where O is taken upon the equator.

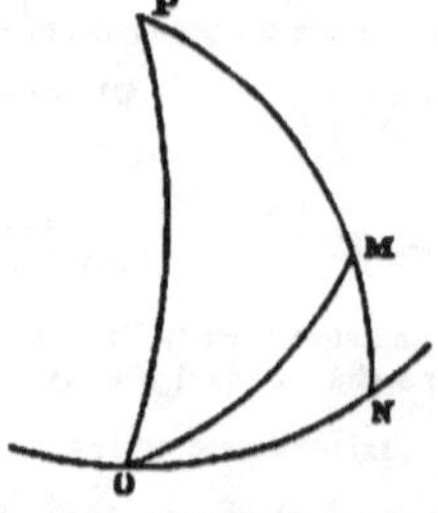

FIG. 35.

Take OP, Fig. 35, for principal meridian; ω is the longitude of M with respect to this meridian; the portion ON of the equator included between O and the point of intersection of the equator with the meridian through M is measured by ω, and the arc MN is measured by θ; the angle MON is the complement of α, and finally OM is $=\lambda$. Now, since N is a right angle, we have in the triangle OMN

$$\cos\lambda = \cos\omega \cos\theta \qquad \tan\alpha = \sin\omega \cot\theta$$

which determine λ and α; ρ is determined by

$$\rho = 2r \sin \tfrac{1}{2}\lambda$$

It is obvious that λ, and consequently ρ, remains the same for all values of ω and θ which give the same value for $\cos\omega \cos\theta$; for example, for the two points of which the latitude of the one equals the longitude of the other.

Take for the axis of ξ and η the right lines representing respectively the equator and the first meridian, and we have, in consequence,

$$\xi=\rho \sin \alpha \qquad \eta=\rho \cos \alpha$$

or

$$\rho=\sqrt{\xi^2+\eta^2} \qquad \tan \alpha=\frac{\xi}{\eta}$$

But

$$\rho=2r \sin \tfrac{1}{2} \lambda \qquad \cos \lambda=\cos \omega \cos \theta \qquad \tan \alpha=\sin \omega \cot \theta$$

The second of these relations gives

$$\sin \tfrac{1}{2} \lambda=\sqrt{\frac{1-\cos \lambda}{2}}=\sqrt{\frac{1-\cos \omega \cos \theta}{2}}$$

so that

$$\xi^2+\eta^2=2r^2(1-\cos \omega \cos \theta)$$

and

$$\xi=\eta \sin \omega \cot \theta$$

These are the formulas of transformation from angular to rectilinear co ordinates. The elimination of θ between these equations gives us the equation of the meridian whose longitude is ω, and the elimination of ω in like manner gives the equation of the parallel of latitude θ.

EQUATION OF THE MERIDIANS.

The result of the elimination of θ is the equation

$$\xi^2+\eta^2=2r^2\left(1-\frac{\xi \cos \omega}{\sqrt{\xi^2+\eta^2 \sin^2 \omega}}\right)$$

By clearing of fractions and radicals this becomes

$$\xi^6+(2+\sin^2 \omega)\xi^4\eta^2+(1+2\sin^2 \omega)\xi^2\eta^4+\sin^2 \omega\,\eta^6-4r^2(\xi^4+\eta^4 \sin^2 \omega)$$
$$-4r^2(1+\sin^2 \omega)\xi^2\eta^2+4r^4(\xi^2+\eta^2)\sin^2 \omega=0$$

This equation of the sixth degree is easily factored into

$$(\xi^2+\eta^2)(\xi^4+[(1+\sin^2 \omega)\eta^2-4r^2]\xi^2+(\eta^4-4r^2\eta^2+4r^4)\sin^2 \omega)=0$$

The factor to be suppressed here is obviously the binomial $\xi^2+\eta^2$, as equating that to zero would only result in giving an imaginary locus (or infinitely small circle), and, in consequence, would be of no practical use. We have, then, remaining a biquadratic equation in ξ and η.

If we write $\xi^2=\xi'$ and $\eta^2=\eta'$, the equation becomes one of the second degree in ξ' and η', viz:

$$\xi'^2+(1+\sin^2 \omega)\xi'\eta'+\eta'^2 \sin^2 \omega-4r^2\xi'-4r^2 \sin^2 \omega\,\eta'+4r^4 \sin^2 \omega=0$$

This last is the equation of an hyperbola whose center is at the intersection of the lines

$$2\xi'+(1+\sin^2 \omega)\eta'-4r^2=0 \qquad (1+\sin^2 \omega)\xi'+2\sin^2 \omega\,\eta'-4r^2 \sin^2 \omega=0$$

or at the point

$$\xi'=-4r^2 \tan^2 \omega \qquad \eta'=+4r^2 \sec^2 \omega$$

Calling m_1 and m_2 the angular coefficients which determine the asymptotes, these quantities are obtained as the roots of the equation

$$m^2 \sin^2 \omega+(1+\sin^2 \omega)m+1=0$$

From which

$$m_1=-\frac{1}{\sin^2 \omega} \qquad m_2=-1$$

Confining ourselves to the region when ξ' and η' are both positive, we can readily construct this hyperbola, on any chosen scale, for each value of ω; then construct the required curve whose co-ordinates, measured on the same scale, are the square roots of ξ' and η', the co-ordinates of each point on the hyperbola. For $\eta=0$ we have

$$\xi=\pm 2r\sin\frac{\omega}{2} \qquad \xi=\pm 2r\cos\frac{\omega}{2}$$

For $\xi=0$ we have

$$\eta=\pm r\sqrt{2}$$

Since $\sin^2\omega=\sin^2(-\omega)$ and the equation of the curve contains only $\sin^2\omega$, the equation represents at the same time the projections of the meridians of longitudes ω and $-\omega$ respectively; these two curves will be symmetrically situated the one to the other with respect to the axis of η. If on the axis of ξ

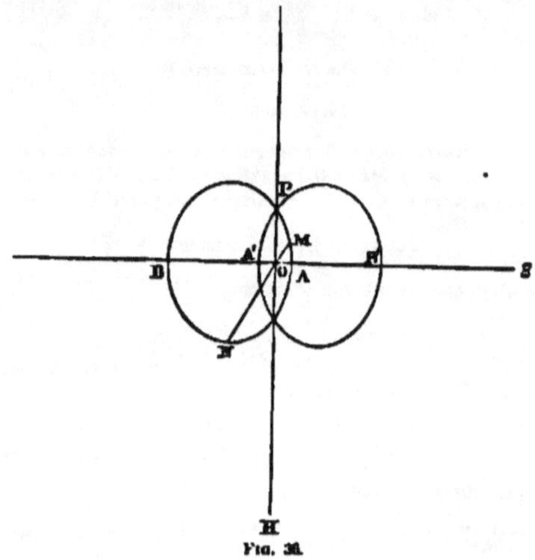

Fig. 36.

we take (Fig. 36) $OA=2r\sin\frac{\omega}{2}$ and $OB=2r\cos\frac{\omega}{2}$, and on the axis of η take $OP=OP^{iv}=r\sqrt{2}$, the curve will pass through the four points A, P, B, P^{iv} and the entire locus will be composed of this curve, and the curve $A'PBP^{iv}$ symmetric to the first with respect to the axis of η.

EQUATION OF A PARALLEL.

To obtain this equation we eliminate ω by the relation

$$\sin\omega=\frac{\xi\tan\theta}{\eta}$$

and obtain

$$\xi^2+\eta^2=2r^2\left(1-\cos\theta\sqrt{\frac{\eta^2-\xi^2\tan^2\theta}{\eta^2}}\right)$$

By clearing this of fractions and radicals, we arrive at an equation of the sixth degree in η and of the fourth in ξ, which will contain only the even powers of the variables. As in the case of the equation of a meridian, this will contain the factor $\xi^2+\eta^2$, and dividing out by this factor we obtain, as the resulting equation of a parallel,

$$\eta^4+(\xi^2-4r^2)\,\eta^2+4r^4\sin^2\theta=0$$

Substitute again $\xi^2=\xi'$ and $\eta^2=\eta'$, and we are conducted to the equation

$$\eta'^2+\xi'\eta'-4r^2\eta'+4r^4\sin^2\theta=0$$

which is of the second degree in ξ', η' and, as in the former case, representing a hyperbola. The center of the hyperbola is on the axis of ξ' and is given $\xi'=4r^2$; one asymptote is parallel to and therefore coincident with the axis of ξ. The same construction being made as before, we obtain for the projection of the parallel of latitude θ the curve ARBS (Fig. 37) and of latitude $-\theta$ the curve

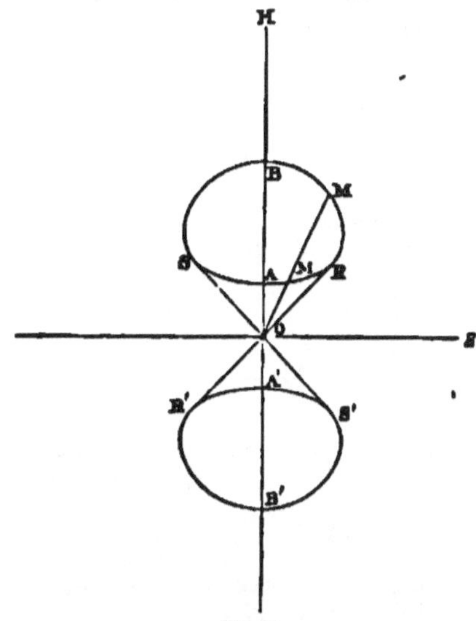

FIG. 37.

A'B'R'S'. These two curves are symmetrically situated with respect to the axis of ξ, and the sum of the squares of the intercepts made by any line OM' with one branch of the curve is constant and equal to the square of the diameter of the sphere, *i. e.*,

$$OM'^2+OM^2=4r^2$$

The truth of this is easily seen if we transform the equation of the parallels into polar co-ordinates, that is, write

$$\xi=\rho\cos\chi \qquad \eta=\rho\sin\chi$$

The equation then becomes

$$\rho^4\sin^4\chi+(\rho^2\cos^2\chi-4r^2)\rho^2\sin^2\chi+4r^4\sin^2\theta$$

Making the obvious reductions, this is

$$\rho^4-4r^2\rho^2+\frac{4r^4\sin^2\theta}{\sin^2\chi}=0$$

Calling the roots of this ρ_1 and ρ_2, we have $\rho_1^2=OM^2$ and $\rho_2^2=OM'^2$; and from the known principles of the theory of equations

$$\rho_1^2+\rho_2^2=4r^2$$

MOLLWEIDE'S PROJECTION.

This projection was invented by Prof. C. B. Mollweide, of Halle, in 1805, and in 1857 a number of applications of it were made by Babinet, whose name thus became attached to it, the projection being known commonly as "Babinet's homalographic projection." The problem proposed for solution here is to represent the entire surface of the earth in an ellipse, the ratio of whose major and minor axes represented by the equator and first meridian respectively shall be 2 : 1; the parallels are to be projected in parallel right lines and the meridians in ellipses, all of which pass through two fixed points, the poles, and each zone of the sphere to be represented upon the chart in its true size.

Let b and $2b$ denote the axes of the limiting ellipse, then the included area will be $=2b^2\pi$; but this is to equal the entire area of the sphere, or $4\pi r^2$; this condition then gives us for the axes of this ellipse

$$b=\sqrt{2}\,r \qquad 2b=2\sqrt{2}\,r$$

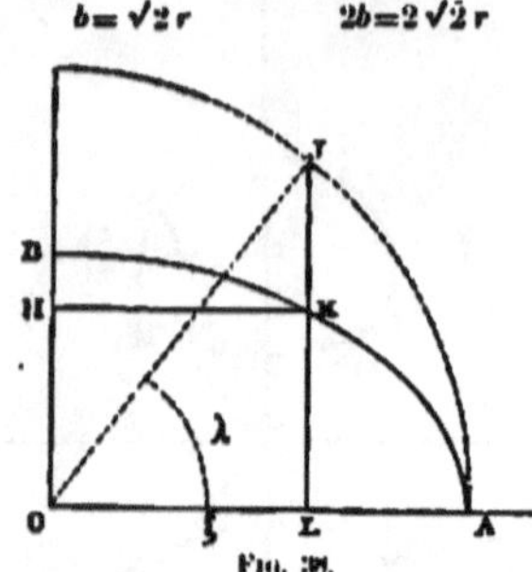

FIG. 38.

The area (Fig. 38) of the elliptic segment ALK = area of circular segment LAJ multiplied by $\frac{OH}{OA}$, that is, by $\frac{1}{2}$. Now, the area of LAJ is equal to the sector OAJ minus the triangle OLJ, or

$$LAJ=\tfrac{1}{2}(2r\sqrt{2})^2\cos^{-1}\frac{\xi}{2r\sqrt{2}}-\frac{2r\sqrt{2}\,\xi\eta}{r\sqrt{2}}=4r^2\cos^{-1}\frac{\xi}{2r\sqrt{2}}-2\xi\eta$$

and then for the elliptic segment we have only to divide this by 2; add to this result the area of the rectangle OLKH or $\xi\eta$, and we obtain, finally,

$$OAKH=2r^2\cos^{-1}\frac{\xi}{2r\sqrt{2}}+\tfrac{1}{2}\xi\eta$$

Assume for the angle AOJ the symbol λ; then follows

$$\cos^{-1}\frac{\xi}{2r\sqrt{2}}=\lambda \qquad \xi=2r\sqrt{2}\cos\lambda \qquad \eta=r\sqrt{2}\sin\lambda$$

and consequently

$$OAKH=2r^2\lambda+r^2\sin 2\lambda$$

This surface is, however, to be equal to the area of the semi-zone between the equator and parallel of φ, or equal to $\pi r^2\sin\varphi$. Equating these, and we have for the fundamental equation of the Mollweide projection

$$\pi\sin\varphi=\sin 2\lambda+2\lambda$$

The values of λ or $\sin\lambda$ have to be obtained from this equation for each given value of φ. Lay off, then, on the semi-minor axis of the ellipse, the lengths $r\sqrt{2}\sin\chi$ measured from the center, and the points so obtained will be the points of intersection of each parallel with the principal meridian or minor axis of the limiting ellipse. Through these points draw parallels to the equator, and they will represent the parallels. For the construction of the meridians by points it is only necessary to divide the equator and parallels in parts which correspond exactly to the points of division of these lines on the sphere. For example, if it is desired to draw the meridians of every ten degrees, we have only to divide the entire equator, and also the meridians of the chart, into 36 equal parts, and through the corresponding points thus obtained draw the ellipses representing the meridians.

For the computation of λ from the above equation, the following method of approximation answers very well. Assume a value λ' such that

$$\sin 2\lambda'+2\lambda'=\pi\sin\varphi'$$

when θ' differs but little from θ; call δ the correction to λ', that is, $\lambda'+\delta=\lambda$; then

$$\sin 2(\lambda'+\delta)+2(\lambda'+\delta)=\pi \sin \theta$$

Subtracting the first of these equations from the second, we have

$$\sin 2(\lambda'+\delta)-\sin 2\lambda'+2\delta=\pi(\sin \theta-\sin \theta')$$

or

$$2\cos(2\lambda'+\delta)+2\delta=\pi(\sin \theta-\sin \theta')$$

As δ will be a very small quantity, we can write

$$\sin \delta=\delta \qquad \cos(2\lambda'+\delta)=\cos 2\lambda'$$

Writing, then, for $\sin \theta$ its value, we obtain for δ the approximate value

$$\delta=\frac{\pi \sin \theta-(\sin 2\lambda'+2\lambda')}{2(1+\cos 2\lambda')}$$

This method of approximation can of course be carried as far as we choose, or until we reach any required degree of exactness. Table VII gives the values of $\sin \lambda$ for values of θ differing by 30′. This was computed by Jules Bourdin, and is more accurate and extended than the one computed by Mollweide himself for the values of λ.

We may just observe before leaving this subject that the equation

$$\pi \sin \theta=\sin 2\lambda+2\lambda$$

is readily derived from the differential equation for equivalent projections. This equation was, in our assumed case of the sphere,

$$\frac{d\xi}{d\omega}\frac{d\eta}{d\theta}-\frac{d\xi}{d\theta}\frac{d\eta}{d\omega}=\cos \theta$$

writing, for convenience, $r=1$. The equation of a meridian whose axes are $\sqrt{2}$ and $\frac{2\omega\sqrt{2}}{\pi}$, is

$$\frac{\pi^2 \xi^2}{8\omega^2}+\frac{\eta^2}{2}=1$$

from which

$$\xi^2=(2-\eta^2)\frac{4\omega^2}{\pi^2}$$

Combining this with the differential equation, we find for the determination of η, since $\frac{d\xi}{d\theta}=0$, the equation

$$2\frac{d\eta}{d\theta}\sqrt{2-\eta^2}=\pi \cos \theta$$

By integration this leads to

$$\eta\sqrt{2-\eta^2}+2\sin^{-1}\frac{\eta}{\sqrt{2}}=\pi \sin \theta+c$$

Since, however, $\eta=0$ and $\theta=0$ at the same time, we must have $c=0$, and so

$$\eta\sqrt{2-\eta^2}+2\sin^{-1}\frac{\eta}{\sqrt{2}}=\pi \sin \theta$$

in which we of course take the smallest arc whose sine is $=\frac{\eta}{\sqrt{2}}$. This equation shows that, since η depends only on θ, all points of the same latitude θ lie on a line parallel to the axis of ξ. Writing $\eta=\sqrt{2}\sin \lambda$, we deduce at once the fundamental equation

$$\pi \sin \theta=\sin 2\lambda+2\lambda$$

In conclusion we will examine briefly a projection proposed by M. Collignon, in which he represents the central equivalent projection in the form of a square. Suppose that, as in Mollweide's projection, the parallels are parallel right lines, and that the meridians are also right lines, parting from a common point, the pole. Let h represent the ordinate of the point taken as pole; then the equation of the meridians will be in the form

$$\xi=(h-\eta)f(\omega)$$

The origin is supposed placed at the foot of the perpendicular from the pole upon the equator; the function $f(\omega)$ is independent of θ. As in the Mollweide projection, η is a function of θ only, or

$$\frac{d\eta}{d\omega}=0$$

and, consequently,

$$\frac{d\xi}{d\omega}=(h-\eta)f'(\omega)$$

The condition for the conservation of surfaces now becomes

$$(h-\eta)f'(\omega)\frac{d\eta}{d\theta}=r^2\cos\theta$$

This would give $f'(\omega)$ as a function of θ, which is contradictory to the previous assumption made concerning $f(\omega)$; the interpretation of this is, since $f'(\omega)$ does not contain ω, that $f'(\omega)=m$, a constant, and so $f(\omega)$ is a linear function of the longitude ω, or

$$f(\omega)=m\omega+n$$

The equation of condition is thus

$$m(h-\eta)d\eta=r^2\cos\theta d\theta$$

from which, by integration, follows

$$m\left(h\eta-\frac{\eta^2}{2}\right)=c+r^2\sin\theta$$

Since for $\theta=\frac{\pi}{2}$ we have $\eta=h$, we find $c=\frac{1}{2}mh^2-r^2$; and again, since $\theta=0$ gives $\eta=0$, $c=0$ or $\frac{1}{2}mh^2=r^2$. Finally, since we wish the extreme meridians limiting the chart to form a square, it will be necessary, since $\theta=0$ and $\omega=\pm\frac{\pi}{2}$, that we have $\xi=\pm h$, the corresponding signs to be taken together; but

$$\xi=(h-\eta)(m\omega+n)$$

In this, making $\xi=\pm h$, and remembering that when $\theta=0$ then $\eta=0$, there follows

$$m\frac{\pi}{2}+n=1 \qquad -m\frac{\pi}{2}+n=-1$$

Solution of these equations gives

$$m=\frac{2}{\pi} \qquad n=0$$

and so, by virtue of the relation $h^2=\frac{2r^2}{m}$,

$$h=r\sqrt{\pi}$$

and finally the equation connecting θ and η is

$$\eta^2-2r\sqrt{\pi}\,\eta+\pi r^2\sin\theta=0$$

The projection need, of course, only be constructed for the positive values of θ, and then repeated symmetrically below the equator for the negative values of θ.

§ VII.

ON THE GENERAL THEORY OF ORTHOMORPHIC PROJECTION.

We have already given some account of the general theory of projections which preserve the angles, or, as we have called them, orthomorphic projections; but as the object in view heretofore has been merely the representation of the sphere, or spheroid, upon a plane, it has not been either necessary or desirable to linger long upon general theories which are ordinarily interesting only from a mathematical point of view. We shall now, however, resume the consideration of orthomorphic projection, give a fuller theoretical account of the subject, and make one or two applications to problems rather more difficult than any yet attempted.

Let the equation of a surface be given in the form

$$f(x, y, z)=0$$

x, y, z denoting rectangular, rectilinear co-ordinates. It is well known that the position of each point on this surface can be given in terms of two independent variables, say u and v, so that in general a definite point of this surface will correspond to certain definite values of u and v, and conversely. For brevity, we write now, as usual in this case,

$$\left(\frac{dx}{du}\right)^2+\left(\frac{dy}{du}\right)^2+\left(\frac{dz}{du}\right)^2=\mathrm{E} \qquad \frac{dx}{du}\frac{dx}{dv}+\frac{dy}{du}\frac{dy}{dv}+\frac{dz}{du}\frac{dz}{dv}=\mathrm{F} \qquad \left(\frac{dx}{dv}\right)^2+\left(\frac{dy}{dv}\right)^2+\left(\frac{dz}{dv}\right)^2=\mathrm{G}$$

The element of length ds^2 is given now by

$$ds^2=dx^2+dy^2+dz^2=\mathrm{E}du^2+2\mathrm{F}du\,dv+\mathrm{G}dv^2$$

Conceive now an elementary triangle on the surface whose vertices are given by u, v; $u+\delta_u$, $v+\delta_v$; $u+\delta'_u$, $v+\delta'_v$,—where δ_u and δ'_u are infinitely small increments of u, and δ_v and δ'_v are infinitely small increments of v. The rectangular co-ordinates of these points are

$$\begin{array}{llll}
\mathrm{A}: & x, & y, & z, \\
\mathrm{B}: & x+\frac{dx}{du}\delta_u+\frac{dx}{dv}\delta_v, & y+\frac{dy}{du}\delta_u+\frac{dy}{dv}\delta_v, & z+\frac{dz}{du}\delta_u+\frac{dz}{dv}\delta_v \\
\mathrm{C}: & x+\frac{dx}{du}\delta'_u+\frac{dx}{dv}\delta'_v, & y+\frac{dy}{du}\delta'_u+\frac{dy}{dv}\delta'_v, & z+\frac{dz}{du}\delta'_u+\frac{dz}{dv}\delta'_v
\end{array}$$

Now we have

$$\overline{\mathrm{AB}}^2=\mathrm{E}\delta_u^2+2\mathrm{F}\delta_u\delta_v+\mathrm{G}\delta_v^2$$

$$\overline{\mathrm{AC}}^2=\mathrm{E}\delta'^2_u+2\mathrm{F}\delta'_u\delta'_v+\mathrm{G}\delta'^2_v$$

and also

$$\overline{\mathrm{BC}}^2=\mathrm{E}(\delta_u-\delta'_u)^2+2\mathrm{F}(\delta_u-\delta'_u)(\delta_v-\delta'_v)+\mathrm{G}(\delta_v-\delta'_v)^2$$

Again

$$\mathrm{AB.AC}\cos\mathrm{BAC}=\left(\frac{dx}{du}\delta_u+\frac{dx}{dv}\delta_v\right)\left(\frac{dx}{du}\delta'_u+\frac{dx}{dv}\delta'_v\right)$$
$$+\left(\frac{dy}{du}\delta_u+\frac{dy}{dv}\delta_v\right)\left(\frac{dy}{du}\delta'_u+\frac{dy}{dv}\delta'_v\right)+\left(\frac{dz}{du}\delta_u+\frac{dz}{dv}\delta_v\right)\left(\frac{dz}{du}\delta'_u+\frac{dz}{dv}\delta'_v\right)$$

and this, by virtue of the last formula, becomes

$$=\mathrm{E}\delta_u\delta'_u+\mathrm{F}(\delta_u\delta'_v+\delta'_u\delta_v)+\mathrm{G}\delta_v\delta'_v$$

It is easy to obtain the equation

$$\overline{\mathrm{AB}}^2.\overline{\mathrm{AC}}^2\sin^2\mathrm{BAC}=4\,\mathrm{ABC}^2=(\mathrm{EG}-\mathrm{F}^2)(\delta_u\delta'_v-\delta'_u\delta_v)^2$$

If we remember that

$$(E\delta^2_u+2F\delta_u\delta_v+G\delta^2_v)(E\delta'^2_u+2F\delta'_u\delta'_v+G\delta'^2_v)$$
$$=(E\delta_u\delta'_u)+2F(\delta_u\delta'_v+\delta'_u\delta_v)+G\delta_v\delta'_v)+(EG-F^2)(\delta_u\delta'_v-\delta'_u\delta_v)^2=\overline{AB}^2.\overline{AC}^2$$

Substituting this in the equation giving $\overline{AB}^2.AC^2\cos^2 BAC$, we find readily the value of $\sin^2 BAC$, and this multiplied by the value of $\overline{AB}^2.\overline{AC}^2$, gives the above equation.

It is now clear that if we have another surface upon which the co-ordinates (ξ, η, ζ) of any point are also functions of u, v, the corresponding elements of both surfaces will be similar if the new functions, which we may designate as E′, F′, G′, are proportional to E, F, G; by corresponding points are meant points (x, y, z) and (ξ, η, ζ) which correspond on each surface to the same values of u, v. If the second surface upon which the given surface is to be projected is a plane, the problem is, of course, very much simplified. Considering ξ, η as the rectangular co-ordinates of a point in a plane, it is clear that they must be determined to satisfy the relation

$$d\xi^2+d\eta^2=m^2(E\,du^2+2\,F\,du\,dv+G\,dv^2)$$

or

$$\left(\frac{d\xi}{du}\right)^2+\left(\frac{d\eta}{du}\right)^2=m^2E \qquad \left(\frac{d\xi}{dv}\right)^2+\left(\frac{d\eta}{dv}\right)^2=m^2G \qquad \frac{d\xi}{du}\frac{d\eta}{du}+\frac{d\xi}{dv}\frac{d\eta}{dv}=m^2F$$

m denotes the ratio of alteration of lengths in the projection.

Since the elementary quantities du and dv are independent of each other, and since $d\xi+id\eta$ and $d\xi-id\eta$ are linear functions of these quantities, we can define the quantities $d\xi+id\eta$ and $d\xi-id\eta$ as linear factors of the quadratic expression $E\,du^2+2\,F\,du\,dv+G\,dv^2$, which are at the same time exact differentials. The same is true of the expressions which we may obtain by multiplying the quantities $d\xi+id\eta$ and $d\xi-id\eta$ by any functions of $\xi+i\eta$ and $\xi-i\eta$ respectively. In order, then, to obtain ξ and η in the most general manner as functions of u and v, divide the given expression $E\,du^2+2\,F\,du\,dv+G\,dv^2$ for the equal of the element of length into its linear factors, multiply each of these factors by the quantity necessary to render them exact differentials, and equate the corresponding integrals to arbitrary functions of $\xi+i\eta$ and $\xi-i\eta$.

SURFACE OF REVOLUTION.

Apply this principle to the simple case of the projection of a surface of revolution upon a plane. plane. For such a surface, if z denote the axis of revolution, we have

$$x^2+y^2=r^2 \qquad \frac{y}{x}=\tan\theta \qquad z=\varphi(r)$$

For one case, then, we can write

$$x=u\cos v \qquad y=u\sin v \qquad z=F(u)$$

and then

$$E=1+\left(\frac{dz}{du}\right)^2 \qquad F=0 \qquad G=u^2$$

and the element of length is

$$=E\,du^2+u^2\,dv^2$$

The integrable factors of this are

$$\sqrt{E}\frac{du}{u}+idv \qquad \sqrt{E}\frac{du}{u}-idv$$

Write

$$U=\int\sqrt{E}\frac{du}{u}$$

then we have for the most general possible relations between ξ, η, and u, v the equations

$$F_1(\xi+i\eta)=U+iv \qquad F_2(\xi-i\eta)=U-iv$$

If we assume that

$$F_1(\xi+i\eta)=\xi+i\eta \qquad F_2(\xi-i\eta)=\xi-i\eta$$

we arrive at once at the relations

$$\xi=\mathrm{U} \qquad \eta=v$$

Now

$$d\xi^2+d\eta^2=d\mathrm{U}^2+dv^2$$

and

$$m^2=\frac{d\mathrm{U}^2+dv^2}{\mathrm{E}du^2+u^2dv^2}=\frac{\frac{\mathrm{E}du^2}{u^2}+dv^2}{\mathrm{E}du^2+u^2dv^2}$$

or

$$m=\frac{1}{u}.$$

This is the Mercator projection. If we assume

$$\mathrm{F}_1(\xi+i\eta)=\log(\xi+i\eta) \qquad \mathrm{F}_2(\xi-i\eta)=\log(\xi-i\eta)$$

we will arrive at the stereographic projection.

PROJECTION OF A CONE.

In this case we have

$$x=u\cos v \qquad y=u\sin v\cos\alpha \qquad z=u\sin v\sin\alpha$$

α being merely a function of v. These equations represent a cone whose vertex is at the origin of co-ordinates. Now,

$$\mathrm{E}=1 \qquad \mathrm{F}=0 \qquad \mathrm{G}=u^2\left[1+v^2\left(\frac{d\alpha}{dv}\right)^2\right]=u^2\mathrm{V}$$

therefore

$$ds^2=du^2+u^2\mathrm{V}\,dv^2$$

The two integrable factors of this are

$$\frac{du}{u}+i\sqrt{\mathrm{V}}dv \qquad \frac{du}{u}-i\sqrt{\mathrm{V}}dv$$

Assume

$$\int\sqrt{\mathrm{V}}\,dv=\int(dv^2+v^2\,d\alpha^2)^{\frac{1}{2}}=\omega$$

then for the most general determination of ξ and η as functions of u and v we have

$$\mathrm{F}_1(\xi+i\eta)=\log u+i\omega \qquad \mathrm{F}_2(\xi-i\eta)=\log u-i\omega$$

If

$$\mathrm{F}_1(\xi+i\eta)=\log(\xi+i\eta) \qquad \mathrm{F}_2(\xi-i\eta)=\log(\xi-i\eta)$$

there results

$$\xi+i\eta=ue^{i\omega} \qquad \xi-i\eta=ue^{-i\omega}$$

and consequently

$$\xi=u\frac{e^{i\omega}+e^{-i\omega}}{2}=u\cos\omega \qquad \eta=u\frac{e^{i\omega}-e^{-i\omega}}{2i}=u\sin\omega$$

From these it is obvious that ξ and η satisfy the equation of a circle whose center is at the origin, and whose radius is $=u$; these are, therefore, the conditions for the projection of a cone by actual development.

If the surface to be projected is a cylinder, y is a function of x, or conversely, and z is independent of both x and y; write then

$$x=u \qquad y=\mathrm{F}(u) \qquad z=v$$

then

$$ds=\mathrm{E}\,du^2+dv^2$$

when

$$\mathrm{E}=1+\left(\frac{dy}{du}\right)^2$$

is only a function of u. The two integrable factors of the square of the linear element are

$$\sqrt{\mathrm{E}}\,du+idv \qquad \sqrt{\mathrm{E}}\,du-idv$$

Write

$$\int \sqrt{E}\,du=\int \sqrt{(dx^2+dy^2)}=s$$

s being the arc cut from the cylinder by a right section, say by the plane of (x,y); we are thus conducted immediately to the desired general equations

$$F_1(\xi+i\eta)=s+iz \qquad F_2(\xi+i\eta)=s-iz$$

Assume

$$F_1(\xi+i\eta)=\xi+i\eta \qquad F_2(\xi-i\eta)=\xi-i\eta$$

then

$$\xi=s \qquad \eta=z=v$$

the equations for the projection by development of the cylinder.

QUINCUNCIAL PROJECTION OF THE SPHERE.

This projection was constructed by Mr. C. S. Peirce, Assistant, United States Coast and Geodetic Survey. The brief description here given of the projection is extracted from the Coast Survey Report for 1877, Appendix No. 15, and was written by Mr. Peirce himself.

For meteorological, magnetological, and other purposes, it is convenient to have a projection of the sphere which shall show the connection of all parts of the surface. It is an orthomorphic or conform projection formed by transforming the stereographic projection, with a pole at infinity, by means of an elliptic function. For that purpose, l being the latitude, and θ the longitude, we put

$$\cos^2\varphi=\frac{\sqrt{1-\cos^2 l\cos^2\theta}-\sin l}{1+\sqrt{1-\cos^2 l\cos^2\theta}}$$

and then $\frac{1}{2}F\varphi$ is the value of one of the rectangular co-ordinates of the point on the new projection. This is the same as taking

$$\cos am\,(x+y\sqrt{-1})\,(\text{angle of mod.}=45^\circ)=\tan\frac{p}{2}(\cos\theta+\sin\theta\sqrt{-1})$$

where x and y are the co-ordinates on the new projection, p is the north polar distance. A table of these co-ordinates is subjoined.

Upon an orthomorphic projection the parallels represent equipotential or level lines for the logarithmic potential, while the meridians are the lines of force. Consequently we may draw these lines by the method used by Maxwell in his Electricity and Magnetism for drawing the corresponding lines for the Newtonian potential. That is to say, let two such projections be drawn upon the same sheet, so that upon both are shown the same meridians at equal angular distances, and the same parallels at such distances that the ratio of successive values of $\tan\frac{p}{2}$ is constant. Then number the meridians and also the parallels. Then draw curves through the intersections of meridians with meridians, the sums of numbers of the intersecting meridians being constant on any one curve. Also do the same thing for the parallels. Then these curves will represent the meridians and parallels of a new projection having north poles and south poles wherever the component projections had such poles.

Functions may, of course, be classified according to the pattern of the projection produced by such a transformation of the stereographic projection with a pole at the tangent points. Thus we shall have:

1. Functions with a finite number of zeroes and infinites (algebraic functions).
2. Striped functions (trigonometric functions). In these the stripes may be equal, or may vary progressively or periodically. The stripes may be simple, or themselves compounded of stripes. Thus, sin (π sin z) will be composed of stripes each consisting of a bundle of parallel stripes (infinite in number) folded over onto itself.
3. Chequered functions (elliptic functions).
4. Functions whose patterns are central or spiral.

PROJECTION OF AN ELLIPSOID.

The position of a point on an ellipsoid is given in terms of the parameters of the two systems of confocal hyperboloids, and in general the position of a point on any one of these three quadric surfaces is given in terms of the parameters of the other two systems of surfaces. Call these parameters λ_1, λ_2, λ_3, with the relation

$$\lambda_1 > \lambda_2 > \lambda_3$$

Now the co-ordinates of a point on the surface $\lambda_1 =$ const. are

$$x = F_1(\lambda_2, \lambda_3) \qquad y = F_2(\lambda_2, \lambda_3) \qquad z = F_3(\lambda_2, \lambda_3)$$

It will be noted that, in the determination of x, y, z, λ will also appear in the values, but as it is constant for the given surface the truth of the general statement is not impaired.

Assume now the system of orthogonal quadric surfaces given by

$$\frac{x^2}{a^2+\lambda_1} + \frac{y^2}{b^2+\lambda_1} + \frac{z^2}{c^2+\lambda_1} = 1, \quad +\infty > \lambda_1 > -c^2$$

$$\frac{x^2}{a^2+\lambda_2} + \frac{y^2}{b^2+\lambda_2} + \frac{z^2}{c^2+\lambda_2} = 1, \quad -c^2 > \lambda_2 > -b^2$$

$$\frac{x^2}{a^2+\lambda_3} + \frac{y^2}{b^2+\lambda_3} + \frac{z^2}{c^2+\lambda_3} = 1, \quad -b^2 > \lambda_3 > -a^2$$

The first of these represents the ellipsoid and the other two the confocal hyperbolids. We may assume, if we choose, $\lambda_1 = 0$, to denote the particular ellipsoid which we wish to project; there would be, however, no particular gain in doing so, but rather a loss by the expressions becoming less symmetrical. The following formulas are too well known to require anything more in this place than the mere statement of them:

$$x^2 = \frac{(a^2+\lambda_1)(a^2+\lambda_2)(a^2+\lambda_3)}{(a^2-b^2)(a^2-c^2)} \qquad y^2 = \frac{(b^2+\lambda_1)(b^2+\lambda_2)(b^2+\lambda_3)}{(b^2-c^2)(b^2-a^2)} \qquad z^2 = \frac{(c^2+\lambda_1)(c^2+\lambda_2)(c^2+\lambda_3)}{(c^2-a^2)(c^2-b^2)}$$

$$ds^2 = \frac{(\lambda_1-\lambda_2)(\lambda_1-\lambda_3)}{(a^2+\lambda_1)(b^2+\lambda_1)(c^2+\lambda_1)} d\lambda_1^2 + \frac{(\lambda_2-\lambda_3)(\lambda_2-\lambda_1)}{(a^2+\lambda_2)(b^2+\lambda_2)(c^2+\lambda_2)} d\lambda_2^2 + \frac{(\lambda_3-\lambda_1)(\lambda_3-\lambda_2)}{(a^2+\lambda_3)(b^2+\lambda_3)(c^2+\lambda_3)} d\lambda_3^2$$

In each of these formulas it is only necessary to place successively λ_1, λ_2, λ_3, equal to constants, in order to obtain the formulas for any particular surface, ellipsoid or hyperboloid. For the element of length we have then the three cases—

Ellipsoid ($\lambda_1 =$ const.):

$$ds^2 = (\lambda_2-\lambda_3)\left(\frac{(\lambda_2-\lambda_1)\,d\lambda_2^2}{(a^2+\lambda_2)(b^2+\lambda_2)(c^2+\lambda_2)} + \frac{(\lambda_1-\lambda_3)\,d\lambda_3^2}{(a^2+\lambda_3)(b^2+\lambda_3)(c^2+\lambda_3)}\right)$$

Hyperboloid of one nappe ($\lambda_2 =$ const.):

$$ds^2 = (\lambda_3-\lambda_1)\left(\frac{(\lambda_2-\lambda_1)\,d\lambda_1^2}{(a^2+\lambda_1)(b^2+\lambda_1)(c^2+\lambda_1)} + \frac{(\lambda_2-\lambda_3)\,d\lambda_3^2}{(a^2+\lambda_3)(b^2+\lambda_3)(c^2+\lambda_3)}\right)$$

Hyperboloid of two nappes ($\lambda_3 =$ const.):

$$ds^2 = (\lambda_1-\lambda_2)\left(\frac{(\lambda_1-\lambda_3)\,d\lambda_1^2}{(a^2+\lambda_1)(b^2+\lambda_1)(c^2+\lambda_1)} + \frac{(\lambda_3-\lambda_2)\,d\lambda_2^2}{(a^2+\lambda_2)(b^2+\lambda_2)(c^2+\lambda_2)}\right)$$

The linear integrable factors of these are

$$\sqrt{\left(\frac{(\lambda_2-\lambda_1)}{(a^2+\lambda_2)(b^2+\lambda_2)(c^2+\lambda_2)}\right)}d\lambda_2 \pm i\sqrt{\left(\frac{(\lambda_1-\lambda_3)}{(a^2+\lambda_3)(b^2+\lambda_3)(c^2+\lambda_3)}\right)}d\lambda_3$$

$$\sqrt{\left(\frac{(\lambda_3-\lambda_1)}{(a^2+\lambda_1)(b^2+\lambda_1)(c^2+\lambda_1)}\right)}d\lambda_1 \pm i\sqrt{\left(\frac{(\lambda_3-\lambda_2)}{(a^2+\lambda_2)(b^2+\lambda_2)(c^2+\lambda_2)}\right)}d\lambda_2$$

$$\sqrt{\left(\frac{(\lambda_1-\lambda_2)}{(a^2+\lambda_1)(b^2+\lambda_1)(c^2+\lambda_1)}\right)}d\lambda_1 \pm i\sqrt{\left(\frac{(\lambda_2-\lambda_3)}{(a^2+\lambda_3)(b^2+\lambda_3)(c^2+\lambda_3)}\right)}d\lambda_3$$

Write L_1, L_2, L_3, for the denominators in the above expressions which contain λ_1, λ_2, λ_3 respectively. Then we have for the general determination of ξ and η in the case of the ellipsoid—

$$\int\frac{\sqrt{\lambda_2-\lambda_1}}{L_2}d\lambda_2+i\int\frac{\sqrt{\lambda_1-\lambda_3}}{L_3}d\lambda_3=F_{1,1}(\xi+i\eta)$$

$$\int\frac{\sqrt{\lambda_2-\lambda_1}}{L_2}d\lambda_2-i\int\frac{\sqrt{\lambda_1-\lambda_3}}{L_3}d\lambda_3=F_{1,2}(\xi-i\eta)$$

for the hyperboloid of one nappe—

$$\int\frac{\sqrt{\lambda_3-\lambda_1}}{L_1}d\lambda_1+i\int\frac{\sqrt{\lambda_3-\lambda_2}}{L_2}d\lambda_2=F_{2,1}(\xi+i\eta)$$

$$\int\frac{\sqrt{\lambda_3-\lambda_1}}{L_1}d\lambda_1-i\int\frac{\sqrt{\lambda_3-\lambda_2}}{L_2}d\lambda_2=F_{2,2}(\xi-i\eta)$$

for the hyperboloid of two nappes—

$$\int\frac{\sqrt{\lambda_1-\lambda_2}}{L_1}d\lambda_1+i\int\frac{\sqrt{\lambda_2-\lambda_3}}{L_3}d\lambda_3=F(\xi-i\eta)$$

$$\int\frac{\sqrt{\lambda_1-\lambda_2}}{L_1}d\lambda_1-i\int\frac{\sqrt{\lambda_2-\lambda_3}}{L_3}d\lambda_3=F_{3,2}(\xi-i\eta)$$

We will confine ourselves now to the case of the ellipsoid, and write, for brevity,

$$U=\int\frac{\sqrt{\lambda_2-\lambda_1}}{L_2}d\lambda_2 \qquad V=\int\frac{\sqrt{\lambda_1-\lambda_3}}{L_3}d\lambda_3$$

The conditions

$$-c^2>\lambda_2>-b^2$$

are easily seen to be fulfilled if we choose for λ_2 the value, in terms of a new variable θ,

$$\lambda_2=\frac{b^2(a^2-c^2)\cos^2\theta+c^2(a^2-b^2)\sin^2\theta}{(a^2-c^2)\cos^2\theta+(a^2-b^2)\sin^2\theta}$$

This gives, for $\theta=0$,

$$\lambda_2=-b^2$$

and for $\theta=\frac{\pi}{2}$,

$$\lambda_2=-c^2$$

For all values of θ lying between these limits the above inequalities are satisfied. We can write λ_2 in a different form, which will be more convenient for the purposes of transformation, viz:

$$\lambda_2=-\frac{b^2+c^2\frac{a^2-b^2}{a^2-c^2}\tan^2\theta}{1+\frac{a^2-b^2}{a^2-c^2}\tan^2\theta}$$

or, briefly,

$$\lambda_2=-\frac{b^2+c^2x^2}{1+x^2}$$

Now, for the determination of U we find the following expressions:

$$a^2+\lambda_2=\frac{(a^2-b^2)+(a^2-c^2)x^2}{1+x^2}$$

$$b^2+\lambda_2=\frac{(b^2-c^2)x^2}{1+x^2}$$

$$c^2+\lambda_2=-\frac{(b^2-c^2)}{1+x^2}$$

$$\lambda_2-\lambda_1=-\frac{(b^2+\lambda_1)+(c^2+\lambda_1)x^2}{1+x^2}$$

$$d\lambda_2=\frac{(b^2-c^2)2x\,dx}{(1+x^2)^2}$$

These substitutions made, give us at once

$$U=\int\sqrt{\frac{(b^2+\lambda_1)+(c^2+\lambda_1)x^2}{(a^2-b^2)+(a^2-c^2)x^2}}\cdot\frac{2\,dx}{1+x^2}$$

Returning now to the angle θ, by means of which we defined x, viz,

$$x^2=\frac{a^2-b^2}{a^2-c^2}\tan^2\theta$$

we find

$$(b^2+\lambda_1)+(c^2+\lambda_1)x^2=\frac{(b^2+\lambda_1)(a^2-c^2)\cos^2\theta+(c^2+\lambda_1)(a^2-b^2)\sin^2\theta}{(a^2-c^2)\cos^2\theta}$$

Writing for $\cos^2\theta$, in the numerator, its value $1-\sin^2\theta$, this reduces to

$$(b^2+\lambda_1)+(c^2+\lambda_1)x^2=\frac{(b^2+\lambda_1)(a^2-c^2)-(a^2+\lambda_1)(b^2-c^2)\sin^2\theta}{(a^2-c^2)\cos^2\theta}$$

The remaining factor under the integral sign is

$$\frac{2\,dx}{1+x^2}\,\frac{1}{\sqrt{(a^2-b^2)+(a^2-c^2)x^2}}$$

Now

$$dx=\sqrt{\frac{a^2-b^2}{a^2-c^2}}\sec^2\theta\,d\theta$$

$$\frac{1}{1+x^2}=\frac{(a^2-c^2)\cos^2\theta}{(a^2-c^2)-(b^2-c^2)\sin^2\theta}$$

$$\sqrt{(a^2-b^2)+(a^2-c^2)x^2}=\frac{\sqrt{a^2-b^2}}{\cos\theta}$$

Multiplying these together gives

$$\frac{2\sqrt{a^2-c^2}\cos\theta\,d\theta}{(a^2-c^2)-(b^2-c^2)\sin^2\theta}$$

The result of the substitution in U of these values of its components is

$$U=2\int\frac{\sqrt{[(b^2+\lambda_1)(a^2-c^2)+(a^2+\lambda_1)(b^2-c^2)\sin^2\theta]}}{(a^2-c^2)-(b^2-c^2)\sin^2\theta}d\theta$$

or

$$U=2\sqrt{\frac{b^2+\lambda_1}{a^2-c^2}}\int\frac{\sqrt{\left[1-\frac{(a^2+\lambda_1)(b^2-c^2)}{(b^2+\lambda_1)(a^2-c^2)}\sin^2\theta\right]}}{1-\frac{b^2-c^2}{a^2-c^2}\sin^2\theta}d\theta$$

The quantity $\frac{b^2-c^2}{a^2-c^2}<1$ occurs in both numerator and denominator; in the numerator we have also the factor $\frac{a^2+\lambda_1}{b^2+\lambda_1}$. This may be written as $1+\frac{a^2-b^2}{b^2+\lambda_1}$, and in this form gives, for $\lambda_1=\infty$,

$$\frac{a^2+\lambda_1}{b^2+\lambda_1}=1$$

for $\lambda_1=c^2$ it gives

$$\frac{a^2+\lambda_1}{b^2+\lambda_1}=\frac{a^2-c^2}{b^2-c^2}$$

In the former of these cases there results

$$\frac{(a^2+\lambda_1)(b^2-c^2)}{(b^2+\lambda_1)(a^2-c^2)}=\frac{b^2-c^2}{a^2-c^2}<1$$

in the latter

$$\frac{(a^2+\lambda_1)(b^2-c^2)}{(b^2+\lambda_1)(a^2-c^2)}=1$$

This quantity being, then, either equal to or less than unity, can be taken as the modulus of an elliptic integral, and we may write

$$\frac{(a^2+\lambda_1)(b^2-c^2)}{(b^2+\lambda_1)(a^2-c^2)}=k$$

and, since the corresponding factor in the denominator is smaller than k^2, we may also write

$$\frac{b^2-c^2}{a^2-c^2}=k^2\sin^2\alpha$$

when

$$\sin^2\alpha=\frac{b^2+\lambda_1}{a^2+\lambda_1}$$

and also

$$\cos^2\alpha=\frac{a^2-b^2}{a^2+\lambda_1} \qquad \sqrt{1-k^2\cos^2\alpha}=\frac{\sqrt{a^2-b^2}}{a^2-c^2}=\Delta\alpha$$

The substitution of these values in U gives again, on writing simply U for $\frac{1}{2}$ U,

$$U=\sqrt{\frac{b^2+\lambda_1}{a^2-c^2}}\int\frac{\sqrt{1-k^2\sin^2\theta}\,d\theta}{1-k^2\sin^2\alpha\sin^2\theta}$$

Now

$$\sqrt{\frac{b^2+\lambda_1}{a^2-c^2}}=\sqrt{\frac{b^2+\lambda_1}{a^2+\lambda_1}\frac{a^2+\lambda_1}{a^2-b^2}\frac{a^2-b^2}{a^2-c^2}}=\frac{\sin\alpha\Delta\alpha}{\cos\alpha}$$

therefore

$$U=\frac{\sin\alpha\Delta\alpha}{\cos\alpha}\int\frac{\Delta\theta d\theta}{1-k^2\sin^2\alpha\sin^2\theta}=\tan\alpha\Delta\alpha\int\frac{d\theta}{\Delta\theta}-k^2\sin\alpha\cos\alpha\Delta\alpha\int\frac{\sin^2\theta d\theta}{(1-k^2\sin^2\alpha\sin^2\theta)\Delta\theta}$$

Write now

$$t=\int\frac{d\theta}{\Delta\theta} \qquad \alpha=\int\frac{d\alpha}{\Delta\alpha}$$

then

$$U=\text{tn}\,\alpha\,\text{dn}\,\alpha\,t-k^2\,\text{sn}\,\alpha\,\text{cn}\,\alpha\,\text{dn}\,\alpha\int\frac{\sin^2\theta d\theta}{(1-k^2\sin^2\alpha\sin^2\theta)\Delta\theta}$$

or

$$U=\text{tn}\,\alpha\,\text{dn}\,\alpha\,t-\int\frac{k^2\,\text{sn}\,\alpha\,\text{cn}\,\alpha\,\text{dn}\,\alpha\,\text{sn}\,\alpha\,\text{sn}^2t\,dt}{1-k^2\,\text{sn}^2\alpha\,\text{sn}^2t}$$

The quantity under the integral sign is the elliptic integral of the third kind, or $\Pi(t,\alpha)$; therefore

$$U=\text{tn}\,\alpha\,\text{dn}\,\alpha\,t-\Pi(t,\alpha)$$

Now we have the formula connecting the three Jacobi functions Z, H, Θ, viz:

$$H(t,a)=tZa+\tfrac{1}{2}\log\frac{\theta(t-a)}{\theta(t+a)}$$

which gives

$$U=\left[(\text{tn}\,a\ \text{dn}\,a-Za)\,t-\tfrac{1}{2}\log\frac{\theta(t-a)}{\theta(t-a)}\right]$$

or placing

$$\text{tn}\,a\ \text{dn}\,a-Za=h \qquad\qquad U=\tfrac{1}{2}\log e^{ht}\frac{\theta(t+a)}{\theta(t+a)}$$

Concerning h, we can readily place it in a different form

$$h=\text{tn}\,a\ \text{dn}\,a-Za=-\frac{d}{da}\log(\theta a\,.\,\text{cn}\,a)$$

since

$$Za=\frac{\theta' a}{\theta a}$$

but, introducing the Π function,

$$\log\Pi(a-K)=\log\sqrt{\frac{K}{K'}}+\log\text{cn}\,a+\log\theta a$$

and consequently

$$h=-\frac{d}{da}\log(\theta a\ \text{cn}\,a)=-\frac{d}{da}\log\Pi(a+K)$$

Introducing the q function defined by the relation

$$q=e^{\frac{-\pi K'}{K}}$$

we have (Cayley's Elliptic Functions, page 295)

$$\theta\left(\frac{2K}{\pi}t'\right)=1-2q\cos 2t'+2q^4\cos 4t'-2q^9\cos 6t'+\ldots$$

$$\Pi\left(\frac{2K}{\pi}t'\right)=2\sqrt[4]{q}(\sin t'-q^2\sin 3t'+q^6\sin 5t'-q^{12}\sin 7t'+\ldots)$$

or writing

$$t=\frac{2K}{\pi}t' \qquad\qquad a=\frac{2K}{\pi}a'$$

these may be expressed in the brief form

$$\theta t=1+2\Sigma(-1)^j q^{j^2}\cos 2jt' \qquad\qquad \Pi t=2\sqrt[4]{q}\,\Sigma(-1)^{j-1}q^{j(j-1)}\sin(2j-1)t'$$

and consequently

$$\frac{\theta(t+a)}{\theta(t-a)}=\frac{1+2\Sigma(-1)^j q^{j^2}\cos 2j(t'+a')}{1+2\Sigma(-1)^j q^{j^2}\cos 2j(t'-a')}$$

and

$$\Pi(a+K)=2\sqrt[4]{q}\,\Sigma q^{j(j-1)}\cos(2j-1)a'$$

From the above expression for a we have

$$\frac{d}{da}=\frac{\pi}{2K}\frac{d}{da'}$$

and therefore

$$h=\frac{d}{da}\log\Pi(a+K)=\frac{\pi}{2K}\frac{d}{da'}\log\Pi(a+K)$$

or

$$h=\frac{\pi}{2K}\frac{\Sigma(2j-1)\,q^{j(j-1)}\sin(2j-1)a'}{\Sigma q^{j(j-1)}\cos(2j-1)a'}$$

By changing λ_1 into λ_3 the expression for U becomes $=i$V. The quantities K, ϵ, do not depend on either λ_1 or λ_3, and in consequence are unaltered by this change; the same is, of course, true of the constants a, a', h which are functions of K, ϵ, and constants. The only quantity, then, which can vary is t, and in this case, on account of the prescribed limits of λ_3, t will become a pure imaginary, say $i(\mathrm{K}'+\tau)$ and so

$$\theta=\mathrm{am}\, i(\mathrm{K}'+\tau)$$

We have now

$$\frac{1}{\mathrm{K}'\mathrm{sn}^2(i\mathrm{K}'+i\tau)}=\frac{a^2+\lambda_3\; b^2+\lambda_1}{a^2+\lambda_1\; b^2+\lambda_3}$$

but

$$\mathrm{sn}^2 i(\mathrm{K}'+\tau)=\frac{1}{\mathrm{K}^2\mathrm{sn}^2 i\tau}$$

and

$$\mathrm{sn}(i\tau,\mathrm{K})=\frac{i\,\mathrm{sn}(\tau,\mathrm{K}')}{\mathrm{cn}(\tau,\mathrm{K}')}=i\,\mathrm{tn}(\tau,\mathrm{K}')$$

the above equation is therefore equivalent to

$$-\mathrm{sn}^2(i\tau)=\mathrm{tn}^2(\tau,\mathrm{K}')=\frac{(a^2+\lambda_3)(b^2+\lambda_1)}{(a^2+\lambda_1)(b^2+\lambda_3)}$$

Writing in this, $\psi=\mathrm{am}(\tau,\mathrm{K}')$, we derive

$$\cos 2\psi=\frac{(a^2+\lambda_1)(b^2+\lambda_3)}{(a^2-b^2)(\lambda_3-\lambda_1)} \qquad \sin 2\psi=\frac{(a^2+\lambda_3)(b^2+\lambda_1)}{(a^2-b^2)(\lambda_3-\lambda_1)}$$

and for the complementary modulus k',

$$k'^2=\frac{(a^2-b^2)(c^2+\lambda_1)}{(a^2-c^2)(b^2+\lambda_1)}$$

from which

$$k'^2\sin^2\psi=\frac{(a^2+\lambda_3)(c^2+\lambda_1)}{(a^2-c^2)(\lambda_1-\lambda_3)}$$

and

$$\Delta^2(\psi\mathrm{K}')=1\mathrm{K}'^2\sin^2\psi=\frac{(a^2+\lambda_1)(c^2+\lambda_3)}{(a^2-c^2)(\lambda_3-\lambda_1)}$$

Now we read for U the value

$$\mathrm{U}=\log\frac{e^{ht}\,\theta(t+a)}{\theta(t-a)}$$

or

$$=ht+\log\sqrt{\frac{\theta(t+a)}{\theta(t-a)}}$$

and we have seen that

$$\mathrm{V}=\frac{1}{i}\mathrm{U}$$

therefore, since $t=i(\mathrm{K}'+\tau)$,

$$\mathrm{V}=h(\mathrm{K}'+\tau)-\frac{i}{2}\log\frac{\theta(i\mathrm{K}'+i\tau+a)}{\theta(i\mathrm{K}'+i\tau-a)}$$

Now (Cayley's Elliptic Functions, page 150)

$$\theta(u+i\mathrm{K}')=ie^{\frac{\pi}{4K}(K'-2iu)}\,\mathrm{H}u$$

and therefore

$$\mathrm{V}=h\mathrm{K}'+h\tau-\frac{i}{2}\log\frac{\mathrm{H}(a+i\tau)}{\mathrm{H}(a-i\tau)}$$

As we are to equate U$+i$V to an arbitrary function of ξ and η, there will be no gain in retaining the pure constant $h\mathrm{K}'$, so we shall omit it and write V in the form

$$\mathrm{V}=h\tau+\frac{1}{2i}\log\frac{\mathrm{H}(a+i\tau)}{\mathrm{H}(a-i\tau)}=h\tau+\tan^{-1}\frac{\cos a'(e^{\tau}-e^{-\tau})-q^2\cos^3 a'(e^{3\tau}-e^{-3\tau})+\ldots}{\sin a'(e^{\tau}+e^{-\tau})-q^2\sin^3 a'(e^{3\tau}+e^{-3\tau})+\ldots}$$

when

$$a' = \frac{\pi a}{2K} \qquad \tau' = \frac{\pi\tau}{2K}$$

Resume now the formulas

$$U + iV = F_1(\xi + i\eta) \qquad U - iV = F_2(\xi + i\eta)$$

and suppose

$$F_1(\xi + i\eta) = \log(\xi + i\eta) \qquad F_2(\xi - i\eta) = \log(\xi - i\eta)$$

We have then, on writing $V = h\tau + \tan^{-1} Q$,

$$\log e^{hu}\sqrt{\frac{\theta(t+a)}{\theta(t-a)}} + ih\tau + i\tan^{-1} Q = \log(\xi + i\eta)$$

$$\log e^{hu}\sqrt{\frac{\theta(t+a)}{\theta(t-a)}} - ih\tau - i\tan\ Q = \log(\xi - i\eta)$$

Observe that

$$\log(\xi + i\eta) + \log(\xi - i\eta) = \log(\xi^2 + \eta^2) = \log\rho^2$$

$$\log(\xi + i\eta) - \log(\xi - i\eta) = \log\frac{\cos\varphi + i\sin\varphi}{\cos\varphi - i\sin\varphi} = \log e^{2i\varphi} = 2i\varphi$$

where

$$\xi = \rho\cos\varphi \qquad \eta = \rho\sin\varphi$$

Now, adding and subtracting the above equations, we obtain

$$e^{hu}\sqrt{\frac{\theta(t+a)}{\theta(t-a)}} = \rho \qquad h\tau + \tan^{-1} Q = \varphi$$

These conditions give the projections of the lines of curvature on the ellipsoid, arising from its intersections with the system of hyperboloids of two nappes, a series of straight lines in the xy plane passing through the origin of co-ordinates. The lines of curvature due to the hyperboloid of one nappe are projected in a series of concentric circles having the origin as center.

The point on the ellipsoid given by the polar co-ordinates (ρ, φ) is of course

$$x^2 = \frac{(a^2+\lambda_1)(a^2+\lambda_2)(a^2+\lambda_3)}{(a^2-b^2)(a^2-c^2)} \qquad y^2 = \frac{(b^2+\lambda_1)(b^2+\lambda_2)(b^2+\lambda_3)}{(b^2-c^2)(b^2-a^2)} \qquad z^2 = \frac{(c^2+\lambda_1)(c^2+\lambda_2)(c^2+\lambda_3)}{(c^2-a^2)(c^2-b^2)}$$

when λ_1 is constant, and

$$a^2 + \lambda_2 = \frac{a^2 - b^2}{1 - k^2\sin^2\alpha\sin^2\theta} \qquad a^2 + \lambda_3 = \frac{(a^2 - b^2)\sin^2\phi}{\sin^2\alpha + \cos^2\alpha\sin^2\phi}$$

$$b^2 + \lambda_2 = \frac{(a^2 - b^2)\sin^2\alpha\sin^2\theta}{1 - k^2\sin^2\alpha\sin^2\theta} \qquad b^2 + \lambda_3 = \frac{(a^2 - b^2)\sin^2\alpha\cos^2\phi}{\sin^2\alpha + \cos^2\alpha\sin^2\phi}$$

$$c^2 + \lambda_2 = \frac{(a^2 - c^2)\sin^2\alpha\cos^2\theta}{1 - k^2\sin^2\alpha\sin^2\theta} \qquad c^2 + \lambda_3 = \frac{(a^2 - c^2)\sin^2\alpha\,\Delta^2(\phi, k')}{\sin^2\alpha + \cos^2\alpha\sin^2\phi}$$

or, introducing the notation of elliptic functions,

$$\theta = \mathrm{am}\, t \qquad \phi = \mathrm{am}(\tau, k')$$

These may be written

$$a^2 + \lambda_2 = (a^2 - b^2) \quad \div a^2 + \lambda_3 = (a^2 - b^2)\,\mathrm{sn}^2(\tau, k') \div$$

$$b^2 + \lambda_2 = (a^2 - b^2)\,\mathrm{sn}^2\alpha\,\mathrm{sn}^2 t \div b^2 + \lambda_3 = (a^2 - b^2)\,\mathrm{sn}^2\alpha\,\mathrm{cn}^2(\tau, k') +$$

$$c^2 + \lambda_2 = (a^2 - c^2)\,\mathrm{sn}^2\alpha\,\mathrm{cn}^2 t \div c^2 + \lambda_3 = (a^2 - c^2)\,\mathrm{sn}^2\alpha\,\mathrm{dn}^2(\tau, k') +$$

where denominator $= 1 - \mathrm{sn}^2\alpha\,\mathrm{sn}^2 t$ for 1st column, and $= 1 - \mathrm{sn}^2\alpha\,\mathrm{sn}^2(\tau, k')$ for 2d column. A number of interesting relations can be obtained from these formulæ, but it is not in the province of this work to take up subjects entirely foreign to projections. Observing, however, the following relations, given in another place,

$$\frac{a^2 - b^2}{a^2 - c^2} = \Delta^2\alpha \qquad \frac{(a^2 - b^2)(b^2 + \lambda_1)}{(b^2 - c^2)} = \frac{(a^2 + \lambda_1)\Delta^2\alpha}{K^2} \qquad \frac{(a^2 - c^2)(c^2 + \lambda_1)}{(b^2 - c^2)} = \frac{(a^2 + \lambda_1)k^2}{K^2\Delta^2\alpha}$$

and substituting in these for $\Delta\epsilon$ its value $\mathrm{dn}\,\sigma$, we can write for x, y, z the following values:

$$x=G\,.\,\mathrm{dn}^2\sigma\,\mathrm{sn}(\tau,k') \qquad y=G\,.\,\mathrm{dn}^2\sigma\,\mathrm{sn}^2\sigma\,\mathrm{sn}\,\theta\,\mathrm{cn}(\tau,k') \qquad z=G\,.\,k'\,\mathrm{sn}^2\sigma\,\mathrm{cn}\,\theta\,\mathrm{dn}(\tau,k')$$

when

$$G=\frac{\sqrt{a^2+\lambda_1}}{\mathrm{dn}\,\sigma\sqrt{(1-k^2\mathrm{sn}^2\sigma\,\mathrm{sn}^2\theta)(\mathrm{sn}^2\sigma+\mathrm{cn}^2\sigma\,\mathrm{sn}^2(\tau,k'))}}$$

The angle ϵ has been defined by the relation

$$\sin^2\epsilon=\frac{b^2+\lambda_1}{a^2+\lambda_1}$$

From this we have

$$b^2+\lambda_1=(a^2+\lambda_1)\sin^2\epsilon$$

and we can also find quite readily

$$c^2+\lambda_1=\frac{(a^2+\lambda_1)k'^2\sin^2\epsilon}{\Delta^2\epsilon}$$

The equation of the ellipsoid can then be written in the form

$$\frac{x^2}{a^2+\lambda_1}+\frac{y^2}{(a^2+\lambda_1)\sin^2\epsilon}+\frac{z^2\Delta^2\epsilon}{(a^2+\lambda_1)k'^2\sin^2\epsilon}=1$$

or

$$x^2+y^2\,\mathrm{cosec}^2\epsilon+z^2\,\mathrm{cosec}^2\epsilon\,\frac{\Delta^2\epsilon}{k'^2}=a^2+\lambda_1$$

or again transforming to elliptic functions

$$x^2+\frac{y^2}{\mathrm{sn}^2\sigma}+\frac{z^2\,\mathrm{dn}^2\sigma}{k'^2\,\mathrm{sn}^2\sigma}=a^2+\lambda_1$$

If the ellipsoid is one of rotation around the axis of x, the following conditions hold:

$$b=c \qquad k=0 \qquad q=0 \qquad K=\frac{\pi}{2} \qquad \theta=t=t' \qquad \sigma=\delta=\sigma' \qquad h=\tan\epsilon$$

Further, since $\phi=\mathrm{am}(\tau,k')$ and $k=1$,

$$\tau'=\tau=\int_0^\psi\frac{d\psi}{\cos\psi}=\log\tan\tfrac{1}{2}(90^\circ+\psi)$$

and

$$\frac{e^{\tau'}-e^{-\tau'}}{e^{\tau'}+e^{-\tau'}}=\sin\phi$$

For the position of a point on the plane corresponding to x, y, z on the ellipsoid of revolution we have, then, the polar co-ordinates

$$\rho=e^{\tan\epsilon\,.\,\theta} \qquad \varphi=\tan\epsilon\log\tan\tfrac{1}{2}(90^\circ+\psi)+\tan^{-1}\cot\epsilon\sin\psi$$

the point on the surface being given by

$$x=G\sin\psi \qquad y=G\sin^2\epsilon\sin\theta\cos\psi \qquad z=G\sin^2\epsilon\cos\theta\cos\psi$$

where

$$G=\frac{\sqrt{a^2+\lambda_1}}{\sqrt{\sin^2\epsilon+\cos^2\epsilon\sin^2\psi}}=\frac{\sqrt{a^2+\lambda_1}}{\sqrt{\sin^2\psi+\sin^2\epsilon\cos^2\psi}}$$

and for the surface

$$\frac{x^2}{a^2+\lambda_1}+\frac{y^2+z^2}{(a^2+\lambda_1)\sin^2\epsilon}=1$$

or

$$x^2+(y^2+z^2)\,\mathrm{cosec}^2\epsilon=a^2+\lambda_1$$

The case of an oblate ellipsoid is not arrived at quite so readily. As the general ellipsoid approaches the form of an oblate ellipsoid, the quantity k' becomes smaller, k gradually approaching its limit of unity. The transformation to this case is as follows: Denote the complementary functions of those that we have been dealing with by the same letters with a suffix, *e. g.*,

$$t_1=\mathrm{K}-t \qquad a_1=\mathrm{K}-a$$

and also write

$$\mathrm{am}\, t_1=\theta_1 \qquad \mathrm{am}\, a_1=\alpha_1$$

then

$$\sin^2\alpha_1=\frac{a^2-c^2}{a^2+\lambda_1} \qquad \cos^2\alpha_1=\frac{c^2+\lambda_1}{a^2+\lambda_1} \qquad \Delta^2\alpha_1=\frac{c^2+\lambda_1}{b^2+\lambda_1}$$

$$\sin^2\theta=-\frac{(b^2+\lambda_1)(c^2+\lambda_2)}{(\lambda_1-\lambda_2)(b^2-c^2)} \qquad \cos^2\theta=\frac{(c^2+\lambda_1)(b^2+\lambda_2)}{(b^2-c^2)(\lambda_1-\lambda_2)} \qquad \Delta^2\theta_1=\frac{(c^2+\lambda_1)(a^2+\lambda_2)}{(a^2-c^2)(\lambda_1-\lambda_2)}$$

and for the equation of the ellipsoid

$$\frac{x^2}{a^2+\lambda_1}+\frac{y^2\Delta^2\alpha_1}{(a^2+\lambda_1)\cos^2\alpha_1}+\frac{z^2}{(a^2+\lambda_1)\cos^2\alpha_1}=1$$

or

$$x^2+y^2\Delta^2\alpha_1\sec^2\alpha_1+z^2\sec^2\alpha_1=a^2+\lambda_1$$

the co-ordinates x, y, z becoming readily

$$x=\mathrm{G}_1\Delta^2\alpha_1\Delta\theta_1\sin\psi \qquad y=\mathrm{G}_1\cos^2\alpha_1\cos\theta_1\cos\psi \qquad z=\mathrm{G}_1\cos^2\alpha\Delta^2\alpha_1\sin\theta_1\Delta(\psi,k')$$

where

$$\mathrm{G}_1=\frac{\sqrt{a^2+\lambda_1}}{\Delta\alpha_1\sqrt{(1-k^2\sin^2\alpha_1\sin^2\theta_1)(\cos^2\alpha_1+k'^2\sin^2\alpha_1\sin^2\varphi)}}$$

The product ht was before given by the equation

$$ht=-\frac{d\log \mathrm{H}(a+\mathrm{K})}{da}t$$

and now becomes, on substituting for t and a their values,

$$ht=\frac{d}{da}\log \mathrm{H}(a_1)(\mathrm{K}-t_1)$$

and also

$$\frac{\theta(t+a)}{\theta(b-a)}=\frac{\theta(t_1+a_1)}{\theta(t_1-a_1)}$$

so that U now assumes the form

$$\mathrm{U}=\mathrm{K}\frac{d}{da_1}\log \mathrm{H}(a_1)+\frac{d}{da_1}\log \mathrm{H}(a_1)\,t_1-\tfrac{1}{2}\log\frac{\theta(t_1+a_1)}{\theta(t_1-a_1)}$$

or merely

$$\mathrm{U}=\frac{d}{da_1}\log \mathrm{H}(a_1)\,t_1-\tfrac{1}{2}\log\frac{\theta(t_1+a_1)}{\theta(t_1-a_1)}$$

since the constant may be neglected without any loss or change in the conditions of the problem. Furthermore, we have for U expression

$$\mathrm{U}=\frac{\sin\alpha\Delta\alpha}{\cos\alpha}\int_0^\theta\frac{\Delta\theta\, d\theta}{1-k^2\sin^2\alpha\sin^2\theta}$$

This becomes by the transformation from (θ,α) to (θ_1,α_1)

$$\mathrm{U}=\frac{k'^2\cos\alpha_1\Delta\alpha_1}{\sin\alpha_1}\int_{\theta_1}^{\frac{\pi}{2}}\frac{d\theta_1}{(\Delta^2\alpha_1\Delta^2\theta_1-k^2\cos^2\alpha_1\cos^2\theta_1)\Delta\theta_1}=\frac{\cos\alpha_1\Delta\alpha_1}{\sin\alpha_1}\int_{\theta_1}^{\frac{\pi}{2}}\frac{d\theta_1}{(1-k^2\sin^2\alpha_1\sin^2\theta_1)\Delta\theta}$$

Since the integration $\int_{\theta_1}^{\frac{\pi}{2}}$ only differs from $\int_0^{\theta_1}$ by a constant, we may write for U the integral

$$\mathrm{U}=\frac{\cos\alpha_1\Delta\alpha_1}{\sin\alpha_1}\int_0^{\theta_1}\frac{d\theta_1}{(1-k^2\sin^2\alpha_1\sin^2\theta)\Delta\theta_1}$$

This expression can be changed into another having an imaginary argument, and for modulus the complementary one to k, that is, k'. We have, however, by immediate reference to the Jacobian notation

$$\theta(t)=\sqrt{\frac{K}{K'}}\,e^{-\frac{\pi t^2}{4KK'}}\,\Pi(K'\pm it,\,k') \qquad \Pi(t)=\sqrt{\frac{K}{K'}}\,\frac{1}{i}\,e^{-\frac{\pi t^2}{4KK'}}\,\Pi(it,\,k')$$

Now

$$h=\frac{d}{da_1}\log\Pi(a_1)=-\frac{\pi a_1}{2KK'}+\frac{d}{da_1}\log\Pi(ia_1,\,k')$$

or simply

$$h=-\frac{\pi a_1}{2KK'}+h_1$$

We can write at once for h_1 its expanded value and have

$$h_1=\frac{\pi}{2K'}\,\frac{e^{a'_1}+e^{-a'_1}-3q'^2(e^{3a'_1}+e^{-3a'_1})+\ldots}{e^{a'_1}-e^{-a'_1}-3q'^2(e^{3a'_1}-e^{-3a'_1})+\ldots}$$

We have, furthermore,

$$U=\frac{\cos a_1\,\Delta a_1}{\sin a_1}\int_0^{\theta_1}\frac{d\theta}{(1-K^2\sin^2 a_1\sin^2\theta_1)\,\Delta\theta_1}=ht_1\Delta-\tfrac{1}{2}\log\frac{\theta(t_1+a_1)}{\theta(t_1-a_1)}=h_1t_1-\tfrac{1}{2}\log\frac{\Pi(K'-i(t_1+a_1),\,K')}{\Pi(K'-i(t_1-a_1),\,K')}$$

and again writing as above

$$t'_1=\frac{\pi t_1}{2K'} \qquad a'_1=\frac{\pi a_1}{2K'}$$

we obtain finally

$$U=h_1t_1-\tfrac{1}{2}\log\frac{e^{(t'_1+a'_1)}+e^{-(t'_1+a'_1)}+q'^2(e^{3(t'_1+a'_1)})+e^{-3(t'_1+a'_1)}+\ldots}{e^{(t'_1-a'_1)}+e^{(t'_1-a'_1)}+q'^2(e^{3(t'_1-a'_1)})+e^{-3(t'_1-a'_1)}+\ldots}$$

For $\theta_1=0$ there results $t_1=0$ and consequently this value for U becomes $=0$; the same is true of the integral expression for U, and these two solutions are identical. If we write again

$$\tau'=\frac{\pi\tau}{2K'}$$

we obtain the transformed expression for V

$$V=h\tau+2\,\frac{1}{i}\log\frac{\Pi(a+i\tau)}{\Pi(a-i\tau)}=h\tau+2\,\frac{1}{i}\log\frac{\Pi(K-a_1+i\tau)}{\Pi(K-a_1-i\tau)}=h_1\tau+2\,\frac{1}{i}\log\frac{\theta(\tau+ia_1,\,k')}{\theta(\tau-ia_1,\,k')}$$

$$=h_1\,\tau+\tan^{-1}\frac{q'(e^{2a'_1}-e^{-2a'_1})\sin 2\tau'-q'^4(e^{4a'_1}-e^{-4a'_1})\sin 4\tau'+\ldots}{1-q'(e^{2a'_1}+e^{-2a'_1})\cos 2\tau'+q'^4(e^{4a'_1}-e^{4a'_1})\cos 4\tau'+\ldots}$$

For the oblate ellipsoid we have now

$$k'=0 \qquad q'=0 \qquad K'=\tfrac{1}{2}\pi \qquad \tau=\tau'=\phi$$

$$a'_1=a_1=\log\tan\tfrac{1}{2}(90^\circ+a_1) \qquad t'_1=t=\log\tan\tfrac{1}{2}(90^\circ+\theta)$$

$$U=\frac{\tan\frac{1}{2}(90^\circ+a_1)+\cot\frac{1}{2}(90^\circ+a_1)}{\tan\frac{1}{2}(90^\circ+a_1)-\cot\frac{1}{2}(90^\circ+a_1)}\log\tan\tfrac{1}{2}(90^\circ+\theta_1)$$

$$-\tfrac{1}{2}\log\frac{\tan\frac{1}{2}(90^\circ+a_1)\tan\frac{1}{2}(90^\circ+\theta_1)+\cot\frac{1}{2}(90^\circ+a_1)\cot\frac{1}{2}(90^\circ+\theta_1)}{\cot\frac{1}{2}(90^\circ+a_1)\tan\frac{1}{2}(90^\circ+\theta_1)+\tan\frac{1}{2}(90^\circ+a_1)\cot\frac{1}{2}(90^\circ+\theta_1)}$$

$$=\frac{1}{\sin a_1}\log\tan\tfrac{1}{2}(90^\circ+\theta_1)-\tfrac{1}{2}\log\frac{\sin^2\frac{1}{2}(\theta_1+a_1)+\cos^2\frac{1}{2}(\theta_1-a_1)}{\cos^2\frac{1}{2}(\theta_1+a_1)+\sin^2\frac{1}{2}(\theta_1-a_1)}$$

$$=\frac{1}{\sin a_1}\log\tan\tfrac{1}{2}(90^\circ+\theta_1)-\tfrac{1}{2}\log\frac{1+\sin a_1\sin\theta_1}{1-\sin a_1\sin\theta_1}$$

$$V=\frac{\phi}{\sin a_1}$$

If we substitute for G_1 (which entered in the expressions for values obtained for the co-ordinates x, y, z, of a point on the oblate ellipsoid) the quantity $G'_1 \cos^2 \epsilon_1$, these values become

$$x = G_1' \cos \theta_1 \sin \phi \qquad y = G_1' \cos \theta \cos \phi \qquad z = G_1' \cos^2 \epsilon_1 \sin^2 \theta_1$$

when

$$G_1' = \frac{\sqrt{a^2+\lambda_1}}{\sqrt{1-\sin^2 \epsilon_1 \sin^2 \theta_1}}$$

and for the equation of the surface there is

$$x^2+y^2+z^2 \sec^2 \epsilon_1 = a^2+\lambda_1$$

The angle ϕ is here the longitude and θ_1 the eccentric anomaly of the meridians.

The preceding projection of the ellipsoid is due to Jacobi, and is to be found in a slightly differing form in Crelle's Journal, vol. 59.

In what has been said up to this point, we have taken the plane as the surface upon which the projection has been made; that supposition, of course, simplifies much the actual forms of the results, but, as we shall see, does not have much effect upon the more general theory, though the steps to be taken in order to project one surface upon any other are more numerous than when one of the surfaces is a plane.

Let x, y, z, as before, denote the co-ordinates of a point upon one of the surfaces, and ξ, η, ζ, the co-ordinates of a point upon the other; the two independent perimeters, in terms of which the co-ordinates of a point on any surface can be given, are u, v for the first surface, and U, V for the the second. We know that it is necessary to determine U, V as functions of u, v, though not as arbitrary functions, since the projection is to fulfill certain assumed conditions; in our case the condition is that the projections of the elements shall be similar to the elements themselves.

Since x, y, z, and ξ, η, ζ are all functions of u, v, we have by differentiation

$$dx = a du + a' dv \qquad dy = b du + b' dv \qquad dz = c du + c' dv$$
$$d\xi = \alpha du + \alpha' dv \qquad d\eta = \beta du + \beta' dv \qquad d\zeta = \gamma du + \gamma' dv$$

$a, b \ldots\ldots \beta', \gamma'$ being determinate functions of u, v.

The condition of this projection is fulfilled, as we already know, first, when all the linear elements that go out from a point of one surface are proportional in length to those that correspond upon the second surface; second, when the corresponding elements make the same angle with each other on both surfaces.

The linear element of the first surface is given by

$$\sqrt{[(a^2+b^2+c^2)\,du^2+2\,(aa'+bb'+cc')\,du\,dv+(a'^2+b'^2+c'^2)\,dv^2]}$$

and of the second by

$$\sqrt{[(\alpha^2+\beta^2+\gamma^2)\,du^2+2\,(\alpha\alpha'+\beta\beta'+\gamma\gamma')\,du\,dv+(\alpha'^2+\beta'^2+\gamma'^2)\,dv^2]}$$

where, as we know,

$$E = a^2+b^2+c^2 \qquad F = aa'+bb'+cc' \qquad G = a'^2+b'^2+c'^2$$

and similarly

$$E' = \alpha^2+\beta^2+\gamma^2 \qquad F' = \alpha\alpha'+\beta\beta'+\gamma\gamma' \qquad G' = \alpha'^2+\beta'^2+\gamma'^2$$

Now the first condition is satisfied when, independently of du and dv, the quantities E, F, G bear to E', F', G', respectively, the same definite ratio, say m; that is, when

$$\frac{E}{E'} = \frac{F}{F'} = \frac{G}{G'} = m$$

This quantity m is then the ratio of the lengths of two corresponding elements on the first and second surfaces; or, if the elements of length are respectively ds and $d\sigma$, we have

$$ds = m\,d\sigma$$

which expresses an increase or diminution of length according as

$$m \lesseqgtr 1$$

This ratio is in general different for different points; in the special case, however, where m is constant, the projections of regions of finite extent will also be similar to the regions projected, and if $m=1$, the areas will be equal, and the first surface will be developed upon the second.

Consider two linear elements through the point given by u, v, their extremities being

$$u,\ v:\ u+du,\ v+dv \qquad u,\ v:\ u+\delta u,\ v+\delta v$$

The cosine of the angle between these two lines is, in rectilinear rectangular co-ordinates,

$$=\frac{dx\ \delta x+dy\ \delta y+dz\ \delta z}{\sqrt{[(dx^2+dy^2+dz^2)\ (\delta x^2+\delta y^2+\delta z^2)]}}$$

or

$$\frac{E\ du\ \delta u+2F\ (du\ \delta v+\delta u\ dv)+G\ dv\ \delta v}{\sqrt{[(E\ du^2+2F\ du\ dv+G\ dv^2)\ (E\ \delta u^2+2F\ \delta u\ \delta v+G\ \delta v^2)]}}$$

The cosine of the angle between the two corresponding elements on the second surface differs from this only by having E, F, G replaced by E', F', G'; we see then that, in order that these two quantities may be equal, E, F, G must be proportional to E', F', G', which is precisely the conclusion arrived at in examining the first condition of the projection; the two conditions are then identical, a fact which is indeed obvious from *a priori* considerations. Write for brevity

$$E du^2+2F du\ dv+G dv^2=\Omega$$

We know that the equation $\Omega=0$ admits of two separate integrations, inasmuch as we can divide the trinomial into two linear factors, either of which equated to zero must satisfy the equation $\Omega=0$, equating the two factors thus to zero and these results to integrations. As we already know the factors will be of the forms $dp+idq$ and $dp-idq$; and these equated to zero give

$$p+iq=\text{const.} \qquad p-iq=\text{const.}$$

where p and q denote real functions of u, v, and consequently

$$\Omega=n\ (dp^2+dq^2)$$

where n is a certain finite function of (u, v).

The same process leads us to

$$P+iQ=\text{const.} \qquad P-iQ=\text{const.}$$

as the two separate integrals of

$$\Omega'=E'du^2+2F'du\ dv+G'dv^2=0$$

and also

$$\Omega'=N\ (dP^2+dQ^2)$$

where P, Q, N denote real functions of U, V.

The difficulties of integration being supposed surmountable, these integrals that we have indicated conduct us to the general and complete solution of the problem. The condition of the projection has already been obtained as

$$\Omega'=m^2\Omega$$

which gives us

$$\frac{(dP+idQ)\ (dP-idQ)}{(dp+idq)\ (dp-idq)}=\frac{m^2n}{N}$$

The numerator of the first member of this equation is divisible by the denominator in two ways; either when

$$dP+idQ \text{ is divisible by } dp+idq$$

and

$$dP-idQ \text{ is divisible by } dp-idq$$

or when

$$dP+idQ \text{ is divisible by } dp-idq$$

and

$$dP-idQ \text{ is divisible by } dp+idq$$

In the first case $dP+idQ$ will vanish with $dp+idq$, or $P+iQ$ will be constant when $p+iq=\text{const.}$, which is equivalent merely to saying that $P+iQ$ is a function of $p+iq$, and also $P-iQ$ a function of $p-iq$. In the second case the converse holds; $P+iQ$ being a function of $p+iq$, and $P-iQ$ a function of $p+iq$. The solutions are then of the form

$$P+iQ=F_1(p+iq) \qquad P-iQ=F_2(p-iq)$$

or

$$P+iQ=\phi_1(p-iq) \qquad P-iQ=\phi_2(p+iq)$$

but the second of these functional signs is not arbitrary; if the function F_1 is real, the function F_2 must be identical with it; if, however, F_1 is imaginary, F_2 differs from it only in being its conjugate function, or function obtained by changing i into $-i$. The same remarks, of course, also hold for the functions ϕ_1 and ϕ_2; thus each of our solutions contains only one arbitrary function, which may be either real or imaginary.

We have by solution of the first two of these equations—

$$P=\tfrac{1}{2}F_1(p+iq)+\tfrac{1}{2}F_2(p-iq) \qquad iQ=\tfrac{1}{2}F_1(p+iq)-\tfrac{1}{2}F_2(p-iq)$$

or P will denote the real and iQ (in the second case $-iQ$) the imaginary part of the function F_1. Solution of these last equations will now afford us the values of U and V as functions of u and v, and so solve completely the problem. Denote the derived functions of F_1 and F_2 by F'_1 and F'_2, so that

$$dF_1(t)=F'_1(t)dt \qquad dF_2(t)=F'_2(t)dt$$

Then we have

$$\frac{dP+idQ}{dp+idq}=F'_1(p+iq) \qquad \frac{dP-idQ}{dp-idq}=F'_2(p-iq)$$

also

$$\frac{m^2 n}{N}=F'_1(p+iq)\,F'_2(p-iq)$$

The ratio of enlargement is therefore determined by the formula—

$$m=\sqrt{\left(\frac{(dp^2+dq^2)}{\Omega}\,\frac{\Omega'}{dP^2+dQ^2}\,F'_1(p+iq)\,F'_2(p-iq)\right)}$$

This will be reverted to in another place.

Assume now that the two surfaces under consideration are planes; then

$$x=u \qquad y=v \qquad z=0 \qquad \xi=U \qquad \eta=V \qquad \zeta=0$$

We have manifestly

$$E=G=1 \qquad F=0$$

and

$$\Omega=du^2+dv^2=0$$

which conducts to the integrals

$$u+iv=\text{const.} \qquad u-iv=\text{const.}$$

In like manner

$$\Omega'=dU^2+dV^2=0$$

gives the integrals

$$U+iV=\text{const.} \qquad U-iV=\text{const.}$$

The two general solutions are now

$$\text{(I)} \qquad U+iV=F_1(u+iv) \qquad U-iV=F_2(u-iv)$$

$$\text{(II)} \qquad U+iV=F_1(u-iv) \qquad U-iV=F_2(u+iv)$$

These results can also be expressed as follows: f denoting an arbitrary function, we equate the real part of $F(x+iy)$ to ξ and the imaginary part either to η or $-\eta$, as the case may be. Introducing the derived functions F'_1 and F'_2, write

$$F'_1(x+iy)=X+iY \qquad F'_2(x-iy)=X-iY$$

when X and Y denote real functions of x and y. We have now for the first solution

$$d\xi+id\eta=(X+iY)(dx+idy) \qquad d\xi-id\eta=(X-iY)(dx-idy)$$

and consequently

$$d\xi=Xdx-Ydy \qquad d\eta=Ydx+Xdy$$

Make now

$$X=S.\cos G \qquad Y=S.\sin G$$

$$dx=ds\cos g \qquad dy=ds\sin g$$

$$d\xi=d\sigma\cos\gamma \qquad d\eta=d\sigma\sin\gamma$$

when ds denotes a linear element of the first plane, and g its inclination to the axis of x; $d\sigma$ denotes in like manner the element of the second plane corresponding to ds on the first, and γ denotes its inclination to the axis of ξ. The above equations give then

$$d\sigma\cos\gamma=Sds\cos(G+g) \qquad d\sigma\sin\gamma=Sds\sin(G+g)$$

from which follows, regarding S as positive,

$$d\sigma=S\,ds \qquad \gamma=G+g$$

It is clear, also, from this that S denotes the ratio of alteration of the element ds to its projection $d\sigma$, and further that S is independent of g; and the independence of the angles G and g shows that all the linear elements proceeding from a point of the first plane and represented on the second plane by elements which cut each other under the same angles measured in the same direction.

If we choose for F a linear function of the form

$$F(p+iq)=A+B(p+iq)$$

when the constants A and B are of the forms

$$A=a+ib \qquad B=C+ic$$

then we shall have

$$F'(p+iq)=B=c+ie \qquad S=\sqrt{c^2+e^2} \qquad G=\tan^{-1}\frac{e}{c}$$

The ratio of alteration is therefore constant in all parts of the plane, and the projection of the first plane is throughout similar to the plane. For any other value of F the similarity would only hold for infinitesimal portions of the plane.

Enough has been said in the previous pages on the projection of the sphere upon a plane, so that we need not allude to that subject here; but we will once more obtain the formulas for the projection of an ellipsoid of revolution upon a plane, solving the problem directly instead of deriving it as a particular case of the more general problem of projecting the ellipsoid of their unequal axes upon a plane.

Denote by a and b the semi-axes of the ellipsoid; then

$$\frac{x^2+y^2}{a^2}+\frac{z^2}{b^2}=1$$

is the equation of the surface, and we can write

$$x=a\cos u\sin v \qquad y=a\sin u\sin v \qquad z=b\cos v$$

Ω now takes the form

$$\Omega=a^2\sin^2 v\,du^2+(a^2\cos^2 v+b^2\sin^2 v)\,dv^2$$

and the differential equation $\Omega=0$ assumes the form

$$du^2+(\cot^2 v+1-\epsilon^2)\,dv^2$$

when (assuming that $b<a$)

$$\epsilon^2=\frac{a^2-b^2}{a^2}$$

This gives

$$du\mp i\,dv\sqrt{(\cot^2 v+1-\epsilon^2)}\,dv=0$$

Assume

$$\sqrt{(1-\epsilon^2)}\tan v=\tan\omega$$

where, in the case of the earth, $90^\circ-\omega$ denotes the geographical latitude of a point and u denotes its longitude; this equation now assumes the form

$$du\mp i\,d\omega\frac{1-\epsilon^2}{(1-\epsilon^2\cos^2\omega)\sin\omega}$$

the integration of which gives

$$u\pm i\log\cot\tfrac{1}{2}\omega\left(\frac{1-\epsilon\cos\omega}{1+\epsilon\cos\omega}\right)^{\frac{\epsilon}{2}}$$

Denoting now by f an arbitrary functional symbol, we must equate ξ to the real part of

$$f\left[u\pm i\log\cot\tfrac{1}{2}\omega\left(\frac{1-\epsilon\cos\omega}{1+\epsilon\cos\omega}\right)^{\frac{\epsilon}{2}}\right]$$

and $i\eta$ to the imaginary part. If we choose for f a linear function, *i. e.*, write

$$f(p+iq)=k\,(p+iq)$$

then we have at once

$$\xi=ku \qquad \eta=\log\left[\cot\frac{\omega}{2}\left(\frac{1-\epsilon\cos\omega}{1+\epsilon\cos\omega}\right)^{\frac{\epsilon}{2}}\right]^k$$

which gives a projection analogous to Mercator's.

Assume now for f an imaginary exponential function, or

$$f(t)=ke^{i\lambda t}$$

then we have at once

$$\xi=k\tan^\lambda\tfrac{1}{2}\omega\left(\frac{1+\epsilon\cos\omega}{1-\epsilon\cos\omega}\right)^{\frac{\epsilon\lambda}{2}}\cos\lambda u \qquad \eta=k\tan^\lambda\tfrac{1}{2}\omega\left(\frac{1+\epsilon\cos\omega}{1-\epsilon\cos\omega}\right)^{\frac{\epsilon\lambda}{2}}\sin\lambda u$$

which, for $\lambda=1$, is analogous to the stereographic projection.

For the case where $b>a$, we have

$$a^2-b^2<0$$

and ϵ consequently imaginary; but

$$\left(\frac{1+\epsilon\cos\omega}{1-\epsilon\cos\omega}\right)^{\frac{\epsilon}{2}}$$

will be real. Write

$$\sqrt{\frac{b^2-a^2}{a^2}}=\epsilon'$$

then, for the determination of ω, we have the equation

$$\sqrt{1+\epsilon'^2}\tan v=\tan\omega$$

and the differential equation of the problem becomes

$$du\mp i\,d\omega\frac{1+\epsilon'^2}{(1+\epsilon'^2\cos^2\omega)\sin\omega}=0$$

giving the integral

$$u \pm i \left(\log \cot \frac{\omega}{2} + \epsilon' \tan^{-1} \epsilon' \cos \omega\right)$$

and this gives for ξ the real and for η the imaginary parts of

$$f\left[u + i\left(\log \cot \frac{\omega}{2} + \epsilon' \tan^{-1} \epsilon' \cos \omega\right)\right]$$

Assume first, f, a linear function, or

$$f(t) = kt$$

We have, then, at once

$$\xi = ku \qquad \eta = k \log \cot \frac{\omega}{2} + \epsilon' \tan^{-1} \epsilon' \cos \omega$$

Secondly, assume

$$f(t) = ke^{\lambda t}$$

Then we find

$$\xi = k \tan^{\lambda} \frac{\omega}{2}\, e^{-\epsilon' \lambda \tan^{-1} \epsilon' \cos \omega} \cos \lambda u \qquad \eta = k \tan^{\lambda} \frac{\omega}{2}\, e^{-\epsilon' \lambda \tan^{-1} \epsilon' \cos \omega} \sin \lambda u$$

Suppose that we have given a sphere of radius A, viz:

$$\xi^2 + \eta^2 + \zeta^2 = A^2$$

and an ellipsoid of revolution

$$\frac{x^2 + y^2}{a^2} + \frac{z^2}{b^2} = 1$$

required to project the latter surface upon the former. The co-ordinates ξ, η, ζ are given in terms of the geographical co-ordinates U, V, by the equations

$$\xi = A \cos U \sin V \qquad \eta = A \sin U \sin V \qquad \zeta = A \cos V$$

The differential equation arrived at is of course precisely the same as in the last example; and so, calling f an arbitrary functional symbol, we have merely to equate U to the real and $i \log \cot \frac{1}{2} V$ to the imaginary parts of

$$f\left(u + i \log\left[\cot \frac{\omega}{2} \left(\frac{1 - \epsilon \cos \omega}{1 + \epsilon \cos \omega}\right)^{\frac{\epsilon}{2}}\right]\right)$$

The simplest solution is, of course, for the case

$$f(t) = t$$

and gives

$$U = u \qquad \tan \tfrac{1}{2} V = \tan \frac{\omega}{2} \left(\frac{1 + \epsilon \cos \omega}{1 - \epsilon \cos \omega}\right)^{\frac{\epsilon}{2}}$$

formulas of great importance in geodesy.

The rectilinear rectangular co-ordinates of the point on the spherical surface corresponding to that denoted by u, v, on the ellipsoid are then

$$\xi = A \cos u \,\frac{2 \tan \frac{\omega}{2} \left(\frac{1 + \epsilon \cos \omega}{1 - \epsilon \cos \omega}\right)^{\frac{\epsilon}{2}}}{1 + \tan^2 \frac{\omega}{2} \left(\frac{1 + \epsilon \cos \omega}{1 - \epsilon \cos \omega}\right)^{\epsilon}}$$

$$\eta = A \sin u \,\frac{2 \tan \frac{\omega}{2} \left(\frac{1 + \epsilon \cos \omega}{1 - \epsilon \cos \omega}\right)^{\frac{\epsilon}{2}}}{1 + \tan^2 \frac{\omega}{2} \left(\frac{1 + \epsilon \cos \omega}{1 - \epsilon \cos \omega}\right)^{\epsilon}}$$

$$\zeta = A \,\frac{1 - \tan^2 \frac{\omega}{2} \left(\frac{1 + \epsilon \cos \omega}{1 - \epsilon \cos \omega}\right)^{\epsilon}}{1 + \tan^2 \frac{\omega}{2} \left(\frac{1 + \epsilon \cos \omega}{1 - \epsilon \cos \omega}\right)^{\epsilon}}$$

when ω is retained for brevity, instead of its value $\tan^{-1}(\tan v\sqrt{1-\epsilon^2})$. By means of the formulas

$$u=\tan^{-1}\frac{y}{x} \qquad v=\cos^{-1}\frac{z}{b}$$

we may transform these into relations giving the rectangular co-ordinates of the point on the sphere corresponding to a point on the ellipsoid in terms of the rectangular co-ordinates of this last point. These results show the great desirability of finding co-ordinates peculiar to the surfaces under consideration, and by means of which the relations between the chosen co-ordinates on the surface of projection and those on the surface projected may be as simple as possible, the ordinary rectangular-rectilinear co-ordinates evidently giving most complicated expressions.

If, instead of assuming

$$f(t)=t$$

we write

$$f(t)=t+\text{const.}$$

there is clearly no gain of generality if the chosen constant be real, as in that case we would have for V the same value as before, and the values of U and u would only differ by this constant. If, however, the constant is taken as imaginary, say, $-i\log k$, the results are quite different. We have in this case

$$U=u \qquad V=k\tan\frac{\omega}{2}\left(\frac{1+\epsilon\cos\omega}{1-\epsilon\cos\omega}\right)^{\frac{\epsilon}{2}}$$

Determine now the ratio m of alteration; we have for that purpose the formulas

$$n=\frac{\Omega}{dp^2+dq^2} \qquad N=\frac{\Omega'}{dP^2+dQ^2} \qquad m=\sqrt{\left(\frac{N}{n}f'_1(p+iq)f'_2(p-iq)\right)}$$

Now

$$\Omega=a^2\sin v\,du^2+(a\cos^2 v+b^2\sin^2 v)\,dv^2$$

and

$$dp^2+dq^2=du^2+\left(\cot^2 v+\frac{b^2}{a^2}\right)dv^2$$

or

$$n=a^2\sin^2 v$$

Similarly we find

$$N=A^2\sin^2 V \qquad f'(t)=1$$

Now we have

$$m=\frac{A\sin V}{a\sin v}=\frac{A\sin V}{a\sin\omega}\sqrt{1-\epsilon^2\cos^2\omega}=\frac{A}{a}\,\frac{k(1-\epsilon^2\cos^2\omega)^{\frac{\epsilon+1}{2}}}{\cos^2\frac{\omega}{2}(1-\epsilon\cos\omega)^{\epsilon}+k^2\sin^2\frac{\omega}{2}(1+\epsilon\cos\omega)^{\epsilon}}$$

a ratio which is dependent merely upon the latitude which is given by $90^\circ-\omega$. The smallest possible deviation from perfect similarity is obtained when k is so determined that m possesses equal values at the extreme limits of latitude of the region to be projected; in this case m will have its greatest or least value at the mean latitude, or nearly so. Calling ω_1 and ω_2 the extreme values of ω, and equating to each other the values of m for these limits we come readily to the expression

$$k=\sqrt{\left(\frac{\dfrac{\cos^2\frac{\omega_1}{2}(1-\epsilon\cos\omega_1)^{\epsilon}}{(1-\epsilon^2\cos^2\omega_1)^{\frac{\epsilon+1}{2}}}-\dfrac{\cos^2\frac{\omega_2}{2}(1-\epsilon\cos\omega_2)^{\epsilon}}{(1-\epsilon^2\cos^2\omega_2)^{\frac{\epsilon+1}{2}}}}{\dfrac{\sin^2\frac{\omega_2}{2}(1+\epsilon\cos\omega_2)^{\epsilon}}{(1-\epsilon^2\cos^2\omega_2)^{\frac{\epsilon+1}{2}}}-\dfrac{\sin^2\frac{\omega_1}{2}(1+\epsilon\cos\omega_1)^{\epsilon}}{(1-\epsilon^2\cos^2\omega_1)^{\frac{\epsilon+1}{2}}}}\right)}$$

In order to determine at what latitude m has its greatest or least value, observe that we have

$$\frac{dm}{m} = \cot V\, dv - \cot \omega\, d\omega + \frac{e^2 \cos \omega \sin \omega\, d\omega}{1 - e^2 \cos^2 \omega}$$

and

$$\frac{dV}{\sin V} = \frac{d\omega}{\sin \omega} - \frac{e^2 \sin \omega\, d\omega}{1 - e^2 \cos^2 \omega} = \frac{(1-e^2)\, d\omega}{(1 - e^2 \cos^2 \omega) \sin \omega}$$

from which follows

$$\frac{dm}{m} = \frac{(1-e^2)\, d\omega}{\sin \omega (1 - e^2 \cos^2 \omega)} (\cos V - \cos \omega)$$

Now we see that, for $V = \omega$,

$$\frac{dm}{d\omega} = 0$$

i. e., for $V = \omega$, m is a maximum or minimum. Denote this value of ω by W, then will

$$k = \left(\frac{1 - e \cos W}{1 + e \cos W}\right)^{\frac{e}{2}}$$

or, expressing W in terms of k,

$$W = \frac{1 - k^{\frac{2}{e}}}{e\left(1 + k^{\frac{2}{e}}\right)}$$

from which W can be determined when k has been computed by means of the above formula. The quantities U, V, and ω are connected by a relation which now becomes

$$\tan \tfrac{1}{2} V = \tan \frac{\omega}{2} \left(\frac{(1 - e \cos W)(1 + e \cos \omega)}{(1 + e \cos W)(1 - e \cos \omega)}\right)^{\frac{e}{2}}$$

It is easy to see that, for $\omega < W$ and $V > \omega$, $\cos V < \cos \omega$, and consequently $\frac{dm}{d\omega}$ will be negative; and for $\omega > W$ and $V < \omega$, $\frac{dm}{d\omega}$ will be positive; so that, for $\omega = V = W$, the value of m will always be a minimum and

$$= \frac{A}{a} \sqrt{(1 - e^2 \cos^2 W)}$$

If we choose the radius of the sphere

$$A = \frac{a}{\sqrt{(1 - e^2 \cos^2 W)}}$$

the representation of the ellipsoid at the latitude of $90° - W$ will be not only similar in infinitesimal portions, but also equal; for other latitudes, however, the projected elements will be greater than the elements themselves. We can expand the logarithm of m in a series according to ascending powers of $\cos V - \cos W$, of which the first terms are

$$\log m = \log\left(\frac{A}{a} (1 - e^2 \cos^2 W)\right) + \frac{e^2}{2(1-e^2)} (\cos V - \cos W)^2 + \frac{2e^4 \cos W}{3(1-e^2)^2} (\cos V - \cos W)^3 + \ldots$$

From what has preceded it will be easy to obtain the formulas for the projection of the general ellipsoid upon a sphere. Denote by R the radius of the sphere; then its equation will be

$$\xi^2 + \eta^2 + \zeta^2 = R^2$$

and for the ellipsoid whose semi-axes are a, b, c,

$$\frac{x^2}{a^2} + \frac{y^2}{b^2} + \frac{z^2}{c^2} = 1$$

The co-ordinates ξ, η, ζ are given in terms of the two independent variables U and V by the relations

$$\xi = R \cos U \sin V \qquad \eta = R \sin U \sin V \qquad \zeta = R \cos V$$

If λ_1 and λ_2 denote the variable parameters belonging to the two hyperboloids, confocal to the given ellipsoid, we have for the equations of these surfaces

$$\frac{x^2}{a^2+\lambda_1}+\frac{y^2}{b^2+\lambda_1}+\frac{z^2}{c^2+\lambda_1}=1 \qquad \frac{x^2}{a^2+\lambda_2}+\frac{y^2}{b^2+\lambda_2}+\frac{z^2}{c^2+\lambda_2}=1$$

and the co-ordinates x, y, z are then given by

$$x^2=\frac{a^2(a^2+\lambda_1)(a^2+\lambda_2)}{(a^2-b^2)(a^2-c^2)} \qquad y^2=\frac{b^2(b^2+\lambda_1)(b^2+\lambda_2)}{(b^2-c^2)(b^2-a^2)} \qquad z^2=\frac{c^2(c^2+\lambda_1)(c^2+\lambda_2)}{(c^2-a^2)(c^2-b^2)}$$

and the element of length on the ellipsoid by

$$ds^2=(\lambda_1-\lambda_2)\left\{\frac{\lambda_1\,d\lambda_1^2}{(a^2+\lambda_1)(b^2+\lambda_1)(c^2+\lambda_1)}+\frac{\lambda_2\,d\lambda_2^2}{(a^2+\lambda_2)(b^2+\lambda_2)(c^2+\lambda_2)}\right\}$$

The equation $ds^2=0$ thus becomes

$$\frac{\lambda_1\,d\lambda_1^2}{L_1^2}+\frac{\lambda_2\,d\lambda_2^2}{L_2^2}=0$$

writing, for convenience,

$$L_1=\sqrt{(a^2+\lambda_1)(b^2+\lambda_1)(c^2+\lambda_1)} \qquad L_2=\sqrt{(a^2+\lambda_2)(b^2+\lambda_2)(c^2+\lambda_2)}$$

and the differential equations are

$$\frac{\sqrt{\lambda_1}\,d\lambda_1}{L_1}\pm i\frac{\sqrt{\lambda_2}\,d\lambda_2}{L_2}=0$$

If we find one integral of these in the form

$$P+iQ=\text{const.}$$

then having in the case of the sphere the integral

$$U+i\log\cot\tfrac{1}{2}V=\text{const.}$$

it is only necessary to equate U to the real, and $i\log\cot\frac{1}{2}V$ to the imaginary part of $f(P+iQ)$, in order to obtain the most general solution of the problem.

The method of transformation here employed is the same as the one previously used in the case of the projection of the ellipsoid upon the plane. The limits of λ_1 and λ_2 are

$$-c^2>\lambda_1>-b^2 \qquad -b^2>\lambda_2>a^2$$

so, as before, λ_1 expressed in terms of the new variable θ is

$$\lambda_1=-\frac{b^2(a^2-c^2)\cos^2\theta+c^2(a^2-b^2)\sin^2\theta}{(a^2-c^2)\cos^2\theta+(a^2-b^2)\sin^2\theta}$$

or

$$\lambda_1=-\frac{b^2+c^2x^2}{1+x^2}$$

The same reductions that have been already employed will conduct to the equation

$$P=\int\frac{\sqrt{b^2+c^2x^2}}{\sqrt{(a^2-b^2)+(a^2-c^2)x^2}}\cdot\frac{2dx}{1+x^2}$$

and, on writing

$$\frac{b^2}{a^2}=\sin^2\alpha \qquad \frac{a^2}{b^2}\,\frac{b^2-c^2}{a^2-c^2}=k^2 \qquad \frac{b^2-c^2}{a^2-c^2}=k^2\sin^2\alpha$$

this is

$$=\frac{2\sin\alpha\,\Delta\alpha}{\cos\alpha}\int\frac{\Delta\theta\,d\theta}{1-k^2\sin^2\alpha\sin^2\theta}$$

since

$$\Delta\alpha=\sqrt{\frac{a^2-b^2}{a^2-c^2}} \text{ and } \cos\alpha=\sqrt{\frac{a^2-b^2}{a^2}}$$

If we call e the eccentricity of the section of the ellipsoid by the plane xy, it is clear that

$$a=\cos^{-1} e$$

It will be a little more convenient to write P for $\frac{1}{2}$P, and thereby drop the factor 2, which will otherwise run all through the work. Introducing elliptic functions by means of the equations

$$t=\int\frac{d\theta}{\Delta\theta}\qquad\qquad a=\int\frac{da}{\Delta a}$$

we can write for P at once the value

$$P=\tfrac{1}{2}\log\frac{e^{2ht}\theta(t+a)}{\theta(t-a)}$$

when

$$h=\operatorname{tn} a\operatorname{dn} a-Za=\frac{d}{da}\log H(a+K)$$

and in terms of the q functions

$$h=\frac{\pi}{2K}\,\frac{\Sigma(2j-1)q^{(j-\frac{1}{2})^2}\sin(2j-1)a'}{\Sigma q^{(j-\frac{1}{2})^2}\cos(2j-1)a'}$$

when

$$a'=\frac{\pi a}{2K}$$

The ratio of the θ functions is also

$$\frac{\theta(t+a)}{\theta(t-a)}=\frac{1+2\,\Sigma(-)^j q^{j^2}\cos 2j(t'+a')}{1+2\,\Sigma(-)^j q^{j^2}\cos 2j(t'-a')}$$

where

$$t'=\frac{\pi t}{2k}$$

In the case of the Q integral we have seen that it is necessary to write

$$\theta=\operatorname{am} i(K+\tau)$$

$i(K+\tau)$ being the new value of t; by change of the modulus k into the complementary k' we introduce the new amplitude ψ, defined by

$$\psi=\operatorname{am}(\tau\,k')$$

giving, finally,

$$Q=h\tau+\frac{1}{2i}\log\frac{H(a+i\tau)}{H(a-i\tau)}$$

or

$$Q=h\tau+\tan^{-1}\frac{\Sigma(-)^{j-1}q^{(j-\frac{1}{2})^2}(e^{(2j-1)\tau'}-e^{-(2j-1)\tau'})\cos(2j-1)a'}{\Sigma(-)^{j-1}q^{(j-\frac{1}{2})^2}(e^{(2j-1)\tau'}+e^{-(2j-1)\tau'})\sin(2j-1)a'}$$

The complex quantity P$+i$Q is now expressed in elliptic functions by

$$\tfrac{1}{2}\log\left[\frac{e^{2ht}\,\theta(t+a)\,H(a+i\tau)}{\theta(t-a)\,H(a-i\tau)}\right]+ih\tau$$

in which the real and imaginary parts are not separated. Writing for P and Q their values in terms of the q functions, the conditions of the projection are satisfied by equating U to the real real and $i\log\cot\frac{1}{2}$V to the imaginary parts of

$$f\left(\tfrac{1}{2}\log e^{2ht}\frac{1+2\,\Sigma(-)^j q^{j^2}\cos(2j-1)(t'+a')}{1+2\,\Sigma(-)^j q^{j^2}\cos(2j-1)(t'-a')}+ih\tau+i\tan^{-1}\frac{\Sigma(-)^{j-1}q^{(j-\frac{1}{2})^2}(e^{(2j-1)\tau'}-e^{-(2j-1)\tau'})\cos(2j-1)a'}{\Sigma(-)^{j-1}q^{(j-\frac{1}{2})^2}(e^{(2j-1)\tau'}+e^{-(2j-1)\tau'})\sin(2j-1)a'}\right)$$

If the function f is taken as linear and of the form $f(\nu)=\nu$, we find

$$U=P\qquad\qquad\cot\tfrac{1}{2}V=e^{Q}$$

By the first of these, all curves dependent only upon t are projected into curves dependent only upon the longitude U, or into the meridians; but

$$t=\int\frac{d\theta}{\Delta\theta}$$

and s is again a function of λ, so that t is a function of λ, and the curves which depend upon t are the lines of curvature cut out by the hyperboloid of two nappes; these lines are then projected into the meridians of the sphere, and the remaining system of lines of curvature is projected into the parallels of latitude. The quantities entering in the solution of this general case are so complicated that, as the problem is scarcely a practical one, it does not seem desirable to continue the research any further. It may be observed, however, in conclusion, that the ratio of alteration m is given by

$$m=-2\frac{\mathrm{R}e^{-u}}{1+e^{-2u}}\left[\frac{b^2\,\mathrm{cn}^2 t+c^2\Delta^2 s\,\mathrm{sn}^2 t}{\mathrm{cn}^2 t+\Delta^2 s\,\mathrm{sn}^2 t}+\frac{b^2\,[\mathrm{cn}^2(\tau,k')-\mathrm{sn}^2(\tau,k)]}{\mathrm{sn}^2(\tau,k')-\mathrm{sn}^2 s\,\mathrm{cn}^2(\tau,k)}\right]^{-1}$$

We have seen that in general there are two solutions to the proposed problem of the orthomorphic projection of one surface upon another, viz,

(I) $\quad \mathrm{P}+i\mathrm{Q}=f_1(p+iq) \qquad \mathrm{P}-i\mathrm{Q}=f_1(p-iq)$

(II) $\quad \mathrm{P}+i\mathrm{Q}=\phi_1(p-iq) \qquad \mathrm{P}-i\mathrm{Q}=\phi_1(p+iq)$

It can now be shown that in one of these solutions the positions of the different parts of the surface are in the projection exactly similar to their positions on the given surface, while from the other solutions results are inverse similarity.

It is to be remarked first, however, that we can speak only of exact and inverse similarity in so far as we may speak of the upper and under sides of the surfaces considered. As, however, by this way of speaking, it is perfectly arbitrary which we call the upper and which the under side of a surface, it is clear that the two projections have no essential points of difference, an exact projection becoming an inverse projection when the side of the surface previously considered as the upper side is made arbitrarily to become the lower side.

The upper and lower sides of the surface will be defined as follows: If $F=0$ is an equation satisfied by one of the surfaces, F is a given function of the co-ordinates x, y, z, which for all points lying on the surface is equal to zero, and for every other point is greater or less than zero; that is, is either positive or negative. By passing through the surface in one direction, F changes from positive to negative, and by passing in the opposite direction the converse takes place. The side of the surface on which F is positive will be called the upper, and the side on which it is negative the lower side of the surface.

Let the equation of the second surface be $\Phi=0$; the same remarks of course apply to this surface as to the surface $F=0$. Differentiating these two equations gives

$$dF=l_1dx_1+m_1dy+n_1dz \qquad d\Phi=\lambda_1d\xi+\mu_1d\eta+\nu_1d\zeta$$

whence l_1, m_1, n_1 are functions of x, y, z, and λ_1, μ_1, ν_1 are functions of ξ, η, ζ.

The projection of $F=0$ upon $\Phi=0$ admits of having six intermediate and simpler projections inserted between the beginning of the operation. These are given in the following table:

Co-ordinates of corresponding points.

(1)	The surface $F=0$	x, y, z
(2)	Representation in the plane	x, y, 0
(3)	Representation in the plane	u, v, 0
(4)	Representation in the plane	p, q, 0
(5)	Representation in the plane	P, Q, 0
(6)	Representation in the plane	U, V, 0
(7)	Representation in the plane	ξ, η, 0
(8)	Projection upon the surface $\Phi=0$	ξ, η, ζ

Leaving to the side the alteration undergone by the projection, we may now consider the relative positions occupied by the infinitesimal linear elements of the surface upon any two representations. We shall call two representations similar when the linear elements that proceed from a point and lie on the right hand in one representation correspond to linear elements lying on the right hand of the corresponding point in the second representation; in the opposite case the projections will be spoken of as inversely-similar. As regards the planes upon which (Nos. 2–7) the intermediate projections are made, we will merely call that side where the positive values of the third co-ordinate are found the upper side; the lower side will then correspond to negative values of this co-ordinate. The surfaces Φ and Ψ have already been mentioned as having their upper sides corresponding to positive, and their lower sides to negative values of Φ and Ψ respectively.

It is quite clear that for any point of the first surface at which we consider x and y invariable and increase z by a positive increment, if we come to the upper side of this surface the representations in (1) and (2) are exactly-similar, or an exactly-similar representation is obtained when n is positive, and an inversely similar representation when n is negative.

In the same way, if ζ be increased by a positive increment, exactly-similar representations are obtained in (7) and (8) when n_1 is positive, and inversely-similar representations when n_1 is negative. In order to compare (2) and (3), consider in (2) a linear element ds, the co-ordinates of whose extremities are $x, y, x+dx, y+dy$; and denote by f its inclination to the axis of x; then

$$dx=ds\cos f \qquad dy=ds\sin f$$

In (3) let $d\sigma$ and φ represent the corresponding quantities; then

$$du=d\sigma\cos\varphi \qquad dv=d\sigma\sin\varphi$$

but

$$dx=a\,du+a'\,dv \qquad dy=b\,du+b'\,dv$$

consequently

$$ds\cos f=d\sigma(a\cos\varphi+a'\sin\varphi) \qquad ds\sin f=d\sigma(b\cos\varphi+b'\sin\varphi)$$

and

$$\tan f=\frac{b\cos\varphi+b'\sin\varphi}{a\cos\varphi+a'\sin\varphi}$$

Regarding x and y as constant and f, φ as variable differentiation gives

$$\frac{df}{d\varphi}=\frac{ab'-a'b}{(a\cos\varphi+a'\sin\varphi)^2+(b\cos\varphi+b'\sin\varphi)^2}=\begin{vmatrix} a, & a' \\ b, & b' \end{vmatrix}\left(\frac{d\sigma}{ds}\right)^2$$

The sign of $\frac{df}{d\varphi}$ manifestly depends only on that of the determinant

$$\begin{vmatrix} a, & a' \\ b, & b' \end{vmatrix}$$

If this is positive, f and φ increase together, and if the determinant is negative these quantities vary in an opposite manner, f increasing while φ decreases. In the first case the representations (2) and (3) are exactly-similar; in the second case they are inversely-similar. The combination of the results now obtained gives that (1) and (3) are exactly-similar or inversely-similar, according as

$$\frac{1}{n}\begin{vmatrix} a, & a' \\ b, & b' \end{vmatrix}$$

is positive or negative.

Upon the surface $\Psi=0$ obtains

$$l_1dx+m_1dy+n_1dz=0$$

or substituting the values of dx, dy, dz as functions of du and dv

$$(l_1a+m_1b+n_1c)du+(l_1a'+m_1b'+n_1c')dv=0$$

but du and dv are independent, and so we must have

$$l_1a+m_1b+n_1c=0 \qquad l_1a'+m_1b'+n_1c'=0$$

and consequently l, m, n are proportional to

$$\begin{vmatrix} b, & b' \\ c, & c' \end{vmatrix} \quad \begin{vmatrix} c, & c' \\ a, & a' \end{vmatrix} \quad \begin{vmatrix} a, & a' \\ b, & b' \end{vmatrix}$$

or

$$\frac{bc'-b'c}{l_1}=\frac{ca'-c'a}{m_1}=\frac{ab'-a'b}{n_1}=k$$

Any one of these can be used as the criterion for the nature of the representations (1) and (3), or better still, replacing these thus by

$$\frac{l_1(bc'-b'c)+m_1(ca'-c'a)+n_1(ab'-a'b)}{l_1^2+m_1^2+n_1^2}=k$$

and then multiplying through by the positive quantity $l_1^2+m_1^2+n_1^2$, we have as the sought criterion the determinant

$$\begin{vmatrix} l_1, & m_1, & n_1 \\ a, & b, & c \\ a', & b', & c' \end{vmatrix}$$

Similarly, the exact or inverse similarity of (6) and (8) will depend upon the positive or negative values of

$$\frac{\beta\gamma'-\gamma\beta'}{\lambda_1}=\frac{\gamma\alpha'-\gamma'\alpha}{\eta_1}=\frac{\alpha\beta'-\alpha'\beta}{\nu_1}$$

or upon the determinant

$$\begin{vmatrix} \lambda_1, & \eta_1, & \nu_1 \\ \alpha_1, & \beta, & \gamma \\ \alpha', & \beta', & \gamma' \end{vmatrix}$$

In like manner the condition for exact or inverse similarity in the representations (3) and (4) depends upon the positive or negative sign of

$$\begin{vmatrix} \frac{dp}{du}, & \frac{dp}{dv} \\ \frac{dq}{du}, & \frac{dq}{dv} \end{vmatrix}$$

and in (5) and (6) upon the sign of

$$\begin{vmatrix} \frac{dP}{dU}, & \frac{dP}{dV} \\ \frac{dQ}{dU}, & \frac{dQ}{dV} \end{vmatrix}$$

In the projection of one plane upon another by the first solution, *i. e.*, by

(I) $\qquad P+iQ=f_1(p+iq) \qquad P-iQ=f_2(p-iq)$

it was found that exact similarity resulted, or, that the elements proceeding from a point in one plane and making certain angles with each other had in the second plane the corresponding elements making the same angles with each other, the angles being measured in the same direction.

The second solution

(II) $\qquad P+iQ=\phi_1(p-iq) \qquad P-iQ=\phi_2(p+iq)$

will only differ from the first in giving us a result that the angles between corresponding elements in the two representations will be equal, but measured in opposite directions. These considerations afford us the means of determining the relation between the representations (4) and (5); these are exactly-similar for the first solution

$$\text{(I)} \qquad P+iQ=f_1(p+iq) \qquad P-iQ=f_2(p-iq)$$

and inversely-similar for the second solution

$$\text{(II)} \qquad (P+iQ)=\phi_1(p-iq) \qquad P-iQ=\phi_2(p+iq)$$

Now, in order to determine whether the projection of Ψ upon Φ has not only its elements similar, but similarly placed, it is necessary to take note of the negative signs of the quantities

$$\frac{1}{n_1}\begin{vmatrix} a, & a' \\ b, & b' \end{vmatrix} \qquad \begin{vmatrix} \frac{dp}{du}, & \frac{dp}{dv} \\ \frac{dq}{du}, & \frac{dq}{dv} \end{vmatrix} \qquad \begin{vmatrix} \frac{dP}{dU}, & \frac{dP}{dV} \\ \frac{dQ}{dU}, & \frac{dQ}{dV} \end{vmatrix} \qquad \frac{1}{\nu_1}\begin{vmatrix} \alpha, & \alpha' \\ \beta, & \beta' \end{vmatrix}$$

If there are none, or an even number of negative signs, the first solution must be chosen to give the desired result; if there is an odd number of negative signs, the second solution must be adopted. The reverse of this method of choice will give an inversely-similar projection.

As we already know, the transition from a stereographic projection to any other orthomorphic projection is merely a particular case of the solution which we have indicated as

$$P+iQ=f_1(p+iq)$$

—it is of course not necessary to write $-iQ=f_2(p-iq)$—so that if x, y denote the co-ordinates of a point upon a stereographic projection, and ξ, η the co-ordinates of the same point upon some other orthomorphic projection, we have

$$(x+iy)=f(\xi+i\eta)$$

f, of course, denoting the arbitrary function derived from the integration of a certain differential equation. If the operation indicated by the functional symbol f is only the addition of a constant, the map is simply shoved along. If it is the multiplication by an imaginary root of unity, the map is merely turned round. If it is the multiplication by a modulus, the scale of the map is changed. If the operation raises the quantity to an integral power, the result is Sir J. Herschel's projection; if to a fractional power, the result is a many-sheeted map, that is, one in which the earth is only covered by a number, finite or infinite, of separate sheets of the map; and on these sheets the whole earth may be represented only once, or several times, or an infinite number of times. When f is an integral algebraic function, the result is a map having a finite number of north and south poles, and the problem to construct a map having north and south poles at given points is resolved by solution of the appropriate algebraic equation. The relation $f(z)=e^z$ is Mercator's projection, and other projections of infinite variety may of course be obtained by a suitable choice of the functional symbol.

Suppose, as a final problem, we take the expression

$$x+iy=(\xi+i\eta)^n$$

the meridians being projected in right lines passing through the point a on the axis of x and the parallels by circles having their centers at this same point. Call θ the angle under which any one meridian cuts x, then

$$(x-a)\tan\theta=y$$

is the equation of this line, or, as it may be written

$$\theta=\tan^{-1}\frac{y}{x-a}$$

The equation of a parallel of radius ρ is in like manner given by

$$(x-a)^2+y^2=\rho^2$$

Consider this latter equation first. It may be written in the form

$$(x+iy-a)(x-iy-a)=\rho^2$$

and, since

$$x+iy=(\xi+i\eta)^n$$

again, as

$$[(\xi+i\eta)^n-a][(\xi-i\eta)^n-a]=\rho^2$$

or, if each factor is divided into its factors,

$$\prod_{j=0}^{j=n-1}\left[\left(\xi-a^{\frac{1}{n}}\cos\frac{2j\pi}{n}\right)+i\left(\eta-a^{\frac{1}{n}}\sin\frac{2j\pi}{n}\right)\right]\left[\left(\xi-a^{\frac{1}{n}}\cos\frac{2j\pi}{n}\right)+i\left(\eta-a^{\frac{1}{n}}\sin\frac{2j\pi}{n}\right)\right]=\rho^2$$

Or, again, this is obviously equivalent to

$$\prod_{j=0}^{j=n-1}\left[\left(\xi-a^{\frac{1}{n}}\cos\frac{2j\pi}{n}\right)^2+\left(\eta-a^{\frac{1}{n}}\sin\frac{2j\pi}{n}\right)^2\right]=\rho^2$$

Π of course denotes the continued product of all the terms following it, obtained by giving to j all of its value from 0 to $n-1$. If we connect each of the points $\sqrt[n]{a}$ to a point $\xi+i\eta$, the length of the connecting line will clearly be given by

$$p=\sqrt{\left(\xi-a^{\frac{1}{n}}\cos\frac{2j\pi}{n}\right)^2+\left(\eta-a^{\frac{1}{n}}\cos\frac{2j\pi}{n}\right)^2}$$

Substituting each value of this in the above equation, it becomes

$$p_1p_2p_3\ldots\ldots p_n=\rho^2$$

Of course we would have obtained the same result had the circle been drawn about any point a, b, or $a+ib$, and so we can state the following

Theorem: When n is real, integral, and positive, the projection $x+iy=(\xi+i\eta)^n$ changes the circle $r=\rho$ around the center $a+ib$ into the curve $p_1p_2p_3\ldots\ldots p_n=\rho^2$, when p denotes a radius vector through each of the points $(a+ib)^{\frac{1}{n}}$.

For the case $n=2$, the system of concentric circles is projected into a system of confocal lemniscates with the foci at $\sqrt{a+ib}$; and in the general case we can say that the circles are projected into lemniscates of the nth order whose foci are at the angles of a regular polygon.

The meridians are transformed in a similar manner; we had for their equation

$$\theta=\tan^{-1}\frac{y}{x-a}$$

It is known that

$$2i\tan^{-1}\frac{y}{x-a}=\log\frac{x+iy-a}{x-iy-a}$$

and so

$$\theta=\frac{1}{2i}\log\frac{(\xi+i\eta)^n-a}{(\xi-i\eta)^n-a}$$

$$=\frac{1}{2i}\log\prod_{j=0}^{j=n-1}\frac{(\xi+i\eta)-a^{\frac{1}{n}}\left(\cos\frac{2j\pi}{n}+i\sin\frac{2j\pi}{n}\right)}{(\xi-i\eta)-a^{\frac{1}{n}}\left(\cos\frac{2j\pi}{n}+i\sin\frac{2j\pi}{n}\right)}$$

$$=\sum_{j=0}^{j=n-1}\frac{1}{2i}\log\frac{\left(\xi-a^{\frac{1}{n}}\cos\frac{2j\pi}{n}\right)+i\left(\eta-a^{\frac{1}{n}}\sin\frac{2j\pi}{n}\right)}{\left(\xi-a^{\frac{1}{n}}\cos\frac{2j\pi}{n}\right)-i\left(\eta-a^{\frac{1}{n}}\sin\frac{2j\pi}{n}\right)}$$

or, finally,

$$\theta=\sum_{j=0}^{j=n-1}\tan^{-1}\frac{\eta-a^{\frac{1}{n}}\sin\frac{2j\pi}{n}}{\xi-a^{\frac{1}{n}}\cos\frac{2j\pi}{n}}$$

If, however, any point $\xi+i\eta$ is connected with the n points given by $\sqrt[n]{a}$ the angle which each of the connecting lines makes with the axis of ξ is given by

$$\theta=\tan^{-1}\frac{\eta-a^{\frac{1}{n}}\sin\frac{2j\pi}{n}}{\xi-a^{\frac{1}{n}}\cos\frac{2j\pi}{n}}$$

and consequently

$$\theta=\theta_1+\theta_2+\;.\;.\;.\;.\;.\;\theta_n$$

and we have the following

Theorem: If from any point $a+ib$, lines are drawn making angles with the radius vector through $a+ib$ of $\theta=\delta$, these lines by the projection $x+iy=(\xi+i\eta)^n$—n positive, real and integral—are transformed into the curves $\theta_1+\theta_2+\theta_3+\;.\;.\;.\;\theta_n=\delta$, where the quantities θ denote the angles made by the radii vectors to the curve from the points $\sqrt[n]{a+ib}$ make with the radius vector of any one of these points.

For $n=2$ these curves are equilateral hyperbolas, and in the general case we can speak of the curves as being hyperbolas of the n order. With these definitions of the preceding systems of curves, we have the theorem that the orthogonal system of confocal lemniscates of the nth order is a group of hyperbolas of the nth order though the n foci of the lemniscates.

The only alterations to be studied in this kind of projection are the alterations of lengths and areas, the former being denoted by the quantity m, the latter by m^2. The value of m has been already shown to be given by the equation

$$m=\sqrt{\left(\frac{\mathrm{N}}{n}f'_1(p+iq)f'_2(p-iq)\right)}$$

where

$$\mathrm{N}=\frac{\Omega^2}{d\mathrm{P}^2+d\mathrm{Q}^2}\qquad\qquad n=\frac{\Omega}{dp^2+dq^2}$$

For the projection of a sphere upon a plane, it has been seen that

$$\frac{\mathrm{N}}{n}=\mathrm{R}^2\cos^2\theta$$

R denoting the radius of the sphere, and θ the latitude of a point.

Tchebychef has based a discussion concerning the most advantageous choice of projections upon the following considerations:

Calling $\frac{n}{N}=G$ and taking the logarithm of the above value of m,

$$\log m=\tfrac{1}{2}\log f'_1(p+iq)+\tfrac{1}{2}\log f'_2(p-iq)-\tfrac{1}{2}\log G$$

Now the equation

$$\frac{d^2F}{dp^2}+\frac{d^2F}{dq^2}=0$$

has for its integral

$$F=\varphi_1(p+iq)+\varphi_2(p-iq)$$

when φ_1 and φ_2 are perfectly arbitrary functional symbols; as then f_1 and f_2 are also quite arbitrary, the above expression for log m varies only as the difference between the integral of

$$\frac{d^2F}{dp^2}+\frac{d^2F}{dq^2}=0$$

and the function $\frac{1}{2}$ log G. The properties of this equation show that this difference is a minimum inside of the space limited by any curve whatever when the value of $F-\frac{1}{2}\log G$ has a constant value over the curve which bounds this region. The integration of

$$\frac{d^2F}{dp^2}+\frac{d^2F}{dq^2}=0$$

gives under this condition

$$F=\tfrac{1}{2}\log f'_1(p+iq)+\tfrac{1}{2}\log f'_2(p-iq)$$

from which, with exception of a constant, the values of $f'_1(p+iq)$ and $f'_2(p-iq)$ may be determined.

Before leaving the subject we may just notice the form of ds employed by Bour in his memoir on the deformation of surfaces. The element of length ds is given by

$$ds^2=Edp^2+2Fdp\,dq+Gdq^2$$

when E, F, G are functions of (p, q) and of the form given in the beginning of this chapter. Suppose that for a certain system of values (p, q) there results

$$F=0 \qquad E=G=\lambda$$

then

$$ds^2=\lambda(dp^2+dq^2)$$

The curves p=constant and q=constant are in this case known as isothermal curves; they are clearly orthogonal, and geometrically they divide the surface into a series of infinitely small squares. It is easy to show geometrically that there exists upon any surface an infinite number of families of orthogonal curves which enjoy this property; that is to say, an infinite number of systems of co-ordinates which conduct to the form

$$ds^2=\lambda(dp^2+dq^2)$$

Suppose, now, that we place

$$p+iq=2\alpha \qquad p-iq=2\beta$$

then

$$dp+idq=2d\alpha \qquad dp-idq=2d\beta$$

and multiplying these together gives

$$ds^2=4\lambda\,d\alpha\,d\beta$$

This equation, like $ds^2=\lambda(dp^2+dq^2)$, has the advantage of being symmetrical with respect to α, β.

It is to be observed here that the expression

$$ds^2 = 4\lambda\, d\alpha\, d\beta$$

is equivalent to

$$ds^2 = Edp^2 + 2Fdpdq + Gdq^2$$

if we make

$$E=0 \qquad F=2\lambda \qquad G=0$$

E, F, G now being given by

$$E=\left(\frac{dx}{d\alpha}\right)^2+\left(\frac{dy}{d\alpha}\right)^2+\left(\frac{dz}{d\alpha}\right)^2=0$$

$$F=\frac{dx}{d\alpha}\frac{dx}{d\beta}+\frac{dy}{d\alpha}\frac{dy}{d\beta}+\frac{dz}{d\alpha}\frac{dz}{d\beta}=2\lambda$$

$$G=\left(\frac{dx}{d\beta}\right)^2+\left(\frac{dy}{d\beta}\right)^2+\left(\frac{dz}{d\beta}\right)^2=0$$

The subject of these isothermal lines is very fully and elegantly treated by M. Haton de la Gonpillière in the Journal de l'École Polytechnique, vol. 22; the same volume also contains Bour's memoir.

The method of geodesic co-ordinates might also be employed in this problem; they are defined by

$$E=1 \qquad F=0 \qquad ds^2=dp^2+Gdq^2$$

The line q=const. is here a geodesic line upon which the lengths dq are measured. The line p=const. is perpendicular to the former, and its element of length is $\sqrt{G}dq$.

---

§ VIII.

GENERAL THEORY OF EQUIVALENT PROJECTIONS.

In this chapter we shall consider principally the alterations that take place in an equivalent projection, and also give the equations for the projection of an ellipsoid of revolution upon a plane. We may first, however, obtain the general condition for the equivalent projection of any surface upon a plane.

Let x, y, z denote the rectangular co-ordinates of any point of a surface and pq the two independent parameters, in terms of which x, y, z can be separately given; then writing, as before,

$$E=\left(\frac{dx}{dp}\right)^2+\left(\frac{dy}{dp}\right)^2+\left(\frac{dz}{dp}\right)^2$$

$$F=\frac{dx}{dp}\frac{dx}{dq}+\frac{dy}{dp}\frac{dy}{dq}+\frac{dz}{dp}\frac{dz}{dq}$$

$$G=\left(\frac{dx}{dq}\right)^2+\left(\frac{dy}{dq}\right)^2+\left(\frac{dz}{dq}\right)^2$$

we have for the element of length

$$ds^2=Edp^2+2Fdp\,dq+Gdq^2$$

and for the element of area

$$d\sigma=Vdpdq$$

where

$$V^2=EG-F^2$$

The quantity V can, of course, be written as a symmetrical determinant, and in fact is given by

$$V=\begin{vmatrix} \left(\frac{dx}{dp}\right)^2+\left(\frac{dy}{dp}\right)^2+\left(\frac{dz}{dp}\right)^2 & \frac{dx}{dp}\frac{dx}{dq}+\frac{dy}{dp}\frac{dy}{dq}+\frac{dz}{dp}\frac{dz}{dq} \\ \frac{dx}{dp}\frac{dx}{dq}+\frac{dy}{dp}\frac{dy}{dq}+\frac{dz}{dp}\frac{dz}{dq} & \left(\frac{dx}{dq}\right)^2+\left(\frac{dy}{dq}\right)^2+\left(\frac{dz}{dq}\right)^2 \end{vmatrix}$$

The co-ordinates of the four points at the angles of this small parallelogram are, upon the surface,

$$(p, q) \qquad (p+dp, q) \qquad (p, q+dq) \qquad (p+dp, q+dq)$$

The corresponding points on the plane are, taking ξ, η as rectangular axes,

$$\begin{array}{ll} \xi & \eta \\ \xi+\frac{d\xi}{dp}dp & \eta+\frac{d\eta}{dp}dp \\ \xi+\frac{d\xi}{dq}dq & \eta+\frac{d\eta}{dq}dq \\ \xi+\frac{d\xi}{dp}dp+\frac{d\xi}{dq}dq & \eta+\frac{d\eta}{dp}dp+\frac{d\eta}{dq}dq \end{array}$$

The area of this projection is

$$\begin{vmatrix} \frac{d\xi}{dp} & \frac{d\xi}{dq} \\ \frac{d\eta}{dp} & \frac{d\eta}{dq} \end{vmatrix} dp\, dq$$

Equating this to the corresponding element on the surface, we obtain as the condition of equivalent projection the differential equation

$$\frac{d\xi}{dp}\frac{d\eta}{dq}-\frac{d\xi}{dq}\frac{d\eta}{dp}=V$$

Nothing of interest can be obtained by attempting to discuss this very general form of the differential equation. It does not seem possible to reduce the question to one of quadratures, except in the case of a surface of revolution, when V will be a function of only one of the variables (p, q). If, in considering surfaces of revolution, we define latitude as the angle made by a normal to the surface with a plane perpendicular to the axis, we can speak of q as the latitude of a point and, longitude being measured in the same manner as upon the sphere, p the longitude of the same point. In this case V is a function of q alone, and one which we may suppose known.

Let PN (Fig. 39) denote the axis of revolution of a surface whose meridional curve PM is supposed known. Let s represent the arc of a meridian measured from P, u the distance PI, v the

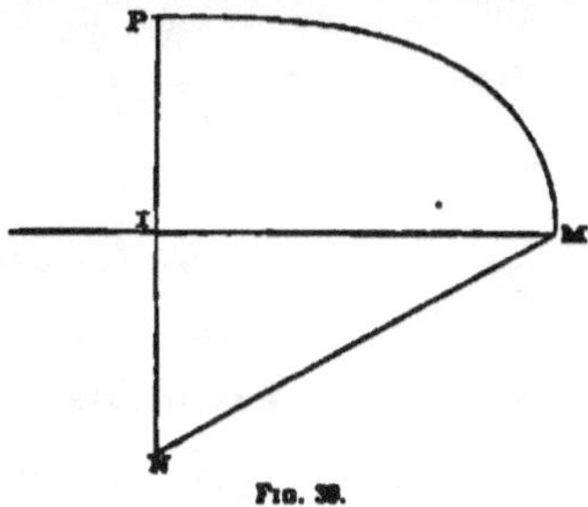

FIG. 39.

distance IM. The curve PM is in general determined by an equation between s and r, and from this equation the values of r and s can be determined as functions of p. It is easy to see that in this case we must have

$$V=-r\frac{ds}{dq}$$

the minus sign being necessary because the arc s increases as the latitude p decreases. Resuming, however, the differential equation

$$\frac{d\xi}{dp}\frac{d\eta}{dq}-\frac{d\xi}{dq}\frac{d\eta}{dp}=V$$

write for brevity

$$\frac{d\xi}{dp}=p_1 \qquad \frac{d\xi}{dq}=q_1$$

then

$$p_1\frac{d\eta}{dq}-p_1\frac{d\eta}{dp}=V$$

This partial differential equation leads by Lagrange's method to the system of ordinary differential equations

$$\frac{dq}{p_1}=-\frac{dp}{q_1}=\frac{d\eta}{V}$$

The first of these equations gives $p_1dp+q_1dq=0$; therefore $f(p, q)=c$ is the result of integration, c being an arbitrary constant. Again

$$d\eta=\frac{Vdq}{p_1}$$

Now if $f(p, q)=c$ is solved for p we will be able to substitute in p_1 the quantity p by its value in terms of q and c; writing this form of p_1 as $p_1(q, c)$, we have by integration

$$\eta=c'+\int\frac{Vdq}{p_1(q, c)}$$

which reduces the problem of finding η to a simple quadrature. In order now to determine the integral of the proposed partial differential equation, it is necessary to establish a relation between the constants c and c'. Remarking that $f(p, q)=\xi$, we have immediately

$$\xi=f(p, q) \qquad \eta=F(\xi)+\int\frac{Vdq}{p_1(q, \xi)}$$

In the quadrature, ξ is, of course, to be regarded as constant. We have already studied this problem at some length in considering the sphere as the surface to be projected, but a remark or two more on the subject will not be out of place in this chapter. For the sphere of radius r we know that $V=r^2\cos q$. That being the case, if it is required to find the equivalent projection upon the tangent cylinder at the equator, we must place $f(p, q)=rp$; then $p_1=r$, or p_1 is constant; again, making $F(\xi)=0$, we find

$$\eta=\int r\cos q dq=r\sin q$$

If we take the pole as center of a central equivalent projection

$$\xi=f(p, q)=2r\sin\tfrac{1}{2}\left(\frac{\pi}{2}-q\right)\cos p$$

$$\eta=F(\xi)-\int\frac{r^2\cos q dq}{\sqrt{4r^2\sin^2\frac{1}{2}\left(\frac{\pi}{2}-q\right)-\xi^2}}=F(\xi)+\sqrt{4r^2\sin^2\tfrac{1}{2}\left(\frac{\pi}{2}-q\right)-\xi^2}$$

If in this projection the ratio $\frac{\eta}{\xi}$ is independent of q, we can place

$$\xi=\varphi(q)\varphi_1(p) \qquad \eta=\varphi(q)\varphi_2(p)$$

The q will be eliminated by division, and differentiation will conduct at once to $\varphi'_1\varphi_2-\varphi_1\varphi'_2=c$ (a constant) and $\varphi\varphi'=\frac{V}{c}$; then

$$\varphi=\sqrt{c+2\int\frac{Vdq}{c}} \qquad \frac{\varphi_1}{\varphi_2}=c_1+c\int\frac{dp}{(\varphi_2)^2}$$

where φ_2 is perfectly arbitrary. The particular equations of the central equivalent projection are obtained by assuming

$$V=r^2\cos q \qquad c=-1 \qquad c_1=2r^2 \qquad c_2=0 \qquad \varphi_2(p)^2=\sin p$$

If, inversely, $\frac{\xi}{\eta}$ is to be only a function of q, the required conditions are obtained by writing

$$\xi=\Psi_1(q)\Psi(p) \qquad \eta=\Psi_2(q)\Psi(p)$$

These conduct to the relation

$$\Psi=\sqrt{cp+c_1}$$

and, Ψ_2 being an arbitrary function,

$$\Psi_1=c_2\Psi_2-\frac{2\Psi_2}{c}\int\frac{Vdq}{\Psi_2^2}$$

Of course, in both of these cases the quantities c, c_1, c_2 are to be regarded as arbitrary constants, of which any desirable disposition can be made.

Designate by a and b the semi-axes, equatorial and polar, of an elliptic meridian, and seek to determine the formulas for the homolographic projection of the spheroid upon which the meridian is found. Suppose that areas upon the spheroid are reduced upon the projection in the ratio $1:k$. The member k is clearly the ratio of the half surface of the spheroid to the area of the limiting ellipse of the map. In the general differential equation it will be only necessary to place $\frac{1}{k}V$ for V in order to take account of the impressed condition. The solution of the problem requires the determination of η as a function of q alone, and the representation of the meridians by ellipses having for semi-principal axes the length on the projection of the polar distance q. Take for axis of η the straight line on the map which joins the poles, and for ξ the perpendicular to this at its middle point, which of course represents the equator. The general form of solution that satisfies all the conditions is given by the group

$$\xi=Ap+B \qquad \eta=F(\xi)+\int\frac{Vdq}{A}$$

and for this case it is clear that

$$B=0 \qquad F(\xi)=0$$

giving

$$\xi=Ap \qquad \eta=\int_0^q\frac{\frac{1}{k}Vdq}{A}$$

Longitude being counted from the meridian represented on the map by the axis of q, it is clear that the quantity B should be made equal to zero. The value of V for an ellipsoid of revolution is

$$V=\frac{b^2\cos q}{(1-e^2\sin^2 q)^2}$$

The meridian of longitude p is represented by an ellipse, of which the principal axis in the direction of η is $=2b$; call the other axis $2a'$; then the equation to the ellipse is

$$\frac{\xi^2}{a'^2}+\frac{\eta^2}{b^2}=1$$

in which a' is an unknown function of p only. Differentiate this for q

$$\frac{\xi}{a'^2}\frac{d\xi}{dq}+\frac{\eta}{b^2}\frac{d\eta}{dq}=0$$

η and $\frac{d\eta}{dq}$ are independent of p, and

$$\xi\frac{d\xi}{dq}=\mathrm{A}p^2\frac{d\mathrm{A}}{dq}$$

The ratio, then, of the only two quantities that contain p is constant, *i. e.*, $\frac{p^2}{a'^2}$ is constant, and we may place

$$a'=mp$$

m being a constant which has to be determined. Substituting mp for a' in the preceding differential equation, and it becomes

$$\mathrm{A}d\mathrm{A}=-\frac{m^2}{b^2}\eta d\eta$$

and on integration

$$\mathrm{A}=\sqrt{c^2-\frac{m^2}{b^2}\eta^2}$$

c being a new constant to be determined again. The equation which gives η becomes on differentiation

$$\mathrm{A}d\eta=\frac{1}{k}\mathrm{V}dq$$

or

$$d\eta\sqrt{c^2-\frac{m^2}{b^2}\eta^2}=\frac{1}{k}\frac{b^2\cos q\,dq}{(1-e^2\sin^2 q)^2}$$

Make

$$\frac{m\eta}{b}=c\sin\lambda$$

Then the first member of this equation becomes

$$=\frac{bc^2}{m}\cos^2\lambda\, d\lambda$$

of which the integral is

$$\frac{1}{4}\cdot\frac{bc^2}{m}(2\lambda+\sin 2\lambda)$$

Now, $q=0$ must give $\lambda=0$ and $\eta=0$, and for $q=\frac{\pi}{2}$ there must arise $\eta=b$; we have then

$$2\lambda+\sin 2\lambda=\frac{4mb}{kc^2}\int_0^q\frac{\cos q\,dq}{(1-e^2\sin^2 q)^2}$$

In the homolographic projection applied to the sphere, the supposition $q=\frac{\pi}{2}$ giving rise to $\eta=b$ gives also $\lambda=\frac{\pi}{2}$. Analogy then leads us to make $c=m$, and thus obtain for the only remaining constant the equation

$$c=\frac{4b}{\pi k}\int_0^{\frac{\pi}{2}}\frac{\cos q\,dq}{(1-e^2\sin^2 q)^2}=\frac{2a}{\pi}$$

since

$$k=\frac{2b}{a}\int_0^{\frac{\pi}{2}}\frac{\cos q\,dq}{(1-e^2\sin^2 q)^2}$$

The result of the substitution of the values of c, m, k is the following equation, in which neither a nor b appears:

$$2\lambda+\sin 2\lambda=\pi\int_0^q\frac{\cos q\,dq}{(1-e^2\sin^2 q)^2}\div\int_0^{\frac{\pi}{2}}\frac{\cos q\,dq}{(1-e^2\sin^2 q)^2}$$

ALTERATIONS.

Upon a surface of revolution the distance between the points whose co-ordinates are (p, q) and $(p+dp, q+dq)$ is given by $\sqrt{P^2dp^2+Q^2dq^2}$, P and Q being known functions of q. Upon the map this distance is represented by $\sqrt{d\xi^2+d\eta^2}$. If k denote the ratio of the first of these distances to the second, k is a variable quantity, being different at different points of the surface, and also dependent upon $\frac{dp}{dq}$. We have in general

$$P^2dp^2+Q^2dq^2=k^2(d\xi^2+d\eta^2)$$

in which

$$d\xi=\frac{d\xi}{dp}dp+\frac{d\xi}{dq}dq \qquad d\eta=\frac{d\eta}{dp}dp+\frac{d\eta}{dq}dq$$

or briefly

$$d\xi=p_1\,dp+q_1\,dq \qquad d\eta=p_2\,dp+q_2\,dq$$

It is required to find the value of the ratio $\frac{dp}{dq}$ which will render k a maximum or minimum in any given point. As the point is absolutely arbitrary in position, we have merely to regard p_1, p_2, q_1, q_2, P and Q as constants, and consider k as a function of $\frac{dq}{dp}$ or, better, of the tangent of the angle which the direction sought makes with the meridian; if we call this angle β, there results

$$\tan\beta=\frac{Pdq}{Qdq}$$

We have for k^2 the expression

$$k^2=\frac{P^2dp^2+Q^2dq^2}{d\xi^2+d\eta^2}=\frac{1}{\frac{p_1^2+p_2^2}{P^2}\sin^2\beta+2\frac{p_1q_1+p_2q_2}{PQ}\sin\beta\cos\beta+\frac{q_1^2+q_2^2}{Q^2}\cos^2\beta}$$

The maximum value of k will occur for the minimum value of the denominator of this section, and conversely. Differentiating the denominator with respect to β, and equating the result to zero, there follows

$$\left[\frac{p_1^2+p_2^2}{P^2}-\frac{q_1^2+q_2^2}{Q^2}\right]\sin 2\beta+2\frac{p_1q_1+p_2q_2}{PQ}\cos 2\beta=0$$

from which

$$\tan 2\beta=-\frac{2(p_1q_1+p_2q_2)PQ}{Q^2(p_1^2+p_2^2)-P^2(q_1^2+q_2^2)}$$

This equation gives a single value of tan 2β, but two values of β, lying between 0° and π, and whose difference is $=\frac{\pi}{2}$.

By differentiating the above expression again for β and substituting the two values of β, and $\beta+\frac{\pi}{2}$, it will be found that the results will have contrary signs, which shows that the two directions upon the map, corresponding to these two values of β, are the directions of the greatest elongation and the greatest diminution, respectively.

The result of these considerations is the following

Theorem: The elements of length which, in their projection, have received the greatest alterations, make right angles with each other upon the surface of revolution.

In order to find the angle between the corresponding elements upon the map, it will be necessary to write for the first

$$d\xi'=p_1dp+q_1dq=\left(\frac{Q}{P}p_1\tan\beta+q_1\right)dq \qquad d\eta'=\left(\frac{Q}{P}p_2\tan\beta+q_2\right)dq$$

For the second, replace β by $\beta+\frac{\pi}{2}$,

$$d\xi''=\left(-\frac{Q}{P}p_1\cot\beta+q_1\right)dq \qquad d\eta''=\left(-\frac{Q}{P}p_2\cot\beta+q_2\right)dq$$

Write also

$$\frac{d\eta'}{d\xi'}=\tan\gamma' \qquad \frac{d\eta''}{d\xi''}=\tan\gamma''$$

then we easily obtain

$$\tan\gamma'\,\tan\gamma''=\frac{\frac{1}{2}(P^2q_2^2-Q^2p_2^2)\sin 2\beta-PQ\,p_2q_2\cos 2\beta}{\frac{1}{2}(P^2q_1^2-Q^2p_1^2)\sin 2\beta-PQ\,p_1q_1\cos 2\beta}$$

$$=\frac{\frac{1}{2}\begin{vmatrix}P^2, & Q^2\\ p_2^2, & q_2^2\end{vmatrix}\tan 2\beta-PQ\,p_2q_2}{\frac{1}{2}\begin{vmatrix}P^2, & Q^2\\ p_1^2, & q_1^2\end{vmatrix}\tan 2\beta-PQ\,p_1q_1}=-1$$

The last is true, as is easily seen by substituting the value of $\tan 2\beta$. The equation

$$\tan\gamma'\,\tan\gamma''=-1$$

shows that the angles γ' and γ'' differ by $\frac{\pi}{2}$; so that we come to the following

Theorem: The elements of length which suffer the greatest alterations have, upon the map as well as upon the sphere, directions at right angles to each other.

The results are of course of a perfectly general nature, and can readily be applied to equivalent projections. At any point of a surface of revolution, and in the directions of the greatest alterations, take two infinitely small lengths equal to unity. The square constructed with these two elements as sides will be an element of the surface of revolution, and will also be unit of area. Now, calculate the two values of h' and h'' from the general formula for h, which correspond to the two elements upon the surface. The square 1×1 will be transformed upon the chart into the rectangle $h'\times h''$. The condition $h'h''=1$ gives then the required equivalent projection. Still, considering h' and and h'' as values of h for two directions at right angles to each other, observe that

$$\frac{1}{h'^2}=\frac{p_1^2+p_2^2}{P^2}\sin^2\beta+2\frac{p_1q_1+p_2q_2}{PQ}\sin\beta\cos\beta+\frac{q_1^2+q_2^2}{Q^2}\cos^2\beta$$

and

$$\frac{1}{h''^2}=\frac{p_1^2+p_2^2}{P^2}\cos^2\beta-2\frac{p_1q_1+p_2q_2}{PQ}\sin\beta\cos\beta+\frac{q_1^2+q_2^2}{Q^2}\sin^2\beta$$

Addition of these gives the remarkable property

$$\frac{1}{h'^2}+\frac{1}{h''^2}=\frac{p_1^2+p_2^2}{P^2}+\frac{q_1^2+q_2^2}{Q^2}$$

a relation independent of the angle β. From this we see that if, in any point, there is no alteration in lengths, *i. e.*, if

$$h'=h''=1$$

the function

$$\frac{p_1^2+p_2^2}{P^2}+\frac{q_1^2+q_2^2}{Q^2}=2$$

Reciprocally, if at a point of an equivalent projection this function equals 2, there is no alteration of lengths around this point, for the equations

$$k'k''=1 \qquad \frac{1}{k'^2}+\frac{1}{k''^2}=2$$

give

$$k'=1 \qquad k''=1$$

The function

$$\frac{p_1^2+p_2^2}{P^2}+\frac{q_1^2+q_2^2}{Q^2}$$

is a characteristic function of equivalent projections; its value computed for different points of the map shows, according to the difference between these numbers and their lower limit 2, how much alteration there is in any given point. Place, for convenience,

$$A=\frac{p_1^2+p_2^2}{P^2} \qquad B=\frac{q_1^2+q_2^2}{Q^2}$$

Then, multiplying these together, we have

$$AB=\frac{(p_1^2+p_2^2)(q_1^2+q_2^2)}{P^2Q^2}=\frac{(p_1q_1+p_2q_2)^2+(p_1q_2-p_2q_1)^2}{P^2Q^2}=\frac{(p_1q_1+p_2q_2)^2+V^2}{P^2Q^2}$$

In this equation P, Q, V are functions of q which, for a given point, have fixed values, so that AB must evidently have the least possible value for $p_1q_1+p_2q_2=0$; that is to say, for the case when the projections of meridians and parallels cut at right angles. For the sphere

$$P^2Q^2=R^4\cos^2 q \qquad V^2=R^4\cos^2 q$$

therefore upon this surface the ratio $\frac{V^2}{P^2Q^2}$ is $=1$. Now at all points of the central projection when the pole is taken as center

$$p_1q_1+p_2q_2=0$$

and, consequently, in this case AB is always equal to unity. We have, then, simultaneously

$$k'k''=1 \qquad \frac{1}{k'^2}+\frac{1}{k''^2}=AB \qquad AB=1$$

consequently, we can write

$$k'=\frac{1}{\sqrt{A}} \qquad k''=\frac{1}{\sqrt{B}}$$

or, taking φ for the co-latitude,

$$k'=\cos\frac{\varphi}{2} \qquad k''=\cos\frac{1}{\frac{\varphi}{2}}$$

In studying the central equivalent projection we called θ the angle on the sphere between the principal meridian and the arc any point M to the center O; the angle upon the chart was designated by Ψ, and the distance OM by r; resuming those symbols, let θ', Ψ', r' denote the corresponding angles for another direction; the angle on the sphere between these two directions is $=\theta'-\theta$; the corresponding angle on the chart is $=\Psi'-\Psi$. *At every point of the sphere and the projection there is an infinite number of groups of two directions which make the same angles between each other on the sphere and on the projection.* Directions which enjoy this property are called *conjugate directions.* The condition for conjugate directions is obviously

$$\theta'-\theta=\Psi'-\Psi$$

or

$$\Psi-\theta=\Psi'-\theta'$$

Then follows

$$\frac{\tan\Psi-\tan\theta}{1+\tan\Psi\tan\theta}=\frac{\tan\Psi'-\tan\theta'}{1+\tan\Psi'\tan\theta'}$$

But φ being the distance from the arbitrary point M to the center O, we have

$$\tan \varphi = \frac{\tan \theta}{\cos^2 \frac{\varphi}{2}} \qquad \tan \varphi' = \frac{\tan \theta'}{\cos^2 \frac{\varphi}{2}}$$

These values substituted in the previous equation give, after easy reductions,

$$\frac{\tan \theta}{\cos^2 \frac{\varphi}{2} + \tan^2 \theta} = \frac{\tan \theta'}{\cos^2 \frac{\varphi}{2} + \tan^2 \theta'}$$

an equation of the second degree for the determination of θ'. It is satisfied obviously by

$$\tan \theta = \tan \theta'$$

which gives

$$\theta = \theta'$$

neglecting negative angles or angles greater than 180°. The other root of the equation is

$$\tan \theta' = \frac{\cos^2 \frac{\varphi}{2}}{\tan \theta}$$

or, upon the sphere, the condition for conjugate directions is

$$\tan \theta \tan \theta' = \cos^2 \frac{\varphi}{2}$$

In like manner is found for the projection the condition

$$\tan \varphi' = \frac{1}{\cos^2 \frac{\varphi}{2} \tan \varphi}$$

or

$$\tan \varphi' \tan \varphi = \frac{1}{\cos^2 \frac{\varphi}{2}}$$

The product $\tan \theta \tan \theta'$ is constant for a given point M, and consequently the corresponding directions belong to the conjugate diameters of an hyperbola which lies in a tangent plane to the sphere and having the point M as center, the tangent to the arc OM as one of its principal axes, and whose asymptotes make with this axis the angle whose tangent is

$$= \cos \frac{\varphi}{2}$$

Upon the chart the conjugate directions are also those of the conjugate diameters of an hyperbola having the radius OM for a principal axis and whose asymptotes make with this direction the angle whose tangent is

$$= \frac{1}{\cos \frac{\varphi}{2}}$$

In other words, upon the projection, as upon the sphere, the asymptotes of the hyperbola which defines the conjugate directions at a point are simply the directions of maximum deviation which have already been determined.

§ IX.

GENERAL THEORY OF PROJECTIONS BY DEVELOPMENT.

The subject of development is so inseparably connected with the higher parts of pure analytic geometry that it seems almost impossible to give much of an account of it in such a treatise as this. An attempt will be made, however, to give the most prominent points in the theory, referring always to the original sources from which the information has been drawn. Considering, as we do here, development in its most general sense to signify the application of one surface to another in such a way that the first shall in every point be made to coincide with the second without either rupture or stretching taking place, we have to find first the differential equation of all surfaces which can be developed or deformed in such a way as to coincide throughout with a given surface. The idea of defining a surface analytically by means of three equations which serve to express the three rectilinear co-ordinates of a point in terms of two independent variables is a very old one and is referred to explicitly in the writings both of Lagrange and Euler; but the glory of perceiving the full importance of the conception is due to Gauss alone, who, in his celebrated "*Disquisitiones generales circa superficies curvas,*" made the whole theory of surfaces, and especially that part of it which pertained to the curvature of surfaces, depend upon these two new parameters. Among the most remarkable of the theorems obtained by Gauss is the one relating to the application of one surface upon another. Gauss, in a certain measure, arrived at this theorem by accident. He was endeavoring to express the measure of curvature of a surface (that is, the reciprocal of the product of the principal radii of curvature at any point) as a function of the quantities p and q (the two independent parameters) when he discovered that this quantity depended only on the functions E, F, G, which serve to express the linear element of the surface in the form

$$ds^2 = Edp^2 + 2Fdp\,dq + Gdq^2$$

From this Gauss was enabled to conclude that if two surfaces are applicable, the one upon the other, that is to say, in such a manner that to each point of the first there corresponds a point of the second, the distance between two infinitely near points on either surface being equal to the distance between the two corresponding points of the other, then the functions E, F, G are to be considered as having the same value for the two surfaces and the measures of curvature will also have the same value for both. This theorem, first stated by Gauss, has led many eminent geometers to undertake the foundation of a theory of surfaces applicable to any given surface; this theory has, of course, its simplest application when the surface upon which the development is to be made is a plane. In the first investigation which follows it is desired *to find a means of ascertaining whether or not two given surfaces are applicable, the one to the other.* We may, for brevity, speak of the two surfaces as S and S'. Let

$$S(x, y, z) = 0$$

denote one surface, and

$$S'(x', y', z') = 0$$

denote the other. Suppose x, y, z to be expressed as functions of two independent variables p and q; for the linear element of this surface we have then the well-known expression

$$ds^2 = E\,dp^2 + 2\,F\,dp\,dq + G\,dq^2$$

where

$$E = \left(\frac{dx}{dp}\right)^2 + \left(\frac{dy}{dp}\right)^2 + \left(\frac{dz}{dp}\right)^2 \qquad F = \frac{dx}{dp}\frac{dx}{dq} + \frac{dy}{dp}\frac{dy}{dq} + \frac{dz}{dp}\frac{dz}{dq} \qquad G = \left(\frac{dx}{dq}\right)^2 + \left(\frac{dy}{dq}\right)^2 + \left(\frac{dz}{dq}\right)^2$$

Denote by p', q' the independent variables which serve to determine x', y', z'; then for the linear element of this surface we have

$$ds'^2 = E'\,dp'^2 + 2\,F'\,dp'\,dq' + G'\,dq'^2$$

where

$$E'=\left(\frac{dx'}{dp'}\right)^2+\left(\frac{dy'}{dp'}\right)^2+\left(\frac{dz'}{dp'}\right)^2$$

and so forth. If, now, the two surfaces are applicable the one to the other in such a way that the distance between two infinitely near points on the first is equal to the distance between the two corresponding points of the second surface, there must exist values of p' and q' expressed in terms of p and q, which will satisfy the equation

$$E\,dp^2+2\,F\,dp\,dq+G\,dq^2=E'\,dp'^2+2\,F'\,dp'\,dq'+G\,dq'^2$$

whatever be the values of p, q, dp, dq. Gauss has shown in his memoir upon orthomorphic projection that there always exists upon a surface particular values of the variables p and q, which make

$$E=G \qquad F=0$$

These have been already referred to in the chapter on orthomorphic projection, so nothing more need be said of them here. Calling them, however, (u, v) and (u', v'), we have as the new form of the above equation of condition

$$\lambda\,(du^2+dv^2)=\lambda'\,(du'^2+dv'^2)$$

λ being a function of (u, v) and λ' of (u', v'). Factor these expressions and write

$$u+iv=\alpha \qquad u-iv=\beta \qquad u'+iv'=\alpha' \qquad u'-iv'=\beta'$$

We have then simply

$$\varphi^2\,d\alpha\,d\beta=\varphi'^2\,d\alpha'\,d\beta'$$

φ^2 denoting the value of λ in terms of α and β, and φ'^2 the corresponding value of λ' in terms of α' and β'.

A very remarkable consequence follows immediately from this equation, viz, that α' depends on only one of the variables α, β, and β' depends upon the other. Write

$$d\alpha'=\frac{d\alpha'}{d\alpha}\,d\alpha+\frac{d\alpha'}{d\beta}\,d\beta \qquad d\beta'=\frac{d\beta'}{d\alpha}\,d\alpha+\frac{d\beta'}{d\beta}\,d\beta$$

The above equality thus becomes

$$\varphi^2\,d\alpha\,d\beta=\varphi'^2\left(\frac{d\alpha'}{d\alpha}\,d\alpha+\frac{d\alpha'}{d\beta}\,d\beta\right)\left(\frac{d\beta'}{d\alpha}\,d\alpha+\frac{d\beta'}{d\beta}\,d\beta\right)$$

Now, $d\alpha$ and $d\beta$ being quite arbitrary, there cannot exist in the second member of this equation any other quantity than $d\alpha d\beta$, with its coefficient; that is, the coefficients of $d\alpha^2$ and $d\beta^2$ must be equal to zero, or

$$\frac{d\alpha'}{d\alpha}\,\frac{d\beta'}{d\alpha}=0 \qquad \frac{d\alpha'}{d\beta}\,\frac{d\beta'}{d\beta}=0$$

From this follows that, α and β being independent variables,

$$\alpha'=f(\alpha) \qquad \beta'=f_1(\beta)$$

or

$$\alpha'=f_1(\beta) \qquad \beta'=f(\alpha)$$

Take the first result, and the equality immediately becomes

$$\varphi^2=\varphi'^2\,f'(\alpha)\,f'_1(\beta)$$

from which

$$\log\varphi^2=\log\varphi'^2+\log f'(\alpha)+\log f'_1(\beta)$$

Differentiation of this function with respect to α and β successively gives at once

$$\frac{d^2\log\varphi^2}{d\alpha\,d\beta}=\frac{d^2\log\varphi'^2}{d\alpha\,d\beta}$$

or

$$\frac{d^2\log\varphi^2}{d\alpha\,d\beta}=\frac{d^2\log\varphi'^2}{d\alpha'\,d\beta'}\,f'(\alpha)f_1'(\beta)$$

and, finally,

$$\frac{1}{\gamma^2}\frac{d^2 \log \gamma^2}{d\alpha\, d\beta}=\frac{1}{\gamma'^2}\frac{d^2 \log \gamma'^2}{d\alpha'\, d\beta'}$$

We have then the remarkable property that if two surfaces are applicable the one to the other the function

$$k=\frac{1}{\gamma^2}\frac{d^2 \log \gamma^2}{d\alpha\, d\beta}$$

has the same value for the two surfaces at the corresponding points. In order to arrive at the geometrical significance of k, we proceed as follows: Resuming the variables u, v, we have

$$4k=\frac{1}{\lambda}\left(\frac{d^2 \log \lambda}{du^2}+\frac{d^2 \log \lambda}{dv^2}\right)$$

Multiply this by the superficial element of the surface, or by

$$ds=\lambda\, du\, dv$$

and integrate throughout the region included by an arbitrary closed curve traced upon the surface, thus

$$4\int\int k\, ds=\int\int\frac{d^2 \log \lambda}{du^2}\, du\, dv+\int\int\frac{d^2 \log \lambda}{dv^2}\, du\, dv$$

Now, denote by the subscript numerals 1, 2, 3 . . . . $2m$ the even number of points in which the line v=constant, produced in a positive direction, meets the closed contour; then, considering the term

$$\int\int\frac{d^2 \log \lambda}{du^2}\, du\, dv$$

and integrating for u alone, we have, if we disregard the second integration for the present,

$$\int\int\frac{d^2 \log \lambda}{du^2}\, du\, dv=dv\left[-\left(\frac{d \log \lambda}{du}\right)_1+\left(\frac{d \log \lambda}{du}\right)_2-\left(\frac{d \log \lambda}{du}\right)_3\right.$$
$$\left.+\ \cdots -\left(\frac{d \log \lambda}{du}\right)_{2m-1}+\left(\frac{d \log \lambda}{du}\right)_{2m}\right]$$

where $\left(\frac{d \log \lambda}{du}\right)_\eta$ denotes the value of this quantity at the point η. Suppose that, in traversing the closed curve, the points of the curve following the points 1, 3, 5 . . . $2m-1$ are on the side of the line v towards which u is counted as positive, and that the points following, 2, 4, 6 . . . $2m$, are on the side of v towards which u is counted as negative; Fig. 40 illustrates what is meant,

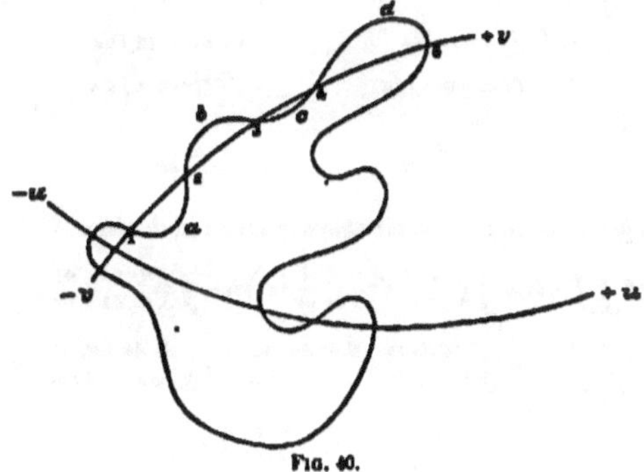

FIG. 40.

the portions $1a2$, $3c4$, &c., lying towards the direction of the positive u, and the portions $2b3$, $4d5$ lying towards the direction of the negative u. Denote by j the positive angle formed at each of the points by the closed curve and the line v, and by ds the element of the closed curve; we have, obviously, for the odd points, 1, 3, 5 . . . $2m-1$,

$$\sqrt{\lambda}\, dv = \sin j\, ds$$

and for the even points, 2, 4, 6 . . . $2m$,

$$\sqrt{\lambda}\, dv = -\sin j\, ds$$

The above expression now becomes

$$= -\sum_{r=1}^{r=2m} \left(\frac{d \log \lambda}{du}\right)_r \sin j_r \frac{ds}{\sqrt{\lambda}}$$

the summation extending over all the points 1, 2 . . . $2m$, where the closed curve is encountered by the line v=const. In effecting the integration with respect to v, which we had for the moment disregarded, we find at once for the reduced value of the term $\int\int \frac{d^2 \log \lambda}{du\, dv}\, du\, dv$ the simple integral

$$-\int \frac{d \log \lambda}{du} \sin j \frac{ds}{\sqrt{\lambda}} = -2 \int \frac{d\sqrt{\lambda}}{du} \sin j \frac{ds}{\lambda}$$

extended all round the closed contour. A similar transformation gives

$$\int\int \frac{d^2 \log \lambda}{dv^2}\, du\, dv = 2 \int \frac{d\sqrt{\lambda}}{dv} \cos j \frac{ds}{\lambda}$$

and so for the equation under consideration

$$2\int\int k\, d\sigma = -\int \left(\frac{d\sqrt{\lambda}}{du} \sin j - \frac{d\sqrt{\lambda}}{dv} \cos j\right) \frac{ds}{\lambda}$$

or again

$$2\int\int k\, d\sigma = \int \left\{ \frac{d \frac{\sin j}{\sqrt{\lambda}}}{du} - \frac{d \frac{\cos j}{\sqrt{\lambda}}}{dv} \right\} ds - \int \left(\frac{\cos j}{\sqrt{\lambda}} \frac{dj}{du} + \frac{\sin j}{\sqrt{\lambda}} \frac{dj}{dv}\right) ds$$

If now du and dv denote the increments, positive or negative, which the quantities u and v receive in passing from the first point of the element ds to its last point, and δu, δv the corresponding increments of u and v for a displacement dn in the direction of the exterior normal to the contour, there will result then, at any point of the contour,

$$(a) \qquad \begin{cases} \sqrt{\lambda}\, du = \cos j\, ds & \sqrt{\lambda}\, dv = \sin j\, ds \\ \sqrt{\lambda}\, \delta u = -\sin j\, dn & \sqrt{\lambda}\, \delta v = \cos j\, dn \end{cases}$$

from which

$$(b) \qquad du = \frac{ds}{dn} \delta v \qquad . \qquad dv = -\frac{ds}{dn} \delta u$$

These relations (a) permit us to express the above relations in the form

$$2\int\int k\, d\sigma = \int \left(\frac{d}{\delta u} \frac{1}{ds} dv - \frac{d}{\delta v} \frac{1}{ds} \delta u\right) ds - \int \left(\frac{dj}{du} \frac{du}{ds} + \frac{dj}{dv} \frac{dv}{ds}\right) ds$$

and (b) give

$$2\int\int k\, d\sigma = \int \left(\frac{d.ds}{\delta u} \frac{\delta u}{\delta s} + \frac{d.ds}{\delta v} \frac{\delta v}{\delta s}\right) \frac{ds}{\delta n} - \int \left(\frac{dj}{\delta u} \frac{du}{ds} + \frac{dj}{\delta v} \frac{dv}{ds}\right) ds$$

or

$$2\int\int k\, d\sigma = \int \frac{\delta.ds}{\delta n\, ds}\, ds - \int dj$$

dj being the increment, positive or negative, which j receives while passing from the first to the last point of ds, and δds the increment of ds for the displacement δn in the direction of the exterior normal to the curve. Furthermore,

$$\int dj = A + B + C + \ldots\ldots - (n-2)\pi$$

A, B, C . . . . . being the interior angles of the contour and n the number of these angles; there results, finally,

$$2\int\int k ds = \int \frac{\delta ds}{\delta n ds} ds - A - B - C - \ldots\ldots + (n-2)\pi$$

We can illustrate this formula by a very simple geometrical construction. Suppose that the closed contour under consideration is only the small parallelogram BACD, formed by the lines p, q, $p+dp$, $q+dq$ (dp and dq to be positive). The expression for element of area is well known to be of the form $\sqrt{EG-F^2}\,dp\,dq$. The integral $\int\int k ds$, of course, reduces in this case to a single element, and is, in fact,

$$=2\sqrt{EG-F^2}\,k\,dp\,dq$$

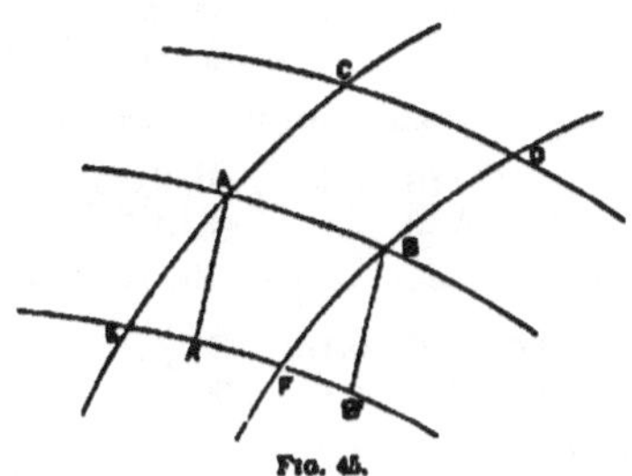

FIG. 45.

In this case, of course, we have $n=4$, and if the angle between the lines p and q be denoted by ω, there results

$$A = \omega$$

$$B = \pi - \left(\omega + \frac{d\omega}{dq} dq\right)$$

$$C = \pi - \left(\omega + \frac{d\omega}{dp} dp\right)$$

$$D = \omega + \frac{d\omega}{dp} dp + \frac{d\omega}{dq} dq + \frac{d^2\omega}{dp\,dq} dp\,dq$$

and consequently

$$-A - B - C\ \ldots\ + (n-2)\pi = -\frac{d^2\omega}{dp\,dq} dp\,dq$$

Now, as is also well known,

$$\cos\omega = \frac{F}{\sqrt{EG}} \qquad \sin\omega = \frac{\sqrt{EG-F^2}}{EG}$$

and forming the expression for $\frac{d\omega}{dp}$, from this we have very easily, for this last quantity,

$$-\frac{d}{dq}\left[\frac{1}{2\sqrt{EG-F^2}}\left(\frac{F}{E}\frac{dE}{dp} + \frac{F}{G}\frac{dG}{dp} - 2\frac{dF}{dp}\right) dp\,dq\right]$$

For future convenience we will write

$$\sqrt{EG-F^2} = V$$

Finally consider the integral of $\frac{\delta ds}{\delta n ds}ds$. This reduces to four elements corresponding to the four sides of the curvilinear parallelogram BACD. The element relative to the side AB has for its value $\frac{(A'B'-AB)AB}{AA'.AB}$ or $\frac{A'B'-AB}{AA'}$, AA′ and BB′ being normal to AB, and EA′FB′ denoting the line $p+dp$; but

$$AA'=AE\sin\omega=\frac{\sqrt{EG-F^2}}{\sqrt{G}}dp$$

$$A'B'-AB=EF-AB+FB'-EA'=\frac{d\sqrt{G}}{dp}dq\,dp-\frac{d}{dq}\frac{F}{\sqrt{G}}dq\,dp$$

We have then for this first element

$$\frac{1}{2V}\left[\frac{dG}{dp}+\frac{F}{G}\frac{dG}{dq}-2\frac{dF}{dq}\right]dq$$

From what precedes we can deduce at once that the element corresponding to the side CD is

$$\frac{1}{2V}\left(\frac{dG}{dp}+\frac{F}{G}\frac{dG}{dq}-2\frac{dF}{dq}\right)dq+\frac{d}{dp}\left[\frac{1}{2V}\left(\frac{dG}{dp}+\frac{F}{G}\frac{dG}{dq}-2\frac{dF}{dq}\right)dp\,dq\right]$$

which gives for the sum of the two elements relative to BA and CD

$$\frac{d}{dp}\left[\frac{1}{2V}\left(\frac{dG}{dp}+\frac{F}{G}\frac{dG}{dq}-2\frac{dF}{dq}\right)\right]dp\,dq$$

In like manner, for the sum of the two elements relative to AC and BD there is found

$$\frac{d}{dq}\left[\frac{1}{2V}\left(\frac{dE}{dq}+\frac{F}{E}\frac{dE}{dp}-2\frac{dF}{dp}\right)\right]dp\,dq$$

The integral $\int\frac{\delta ds}{\delta n ds}ds$ has then for its value

$$\frac{d}{dp}\left[\frac{1}{2V}\left(\frac{dG}{dp}+\frac{F}{G}\frac{dG}{dq}-2\frac{dF}{dq}\right)\right]dp\,dq+\frac{d}{dq}\left[\frac{1}{2V}\left(\frac{dE}{dq}+\frac{F}{E}\frac{dE}{dp}-2\frac{dF}{dp}\right)\right]dp\,dq$$

The equation

$$2\int\int k\,ds=\int\frac{\delta ds}{\delta n ds}ds-A-B\;.\;.\;.\;.\;.\;+(n-2)\pi$$

then becomes

$$4VK=\frac{d}{dp}\left[\frac{1}{V}\left(\frac{dG}{dp}+\frac{F}{G}\frac{dG}{dq}-2\frac{dF}{dq}\right)\right]+\frac{d}{dq}\left[\frac{1}{V}\left(\frac{dE}{dq}-\frac{F}{G}\frac{dG}{dp}\right)\right]$$

This formula gives the value of k as a function of the arbitrary parameters p and q.

It is now quite easy to determine the geometrical significance of k, and to show that, disregarding the sign, the double of this function expresses the measure of curvature. If p and q denote the two rectangular co-ordinates x and y, we shall have, since z is the third co-ordinate,

$$E=1+\left(\frac{dz}{dx}\right)^2\qquad F=\frac{dz}{dx}\frac{dz}{dy}\qquad G=1+\left(\frac{dz}{dy}\right)^2$$

and then

$$\frac{1}{V}\left(\frac{dG}{dp}+\frac{F}{G}\frac{dG}{dq}-2\frac{dF}{dq}\right)=\frac{-2\frac{dz\,d^2z}{dx\,dy^2}}{\left[1+\left(\frac{dz}{dy}\right)^2\right]\sqrt{1+\left(\frac{dz}{dx}\right)^2+\left(\frac{dz}{dy}\right)^2}}$$

$$\frac{1}{V}\left(\frac{dE}{dq}-\frac{F}{G}\frac{dG}{dp}\right)=\frac{2\frac{dz\,d^2z}{dx\,dy^2}}{\left[1+\left(\frac{dz}{dy}\right)^2\right]\sqrt{1+\left(\frac{dz}{dx}\right)^2+\left(\frac{dz}{dy}\right)^2}}$$

and consequently

$$2k=\frac{1}{\sqrt{1+\left(\frac{dz}{dx}\right)^2+\left(\frac{dz}{dy}\right)^2}}\left\{\frac{d}{dx}\frac{-\frac{dz}{dx}\frac{d^2z}{dy^2}}{\left[1+\left(\frac{dz}{dy}\right)^2\right]\sqrt{1+\left(\frac{dz}{dx}\right)^2+\left(\frac{dz}{dy}\right)^2}}+\frac{d}{dy}\frac{\frac{dz}{dx}\frac{d^2z}{dxdy}}{\left[1+\left(\frac{dz}{dy}\right)^2\right]\sqrt{1+\left(\frac{dz}{dx}\right)^2+\left(\frac{dz}{dy}\right)^2}}\right\}$$

$$=\frac{1}{\sqrt{1+\left(\frac{dz}{dx}\right)^2+\left(\frac{dz}{dy}\right)^2}}\left\{-\frac{d^2z}{dy^2}\frac{d}{dx}\frac{\frac{dz}{dx}}{\left[1+\left(\frac{dz}{dy}\right)^2\right]\left[1+\left(\frac{dz}{dx}\right)^2+\left(\frac{dz}{dy}\right)^2\right]}+\frac{d^2z}{dxdy}\frac{d}{dy}\frac{\frac{dz}{dx}}{\left[1+\left(\frac{dz}{dy}\right)^2\right]\left[1+\left(\frac{dz}{dx}\right)^2+\left(\frac{dz}{dy}\right)^2\right]}\right\}$$

which becomes, on developing and reducing,

$$2k=\frac{\left(\frac{d^2z}{dxdy}\right)^2-\frac{d^2z}{dx^2}\frac{d^2z}{dy^2}}{\left[1+\left(\frac{dz}{dx}\right)^2+\left(\frac{dz}{dy}\right)^2\right]^2}$$

which, apart from the sign, is the expression in terms of x, y, z, for the reciprocal of the product of the principal radii of curvature.

The result of what precedes is expressed by saying that when two surfaces are applicable, the one upon the other, their measures of curvature at the corresponding points are equal.

This theorem of Gauss's constitutes a necessary but not sufficient condition of the applicability of the two surfaces considered. It is to be observed, however, that when a first relation has been obtained between the corresponding points, it is always easy to find a second; we can then calculate the values of p' and q', which alone are admissible, and on substituting them in

$$Edp^2+2Fdpdq+Gdq^2=E'dp'^2+2F'dp'dq'+G'dq'^2$$

determine whether or not the surfaces are applicable to one another. If k is a function of p and q, and k' of p' and q', we must have, if the surfaces S and S' are applicable, $k=k'$. By Gauss's theorem we know that the quantities k and k' are the respective measures of curvature for S and S'. Differentiating, we have

$$\frac{dk}{dp}dp+\frac{dk}{dq}dq=\frac{dk'}{dp'}dp'+\frac{dk'}{dq'}dq'$$

or, as we may write for simplification,

$$(\alpha)\qquad mdp+ndq=m'dp'+n'dq$$

This equation combined with

$$(\beta)\qquad Edp^2+2Fdpdq+Gdq^2=E'dp'^2+2F'dp'dq'+G'dq'^2$$

serves to determine dp' and dq' as functions of dp and dq; but the values of dp' and dq' should be expressed as linear functions of dp and dq, since p' and q' are expressible as functions of p and q; it is evident, from the forms of the two equations from which the determination of dp' and dq' as functions of dp and dq is to be made, that this can only happen where there exist certain determinate relations between the quantities E, F, G and E', F', G', and the quantities m, n and m', n'. In order to determine this relation, square equation (α) and add it to equation (β), previously multiplied by an indeterminate quantity λ; we will thus have

$$(m^2+\lambda E)dp^2+2(mn+\lambda F)dpdq+(n^2+\lambda G)dq^2=(m'^2+\lambda E')dp'^2+2(m'n'+\lambda F')dp'dq'+(n'^2+\lambda G')dq'^2$$

If now the indeterminate λ is determined by the condition that the first member of this equation shall be the square of a binomial of the first degree in dp and dq, it will be necessary that the second member be also the square of a binomial of the first degree in dp' and dq'; or, in other words, the values of λ which render the two members perfect squares must be equal; we have, then, by simple algebra

$$\frac{1}{V^2}\begin{vmatrix} E, & 2F, & G \\ m, & n, & o \\ o, & m, & n \end{vmatrix} = \frac{1}{V'^2}\begin{vmatrix} E', & 2F', & G' \\ m', & n', & o \\ o, & m', & n' \end{vmatrix}$$

This is the sought relation, which, for brevity, we shall write in the form

$$H = H'$$

and we have then the two differential equations

$$m\,dp + n\,dq = m'dp' + n'dq' \qquad \frac{dH}{dp}dp + \frac{dH}{dq}dq = \frac{dH'}{dp'}dp' + \frac{dH'}{dq'}dq'$$

corresponding to

$$K = K' \qquad H = H'$$

and between these four equations we can determine p', q', dp', dq' as functions of p, q, dp, dq, in such a way that whatever be the values of p, q, dp, dq, if these are substituted in equation (β) we will have the final necessary and sufficient conditions which must be fulfilled in order that the two surfaces S and S' may be applicable the one upon the other. If we subtract from the square of equation (α) the product, member by member, of (β) and the equation $H = H'$, we have

$$\frac{1}{V^2}[(En - Fm)\,dp + (Fn - Gm)\,dq]^2 = \frac{1}{V'^2}[(E'n' - F'm')\,dp' + (F'n' - G'm')\,dq']^2$$

or, writing

$$\frac{E}{V} = e \qquad \frac{F}{V} = f \qquad \frac{G}{V} = g$$

$$\frac{E'}{V'} = e' \qquad \frac{F'}{V'} = f' \qquad \frac{G'}{V'} = g'$$

$$\begin{vmatrix} e, & f \\ m, & n \end{vmatrix} dp + \begin{vmatrix} f, & g \\ m, & n \end{vmatrix} dq = \begin{vmatrix} e', & f' \\ m', & n' \end{vmatrix} dp' + \begin{vmatrix} f', & g' \\ m', & n' \end{vmatrix} dq'$$

This equation combined with

$$m\,dp + n\,dq = m'dp' + n'dq' \qquad \frac{dH}{dp}dp + \frac{dH}{dq}dq = \frac{dH}{dp'}dp' + \frac{dH}{dq'}dq'$$

affords us the means of eliminating dp' and dq'; we have, in fact,

$$\frac{\frac{dH}{dp}[(e'n' - f'm')n - (fn - gm)m'] + \frac{dH}{dq}[-(e'n' - f'm')m + (en - fm)m']}{en^2 - 2fmn + gm^2} = \frac{dH'}{dp'}$$

$$\frac{\frac{dH}{dp}[(f'n' - g'm')n - (fn - gm)n'] + \frac{dH}{dq}[-(f'n' - g'm')m + (en - fm)n']}{en^2 - 2fmn + gm^2} = \frac{dH'}{dq'}$$

from which is readily deduced

$$\frac{n\frac{dH}{dp} - m\frac{dH}{dq}}{en^2 - 2fmn + gm^2} = \frac{n'\frac{dH'}{dp'} - m'\frac{dH'}{dq'}}{e'n'^2 - 2f'm'n' + g'm'^2}$$

$$\frac{\begin{vmatrix} f, & g, \\ m, & n, \end{vmatrix}\frac{dH}{dp} - \begin{vmatrix} e, & f, \\ n, & n, \end{vmatrix}\frac{dH}{dq}}{en^2 - 2fmn + gm^2} = \frac{\begin{vmatrix} f', & g' \\ m', & n' \end{vmatrix}\frac{dH'}{dp'} - \begin{vmatrix} e', & f' \\ m', & n' \end{vmatrix}\frac{dH'}{dq'}}{e'n'^2 - 2f'm'n' + g'm'^2}$$

or, on account of the equality $H=H'$,

$$\text{(I)}\qquad \frac{1}{V}\begin{vmatrix} \frac{dH}{dp}, & \frac{dH}{dq} \\ m, & n \end{vmatrix} = \frac{1}{V'}\begin{vmatrix} \frac{dH'}{dp'}, & \frac{dH'}{dq'} \\ m', & n' \end{vmatrix}$$

$$\text{(II)}\qquad \frac{1}{V}\left[(Fn-Gm)\frac{dH}{dp}-(En-Fm)\frac{dH}{dq}\right]=\frac{1}{V'}\left[(F'n'-G'm')\frac{dH'}{dp'}-(E'n'-F'm')\frac{dH'}{dq'}\right]$$

(I) and (II) combined with (III)

$$K=K'$$

and (IV)

$$H=H'$$

afford the means of completely solving the problem; for the values of p' and q' obtained by solving any two of these four equations must satisfy the other two, whatever be the values of p and q. It will not be necessary to carry further these general considerations, as the mathematical reader who is interested in this most beautiful branch of geometry will naturally seek the original memoirs for full information.

A brief account of the method of determining all the surfaces applicable to a given surface will now be given, under the supposition that the linear element ds of the surface s can be represented by

$$ds^2=4\varphi^2 d\alpha\, d\beta$$

α and β being the imaginary variables already defined, and φ a known function of α and β. Represent by ξ, η, ζ the unknown functions of α and β which express the rectangular co-ordinates of the points of any surface S' applicable to S; the square of the linear element of this second surface is

$$ds'^2=\left[\left(\frac{d\xi}{d\alpha}\right)^2+\left(\frac{d\eta}{d\alpha}\right)^2+\left(\frac{d\zeta}{d\alpha}\right)^2\right]d\alpha^2+2\left(\frac{d\xi}{d\alpha}\frac{d\xi}{d\beta}+\frac{d\eta}{d\alpha}\frac{d\eta}{d\beta}+\frac{d\zeta}{d\alpha}\frac{d\zeta}{d\beta}\right)d\alpha\, d\beta+\left[\left(\frac{d\xi}{d\beta}\right)^2+\left(\frac{d\eta}{d\beta}\right)^2+\left(\frac{d\zeta}{d\beta}\right)^2\right]d\beta^2$$

Equating the second member of this last equation to the second member of the preceding, it is clear that we must have the conditions

$$\text{(A)}\qquad \begin{cases} \left(\frac{d\xi}{d\alpha}\right)^2+\left(\frac{d\eta}{d\alpha}\right)^2+\left(\frac{d\zeta}{d\alpha}\right)^2=0 \\ \left(\frac{d\xi}{d\beta}\right)^2+\left(\frac{d\eta}{d\beta}\right)^2+\left(\frac{d\zeta}{d\beta}\right)^2=0 \\ \frac{d\xi}{d\alpha}\frac{d\xi}{d\beta}+\frac{d\eta}{d\alpha}\frac{d\eta}{d\beta}+\frac{d\zeta}{d\alpha}\frac{d\zeta}{d\beta}=2\varphi^2 \end{cases}$$

write

$$\text{(B)}\qquad \begin{cases} \frac{d\xi}{d\alpha}=i(m^2+n^2) & \frac{d\eta}{d\alpha}=m^2-n^2 & \frac{d\zeta}{d\alpha}=2mn \\ \frac{d\xi}{d\beta}=i(m'^2+n'^2) & \frac{d\eta}{d\beta}=m'^2-n'^2 & \frac{d\zeta}{d\beta}=2m'n' \end{cases}$$

These values satisfy identically the first two equations (A), and for the satisfaction of the third we have to insert the condition

$$\text{(C)}\qquad \begin{vmatrix} m, & n \\ m', & n' \end{vmatrix}=i\varphi$$

Differentiating the first row of equations (B) for β and the second for α, we see that the following conditions must also exist:

$$\frac{d\,(m^2+n^2)}{d\beta}=\frac{d\,(m'^2+n'^2)}{d\alpha}\qquad \frac{d\,(m^2-n^2)}{d\beta}=\frac{d\,(m'^2-n'^2)}{d\alpha}\qquad \frac{d\,(mn)}{d\beta}=\frac{d\,(m'n')}{d\alpha}$$

which conduct to the following:

$$m\frac{dm}{d\beta}=m'\frac{dm'}{d\alpha}\qquad n\frac{dn}{d\beta}=n'\frac{dn'}{d\alpha}\qquad n\frac{dm}{d\beta}+m\frac{dn}{d\beta}=n'\frac{dm'}{d\alpha}+m'\frac{dn'}{d\alpha}$$

and, with reference to C, these give

$$\text{(D)}\qquad \frac{1}{m'}\frac{dm}{d\beta}=\frac{1}{m}\frac{dm'}{d\alpha}=\frac{1}{n'}\frac{dn}{d\beta}=\frac{1}{n}\frac{dn'}{d\alpha}=\frac{m^2}{i\varphi}\frac{d}{d\beta}\frac{n}{m}=-\frac{m'^2}{i\varphi}\frac{d}{d\alpha}\frac{n'}{m'}$$

Eliminating n' by means of C,

$$\text{(E)}\qquad m\frac{dm}{d\beta}=m'\frac{dm'}{d\alpha}$$

$$\text{(F)}\qquad \frac{d}{d\alpha}\frac{n}{m}=-\frac{i}{m}\frac{d}{d\alpha}\frac{\varphi}{m}$$

$$\text{(G)}\qquad \frac{d}{d\beta}\frac{n}{m}=-\frac{i\varphi}{m'}\frac{d}{d\beta}\frac{1}{m}$$

Equation (E) shows that m^2 and m'^2 are the partial derivatives with respect to α and β of some one function of (α, β); we can then write

$$\text{(H)}\qquad m^2=\frac{dz}{d\alpha}=p_1\qquad m'^2=\frac{dz}{d\beta}=q_1$$

(F) and (G) then become

$$\text{(I)}\qquad \frac{d}{d\alpha}\frac{n}{m}=\frac{i}{\sqrt{q_1}}\frac{d}{d\alpha}\frac{\varphi}{\sqrt{p_1}}=-\frac{i\varphi}{\sqrt{q_1}}\frac{d}{d\alpha}\frac{1}{\sqrt{p_1}}-\frac{i}{\sqrt{pq}}\frac{d\varphi}{d\alpha}$$

$$\text{(J)}\qquad \frac{d}{d\beta}\frac{n}{m}=-\frac{i\varphi}{\sqrt{q_1}}\frac{d}{d\beta}\frac{1}{\sqrt{p_1}}$$

Differentiating the first of these with respect to β and the second with respect to α, and equating the results, we obtain

$$\frac{d\frac{1}{\sqrt{p_1}}}{d\alpha}\frac{d\frac{\varphi}{\sqrt{q_1}}}{d\beta}+\frac{1}{\sqrt{p_1q_1}}\frac{d^2\varphi}{d\alpha d\beta}+\frac{d\varphi}{d\alpha}\frac{d\frac{1}{\sqrt{p_1q_1}}}{d\beta}=-\frac{d\frac{1}{\sqrt{p_1}}}{d\beta}\frac{d\frac{\varphi}{\sqrt{q_1}}}{d\alpha}$$

or

$$\text{(K)}\qquad \varphi(rt-s^2)-2\frac{d\varphi}{d\beta}q_1r_1-2\frac{d\varphi}{d\alpha}p_1t_1+4p_1q_1\frac{d^2\varphi}{d\alpha d\beta}=0$$

where

$$r=\frac{d^2z}{d\alpha^2}\qquad s=\frac{d^2z}{d\alpha d\beta}\qquad t=\frac{d^2z}{d\beta^2}$$

The solution of the proposed problem is thus made to depend upon the integration of equation (K); for, if we know z as a function of α and β, equations (H) will determine m^2 and m'^2 as functions of the same variables; (F) and (G) will determine $\frac{n}{m}$ and n, and n' will be found from (C); finally, then, m, m', n, n' being known, simple quadratures will suffice to determine ξ, η, ζ, for we have

$$d\xi=i(m^2+n^2)d\alpha+i(m'^2+n'^2)d\beta$$

$$d\eta=(m^2-n^2)d\alpha+(m'^2-n'^2)d\beta$$

$$d\zeta=2mnd\alpha+2m'n'd\beta$$

Another more general investigation may be made which shall lay no restrictions whatever upon the original parameters p and q. If p' and q' are the corresponding parameters for the second surface, we know that the relations

$$\varphi(p, q, p', q')=0\qquad \varphi_1(p, q, p', q')=0$$

can be established, which will enable us to write an expression for the element of length upon this surface, which shall depend only on p and q. Let the equation of one surface be

$$f(x, y, z)=0$$

and let p and q denote the independent parameters in terms of which the values of x, y, z may be expressed. We may, of course, consider p and q so determined that p=const. and q=const. shall be the equations of the lines of curvature, in which case

$$f=\text{const.} \qquad p=\text{const.} \qquad q=\text{const.}$$

will denote the equations of the orthogonal surfaces. The expression for the element of length on f is

(1)
$$ds^2=Edp^2+2Fdpdq+Gdq^2$$

where

$$E=\left(\frac{dx}{dp}\right)^2+\left(\frac{dy}{dp}\right)^2+\left(\frac{dz}{dp}\right)^2$$

(2)
$$F=\frac{dx}{dp}\frac{dx}{dq}+\frac{dy}{dp}\frac{dy}{dq}+\frac{dz}{dp}\frac{dz}{dq}$$

$$G=\left(\frac{dx}{dq}\right)^2+\left(\frac{dy}{dq}\right)^2+\left(\frac{dz}{dq}\right)^2$$

Now, suppose a second surface to exist which is developable upon the first. Call ξ, η, ζ the co-ordinates in this case and $d\sigma$ the element of length. If this second surface can be developed upon the first, it will be necessary and sufficient that the points of the one be made to correspond to those of the other—that we shall have $d\sigma=ds$ in every direction around two corresponding points. This equality must hold, then, whatever be the values of dp and dq, which define these different directions. Now, for $d\sigma$ we have

$$d\sigma^2=E'dp^2+2F'dpdq+G'dq^2$$

and, for $d\sigma=ds$, we must have

$$E=E' \qquad F=F' \qquad G=G'$$

It follows from this that the three new variables ξ, η, ζ are three functions of p and q, such that being substituted for x, y, z in equations (2), these equations shall be identically satisfied. Conversely, every solution of equations (2) will furnish a surface which may be developed upon the given surface. It is only necessary to eliminate from these equations any two of the quantities x, y, z in order to find the desired equation, which will be the resulting differential equation satisfied by the remaining quantity, say by z. We have, now, from the first and third of (2)

(3)
$$\left(\frac{dx}{dp}\right)^2+\left(\frac{dy}{dp}\right)^2=E-\left(\frac{dz}{dp}\right)^2 \qquad \left(\frac{dx}{dq}\right)^2+\left(\frac{dy}{dq}\right)^2=G-\left(\frac{dz}{dq}\right)^2$$

For convenience, write

(4)
$$\alpha^2=E-\left(\frac{dz}{dp}\right)^2 \qquad \beta^2=G-\left(\frac{dz}{dq}\right)^2$$

then we may replace these two equations by the four

(5)
$$\frac{dx}{dp}=\alpha\cos\theta \qquad \frac{dx}{dq}=\beta\cos\varphi \qquad \frac{dy}{dp}=\alpha\sin\theta \qquad \frac{dy}{dq}=\beta\sin\varphi$$

The second of (2) now becomes

$$\alpha\beta(\cos\theta\cos\varphi+\sin\theta\sin\varphi)=F-\frac{dz}{dp}\frac{dz}{dq}=\gamma\alpha\beta$$

or

(6)
$$\cos(\varphi-\theta)=\gamma$$

Differentiating the equations of the first row of (5) for q and p respectively, and subtracting, we eliminate x; the same operations performed upon the second row eliminate y; we have then

$$(7)\qquad \frac{d\alpha}{dq}\cos\theta-\alpha\sin\theta\frac{d\theta}{dq}-\frac{d\beta}{dp}\cos\varphi+\beta\sin\varphi\frac{d\varphi}{dp}=0$$

$$\frac{d\alpha}{dq}\sin\theta+\alpha\cos\theta\frac{d\theta}{dq}-\frac{d\beta}{dp}\sin\varphi-\beta\cos\varphi\frac{d\varphi}{dp}=0$$

Multiply the first of these by $\cos\theta$, the second by $\sin\theta$, and add; then multiply the first by $\cos\varphi$ and the second by $\sin\varphi$, and add; we have then

$$(8)\qquad \frac{d\alpha}{dq}-\frac{d\beta}{dp}\cos(\varphi-\theta)+\beta\frac{d\varphi}{dp}\sin(\varphi-\theta)=0 \qquad \frac{d\alpha}{dq}\cos(\varphi-\theta)+\alpha\frac{d\theta}{dq}\sin(\varphi-\theta)-\frac{d\beta}{dp}=0$$

or

$$(9)\qquad \sin(\varphi-\theta)\frac{d\varphi}{dp}=\frac{1}{\beta}\left[\gamma\frac{d\beta}{dp}-\frac{d\alpha}{dq}\right]=\lambda \qquad \sin(\varphi-\theta)\frac{d\theta}{dq}=\frac{1}{\alpha}\left[\frac{d\beta}{dp}-\gamma\frac{d\alpha}{dq}\right]=\mu$$

For brevity we may also write

$$\sin(\varphi-\theta)=\tau$$

Then, since $\cos(\varphi-\theta)=\gamma$,

$$\tau^2=1-\gamma^2$$

Equations (9) are now written

$$(10)\qquad \frac{d\varphi}{dq}=\frac{\lambda}{\tau} \qquad \frac{d\theta}{dq}=\frac{\mu}{\tau}$$

From the expression $\cos(\varphi-\theta)=\gamma$ we have, by differentiating,

$$(11)\qquad \frac{d\varphi}{dq}=\frac{1}{\tau}\left(\mu-\frac{d\gamma}{dq}\right) \qquad \frac{d\theta}{dp}=\frac{1}{\tau}\left(\lambda-\frac{d\gamma}{dp}\right)$$

Finally, form the expressions for $\frac{d^2\varphi}{dpdq}$ from 10 and 11, and equate the results.

$$(12)\qquad \frac{1}{\tau}\frac{d\gamma}{dq}-\frac{\gamma}{\tau^2}\frac{d\tau}{dq}=\frac{1}{\tau}\left(\frac{d\mu}{dp}-\frac{d^2\gamma}{dpdq}\right)-\frac{1}{\tau^2}\left(\mu-\frac{d\gamma}{dq}\right)\frac{d\tau}{dq}$$

From the relation $1-\gamma^2=\tau^2$, we have

$$(13)\qquad \tau\frac{d\tau}{dp}=-\gamma\frac{d\gamma}{dp} \qquad \tau\frac{d\tau}{dq}=-\gamma\frac{d\gamma}{dq}$$

Equation (12) now becomes

$$(14)\qquad \left(\frac{d\gamma}{dq}-\frac{d\mu}{dp}+\frac{d^2\gamma}{dpdq}\right)(1-\gamma^2)+\gamma\left(\lambda\frac{d\gamma}{dq}-\mu\frac{d\gamma}{dp}+\frac{d\gamma}{dp}\frac{d\gamma}{dq}\right)=0$$

which contains only E, F, G, and the differential coefficients of these quantities and z with respect to p and q, and is consequently the differential equation sought.

Denote now by k the Gaussian measure of curvature given by the expression

$$(15)\qquad -2V^4k=V^2\left\{\frac{d}{dp}\left[\frac{1}{V}\left(\frac{dF}{dq}-\frac{dG}{dp}\right)\right]+\frac{d}{dq}\left[\frac{1}{V}\left(\frac{dF}{dp}-\frac{dE}{dq}\right)\right]\right\}-V^2\rho$$

when $V^2=EG-F^2$, and ρ denotes the determinant

$$\begin{vmatrix} E, & G, & F \\ \frac{dE}{dp}, & \frac{dG}{dp}, & \frac{dF}{dp} \\ \frac{dE}{dq}, & \frac{dG}{dq}, & \frac{dF}{dq} \end{vmatrix}$$

Again, denoting for brevity the minors of this determinant corresponding to any element by placing that element in brackets, thus:

$$[E]\ \frac{dG}{dp}\frac{dF}{dq}-\frac{dG}{dq}\frac{dF}{dp},\ \&c.$$

also, write

$$A=\frac{dG}{dp}-2\frac{dF}{dq}\qquad\qquad B=\frac{dE}{dq}-2\frac{dF}{dp}$$

Substituting now in (14) the values of $\lambda, \mu, \nu, \gamma, \alpha, \beta$, we find, after some rather tedious but not difficult reductions, the following form for this equation:

$$(16)\qquad 4V^2(rt-s^2)\left\{4G\sqrt{G}\left(\frac{d\sqrt{G}}{dp}-\frac{dF}{dq\sqrt{G}}\right)\frac{dz}{dp}\left(-E\frac{dG}{dq}+AF\right)\frac{dz}{dq}\right\}r$$
$$+4\left(\frac{GdE}{dq}\frac{dz}{dp}+E\frac{dG}{dp}\frac{dz}{dq}\right)s+\left\{4E\sqrt{E}\left(\frac{d\sqrt{E}}{dq}-\frac{d}{dp}\frac{F}{\sqrt{E}}\right)\frac{dz}{dq}-\left(G\frac{dE}{dp}+BF\right)\frac{dz}{dp}\right\}t$$
$$+\left\{2\sqrt{G}\left[\frac{d}{dq}\sqrt{G}\frac{dE}{dq}-G\frac{d}{dp}\frac{A}{\sqrt{G}}\right]+2[E]\right\}\left(\frac{dz}{dp}\right)^2$$
$$+\left\{F\sqrt{F}\left[\frac{d}{dp}\frac{4A}{\sqrt{F}}+\frac{d}{dq}\frac{4}{\sqrt{F}}\frac{dE}{dq}\right]+2[F]\right\}\frac{dz}{dp}\frac{dz}{dq}$$
$$+\left\{2\sqrt{E}\left[\frac{d}{dp}\sqrt{E}\frac{dG}{dp}-E\frac{d}{dq}\frac{B}{\sqrt{E}}\right]+2[G]\right\}\left(\frac{dz}{dq}\right)^2-2V^4k=0$$

where r, s, t denote respectively the second derivatives

$$\frac{d^2z}{dp^2}\qquad\qquad\frac{d^2z}{dpdq}\qquad\qquad\frac{d^2z}{dq^2}$$

It is to be observed that this equation is linear in $(rs-t^2)$ and in r, s, and t, and is in fact a partial differential equation of the second order, of the form

$$Rr+Ss+Tt+U(rs-t^2)=W$$

where R, S, T, U, and W are given functions of $x, y, z, \frac{dz}{dp}, \frac{dz}{dq}$, or of $p, q, \frac{dz}{dp}, \frac{dz}{dq}$. A general integration is of course impossible, so we will note only a few special cases. If we consider now that the three surfaces f, p, q are orthogonal, then the curves given briefly as p=const. and q=const. cut at right angles, since they belong to the two different sets of lines of curvature, and for this case F=0, and equation (10) takes the form

$$(17)\qquad 4EG(rt-s^2)+\left(\frac{dG^2}{dp}\frac{dz}{dp}-E\frac{dG}{dq}\frac{dz}{dq}\right)r+4\left(G\frac{dE}{dq}\frac{dz}{dp}+E\frac{dG}{dp}\frac{dz}{dq}\right)s$$
$$+\left(\frac{d.E^2}{dq}\frac{dz}{dq}-G\frac{dE}{dp}\frac{dz}{dp}\right)t+2\sqrt{G}\left[\frac{d}{dq}\sqrt{G}\frac{dE}{dq}-G\frac{d}{dq}\frac{1}{\sqrt{G}}\frac{dG}{dp}\right]\left(\frac{dz}{dp}\right)^2$$
$$+[F]\frac{dz}{dp}\frac{dz}{dq}+2\sqrt{E}\left[\frac{d}{dp}\sqrt{E}\frac{dG}{dp}-E\frac{d}{dp}\frac{1}{\sqrt{E}}\frac{dE}{dq}\right]\left(\frac{dz}{dq}\right)^2-2E^2G^2K_0=0$$

where

$$k_0=-\tfrac{1}{2}\left\{\frac{d}{dp}\frac{1}{\sqrt{EG}}\frac{dG}{dp}-\frac{d}{dq}\frac{1}{\sqrt{EG}}\frac{dE}{dq}\right\}$$

For E=G this becomes

$$(18)\qquad 4E^2(rt-s^2)+\frac{dE^2}{dp}\left[\frac{dz}{dp}(r-\tfrac{1}{2}t)+2\frac{dz}{dq}s\right]+\frac{d.E^2}{dq}\left[\frac{dz}{dq}(t-\tfrac{1}{2}r)+2\frac{dz}{dp}s\right]$$
$$+1\left\{\tfrac{1}{2}\frac{d^2E^2}{dq^2}+E\frac{d^2E}{dq^2}-E^2\frac{d^2\log E}{dp^2}\right\}\left(\frac{dz}{dp}\right)^2+\left\{\tfrac{1}{2}\frac{d^2E^2}{dp^2}+E\frac{d^2E}{dp^2}-E^2\frac{d^2\log E}{dq^2}\right\}\left(\frac{dz}{dq}\right)^2$$
$$-2E^4\left(\frac{d^2}{dp^2}-\frac{d^2}{dq^2}\right)\log E=0$$

Suppose in equation (16) that $E=G=0$; this requires that the quantities denoted by α and β shall be imaginary, or

$$\alpha=i\alpha' \qquad \beta=i\beta'$$

It would be convenient in starting from this hypothesis to use the imaginary variables defined by

$$\alpha=p+iq \qquad \beta=p-iq$$

Of course these letters, α, β, have no connection with those previously employed. Using these variables, Bour has treated the problem very fully in the Journal de l'École Polytechnique, vol. 22. We will return to that point, however, but may observe, retaining our variables p and q, the form assumed by the equation under the assumed hypothesis of $E=G=0$. Write $F=2\phi$; then we have

$$(rt-s^2)-\frac{1}{\phi}\frac{d\phi}{dq}\frac{dz}{dq}r-\frac{1}{\phi}\frac{d\phi}{dp}\frac{dz}{dp}t+2\left[\frac{1}{\phi}\left(\frac{d^2\phi}{dpdq}-\frac{1}{\phi}\frac{d\phi}{dp}\frac{d\phi}{dq}\right)\frac{dz}{dp}\frac{dz}{dq}+\left(\frac{1}{\phi}\frac{d\phi}{dp}\frac{d\phi}{dq}-\frac{d^2\phi}{dpdq}\right)\right]=0$$

Add and subtract

$$\frac{1}{\phi^2}\frac{d\phi}{dp}\frac{d\phi}{dq}\frac{dz}{dp}\frac{dz}{dq}$$

and this becomes readily

$$0=2\left(\frac{dz}{dp}\frac{dz}{dq}-\phi\right)\frac{d^2\log\phi}{dpdq}-s^2+\left(r-\frac{dz}{dp}\frac{d\log\phi}{dp}\right)\left(t-\frac{dz}{dp}\frac{d\log\phi}{dq}\right)$$

Make in this equation

$$\frac{dz}{dp}\frac{dz}{dq}=\phi$$

and we find that the equation is satisfied; therefore

$$\frac{dz}{dp}\frac{dz}{dq}=\phi$$

is a singular solution of the differential equation.

If we make $E=1$, $F=0$, and G a function of p only, we come to the case of the development upon surfaces of revolution, a particular case of which has been studied by Weingarten in vol. 59 of Crelle's Journal. There are several other suppositions which might be made, and which would conduct to interesting results, but the object of the investigation has been attained, and so we may leave the subject here. It is quite possible that a singular solution might be found for equation (16) or equation (17), which is a sufficiently general form, which would prove valuable in studying the general geometric properties of this class of surfaces.

It is to be observed that for E, F, G all constants (of course including the case $F=0$), the general equation reduces to

$$rt-s^2=0$$

the simplest class of developable surfaces, viz, those which can be developed upon a plane.

This equation is deduced from K by the supposition γ=const., and as we know has for its general integral the result of elimination of a between the equations

$$(a) \qquad z-a\alpha-f(a)\beta=f(a) \qquad -\alpha-f'(a)\beta=f'_1(a)$$

where a is a function of α and β, and f and f' are arbitrary functional symbols. By successively differentiating for α and β, we find

$$\frac{dz}{d\alpha}=a=m' \qquad \frac{dz}{d\beta}=f(a)=m''$$

Also

$$\frac{n}{m}=-i\varphi\int\frac{1}{\sqrt{f(a)}}d\frac{1}{\sqrt{a}}$$

from which

$$n=-i\varphi\sqrt{a}\int\frac{1}{\sqrt{f(a)}}d\frac{1}{\sqrt{a}}$$

and further

$$n'=\frac{i\varphi}{\sqrt{a}}-i\varphi\sqrt{f(a)}\int\frac{1}{\sqrt{f(a)}}d\frac{1}{\sqrt{a}}$$

Finally, m, n, m' n' being thus known as functions of a and also of α and β, by virtue of

$$\alpha=f'(a) \qquad \beta=-f'(a)$$

we can obtain the values of ξ, η, ζ by the equations

$$d\xi=i(m^2+n^2)d\alpha+i(m'^2+n'^2)d\beta$$

$$d\eta=(m^2-n^2)d\alpha+(m'^2-n'^2)d\beta$$

$$d\zeta=3mnd\alpha+2m'n'd\beta$$

The quantities m, n, m', n' being functions of the same quantity a, a function of α and β, we deduce readily the equations

$$mm'dn^2=nn'dm^2 \qquad mm'dn'^2=nn'dm'^2$$

$$mm'd.2mn=(nm'+n'm)d.m^2 \qquad mm'd.2m'n'=(nm'+n'm)d.m'^2$$

and consequently

$$d.i(m^2+n^2=i\frac{(mm'+nn')}{mm'}da \qquad d.i(m'^2+n'^2)=i\frac{(mm'+nn')}{mm'}f'_1(a)da$$

$$d.(m^2-n^2)=\frac{mm'-nn'}{mm'}da \qquad d.(m'^2n'^2)=\frac{mm'-nn'}{mm'}f'(a)da$$

$$d.2mn=\frac{nm'+n'm}{mm'}da \qquad d.2m'n'=\frac{nm'+n'm}{mm'}f'(a)da$$

The equation

$$\alpha+\beta f'(a)=-f'_1(a)$$

now can be placed in any one of the three forms

$$\alpha d.i(m^2+n^2)+\beta d.i(m'^2+n'^2)=-i\frac{(mm'+nn')}{mm'}f'_1(a)\,da=du$$

$$\alpha d.i(m^2-n^2)+\beta d.i(m'^2-n'^2)=-\frac{mm'-nn'}{mm'}f'(a)\,da=dv$$

$$\alpha d.2mn+\beta d.2m'n'=-\frac{nm'+n'm}{mm'}f'_1(a)\,da=dw$$

u, v, w being functions of a, which are determinable by simple quadratures.

The equations giving $d\xi$, $d\eta$, $d\zeta$ can now be integrated by parts giving with reference to these last three relations,

$$(b)\qquad\begin{cases}\xi=i(m^2+n^2)\alpha+i(m'^2+n'^2)\beta-u\\ \eta=1(m^2-n^2)\alpha+(m'^2-n'^2)\beta-v\\ \zeta=2mn\alpha+2m'n'\beta-w\end{cases}$$

We have thus only to eliminate α, β, a between equations (a) and (b) in order to obtain the equation of the required surfaces. The elimination of α and β is readily effected. Take three quantities, λ, μ, ν, functions of a, multiply the equations of group (b) by these quantities respectively, and add the results.

$$(c)\qquad \lambda\xi+\mu\eta+\nu(nu)\zeta=[i\lambda(m^2+n^2)+\mu(m^2-n)^2+2\nu mn]\alpha$$
$$+[i\lambda(m'^2+n'^2)+\mu(m'^2-n'^2)+2\nu m'n']\beta-\lambda u-\mu u-\nu w$$

Multiply the same equations by $d\lambda$, $d\mu$, and $d\nu$, respectively, and the set giving du, dv, and dw by $-\lambda$, $-\mu$, $-\nu$, respectively, and add the resulting six equations giving

$$(d)\qquad \xi d\lambda+\eta d\mu+\zeta d\nu=[d.i\lambda(m^2+n^2)+d.\mu(m^2-n^2)+d.\nu 2mn]\alpha$$
$$+[d.i\lambda(m'^2+n'^2)+d.\mu(m'^2-n'^2)+d.\nu 2m'n']\beta-d.\lambda u-d.\mu v-d.\nu w$$

Now since λ, μ, ν are indeterminate, we may write

$$i\lambda(m^2+n^2)+\mu(m^2-n^2)+2\nu mn=0 \qquad i\lambda(m'^2+n'^2)+\mu(m'^2-n'^2)+2\nu m'n'=0$$

which gives

$$\lambda=ik(mm'+nn') \qquad \mu=k(mm'-nn') \qquad \nu=k(nm'+n'm)$$

k being any quantity whatever; α and β then disappear from equations (c) and (d), and these equations reduce to

$$(e)\qquad \lambda\xi+\mu\eta+\nu\zeta=-(\lambda u+\mu v+\nu w) \qquad \xi d\lambda+\eta d\mu+\zeta d\nu=-(d\lambda u+d\mu v+d\nu w)$$

If we wished to determine the equation of the required surface in rectangular co-ordinates, it would be necessary to eliminate α between equations (e); but since the first of these is linear in ξ, η, ζ, and the second is the derivative of the first with respect to the parameter α, it is obvious at once that the surface is the envelope of a moving plane—that is to say, it is a *developable surface*.

Bonnet has given a very elaborate discussion of the surfaces which are developable upon surfaces of revolution, and in volume 59 of Crelle, Weingarten has given an investigation of a very curious class of surfaces applicable to one another. The principal theorem which he proves is, that *the surfaces of centers of all surfaces for which at any point one principal radius of curvature is a function solely of the other, form a system of surfaces which are all developable upon one another.*

As the intention of these last three chapters is merely to give the reader a slight idea of the more general theory of projection by different methods, it is not at all necessary to go into the subject with any more fullness, either for the purpose of deducing other new principles or of applying any further those already obtained. Enough has been said to meet all the requirements of the reader who merely wishes a slight acquaintance with this subject; all others would naturally go to the original memoirs and discover for themselves the geometrical gems which abound in the writings of Gauss, Jacobi, Liouville, Bonnet, Bour, Codazzi and a host of others.

PART II.

CONSTRUCTION OF PROJECTIONS.

CONSTRUCTION OF PROJECTIONS.

STEREOGRAPHIC PROJECTION.

The stereographic projection is one in which the eye is supposed to be placed at the surface of the sphere, and in the hemisphere opposite to that which it is desirable to project. The exact position of the eye is at the extremity of the diameter, passing through the point assumed as the center of the map.

It has been shown in the first part of this paper that only the scale of the perspective projection is altered by an alteration of the position of the plane of projection; this being the case, and it being more convenient to take the plane as passing through the center of the sphere, we will hereafter assume the plane of projection to be such a diametral plane.

The stereographic projection has been also found among the possible orthomorphic projections, *i. e.*, projections which preserve the angles; so we need here merely state this property, leaving it for those who wish a proof of it to refer to Part I.

In Fig. I, Part I, let C denote the center of the sphere, V the point of sight, Op the trace of the plane of projection upon the plane of the paper, P the pole of the equator, and M any other point in the sphere whose latitude is θ and longitude ω; PZZ′ denotes the first meridian. We have now

$$\omega=ZPM \qquad 90^\circ-\theta=PM \qquad r=CZ \qquad VC=c$$

to these add

$$PZ=\frac{\pi}{2}-\alpha \qquad MZ=\varphi \qquad PZM=\psi \qquad VO=c'$$

Taking O, the projection of Z, as the origin of co-ordinates, and p, the projection of P, as another point of the axes of x, draw OY perpendicular to Op, and we have the axes to which it is most convenient to refer the projection. Assume m as the projection of M; then O$n=x$, $mn=y$. We have already found for x and y the values

$$x=\frac{cr\,(\sin\alpha\cos\theta\cos\omega-\cos\alpha\sin\theta)}{c'+r\,(\cos\alpha\cos\theta\cos\omega+\sin\alpha\sin\theta)} \qquad y=\frac{cr\cos\theta\sin\omega}{c'+r\,(\cos\alpha\cos\theta\cos\omega+\sin\alpha\sin\theta)}$$

For the case of stereographic projection, we have $c=r$, and, since the plane of projection passes through the center of the sphere $c'=r$, these formulas become

$$x=\frac{r\,(\sin\alpha\cos\theta\cos\omega-\cos\alpha\sin\theta)}{1+\cos\alpha\cos\theta\cos\omega+\sin\alpha\sin\theta)} \qquad y=\frac{r\cos\theta\sin\omega}{1+\cos\alpha\cos\theta\cos\omega+\sin\alpha\sin\theta)}$$

By elimination of θ and ω from these equations, we would arrive at the equations of the meridians and parallels which would be found to be the equations of circles (see Part I). By varying the angle α, we can make the plane of projection assume any position that we please, and as the above equations are true for any value of α, we are enabled to say that *all circles of the sphere are, in stereographic projection, represented by circles.* Two particular forms of this projection are of special interest and value, and we shall now take them up.

STEREOGRAPHIC EQUATORIAL PROJECTION.

The plane of the equator is here taken for the plane of projection, and, therefore, in our formulas we must write $\alpha=\frac{\pi}{2}$; this gives

$$x=\frac{r\cos\theta\cos\omega}{1+\sin\theta} \qquad y=\frac{r\cos\theta\sin\omega}{1+\sin\theta}$$

Write here

$$\zeta=90^\circ-\theta$$

then, from these equations, we can obtain for the meridians

$$\frac{y}{x}=\tan\omega$$

and for the parallels

$$x^2+y^2=r^2\tan^2\frac{\zeta}{2}$$

The meridians are thus seen to be projected into right lines passing through the origin, and the parallels are projected into concentric circles, whose center is at the origin, and the radius of each of them is equal to the radius of the sphere multiplied by the tangent of half the co-latitude of the point. The radius of the equator or bounding circle of the projection is found by making $\theta=0$, *i. e.*, $\zeta=90$, and is consequently $=r$. The construction of the projection is now very simple. Take any point as center, and with a radius (on the proper scale) equal to the radius of the sphere describe a circle; this will be the bounding circle of the map. Now, divide the circumference of this into equal parts of 5°, or 10°, or whatever subdivision may be most desirable; the diameters drawn through these points of division will be the meridians. The parallels are all circles concentric with the one already drawn, and can be constructed for any latitude by multiplying the radius of the sphere by the tangent of one-half the complement of the latitude, and taking this quantity as the radius of the parallel required. Table I is constructed by means of the formula

$$\rho=r\tan\frac{\zeta}{2}$$

ρ denoting the radius of the projected parallel. The values of ρ are given for every 5° of latitude on the assumption of $r=1$. This projection, and others closely allied to it, have been very fully worked out in Part I; and as it is designed in Part II to give only the most elementary and necessary principles connected with the construction of the various projections, it will not be necessary to say any more upon this particular case.

STEREOGRAPHIC MERIDIAN PROJECTION.

This is the projection generally employed when it is desired to represent an entire hemisphere on the map. The eye is supposed to be placed at some point of the equator, and the plane of the meridian 90° distant from this point is taken as the plane of projection. For terrestrial charts the plane of the meridian of Greenwich is usually taken as the plane of projection, the eye being then situated at the point on the equator whose longitude is 90°, or 270°. The meridian passing through the eye is taken as the first meridian in reckoning longitude. For maps of polar regions the stereographic equatorial projection is obviously to be preferred to this, but this gives an excellent and simple method of representing the two hemispheres on two separate charts. We have, in this case, $\epsilon=0$, and therefore

$$x=\frac{-r\sin\theta}{1+\cos\theta\cos\omega}\qquad y=\frac{r\cos\theta\sin\omega}{1+\cos\theta\cos\omega}$$

The meridians are circles given by the equation

$$x^2+y^2+2yr\cot\omega-r^2=0$$

The centers of these circles lie on the axis of y at the points given by

$$\xi=0\qquad \eta=-r\cot\omega$$

and their radii are given by

$$R=r\operatorname{cosec}\omega$$

For the bounding meridian $\omega=90°$ and $R=r$. We have then merely to draw a circle from any assumed point as center, whose radius $=r$ (on the chosen scale), and this will be the bounding circle of the map. Draw two diameters of this circle at right angles to each other and they will denote the equator and first meridian respectively. The distance from the center of the map to the intersection of any meridian with the equator is given by the formula

$$s=r\tan\frac{\omega}{2}$$

If now on the line representing the equator we lay off the distances $\pm r \cot \omega$ to the right and left of the first meridian we will find the centers of the projections of the meridians; it is then only necessary to draw circles from these points as centers with radii $=r \operatorname{cosec} \omega$, and the meridians will be constructed on the map. For the parallels we have the equation

$$x^2+y^2+2rx \operatorname{cosec} \theta+r^2=0$$

which represents circles having their centers on the axis of x at the points given by

$$\xi'=-r \operatorname{cosec} \theta \qquad \eta'=0$$

and whose radii are given by

$$R=r \cot \theta$$

For the distance from the center of the map to the intersection of any particular parallel with the first meridian, we have

$$d'=r \tan \frac{\theta}{2}$$

The construction of the parallels is similar to that of the meridians, the centers merely being taken on the projection of the first meridian. One thing is, however, to be observed in constructing these curves: For the meridians it is to be noticed that the formula $\eta=-r \cot \omega$ gives, when the real sign of this is *minus*, the centers of meridians that lie on the + side of the first meridian; and when the sign of η is positive this formula gives the centers of the meridians lying on the negative side of the first meridian. In the case of the parallels, however, the formula $\xi'=\pm r \operatorname{cosec} \theta$ gives the centers of those parallels which lie on the $\pm$ sides of the equator, respectively. Germain has given a table which facilitates the construction of this projection. In using it the following points are to be observed: Calling ρ_m the radius of the projection of a meridian; ρ_p the radius of the projection of a parallel; (ξ_m, η_m) and (ξ_p, η_p) the co-ordinates of the centers of the meridians and parallels respectively, and δ_m and δ_p the distances from the center of the chart to the intersections of the meridians with the equator and of the parallels with the first meridian, we have, for all these quantities, the formulas

$$\rho_m=r \cot \theta \qquad \xi_m=0 \qquad \eta_m=r \cot \omega \qquad \delta_m=r \tan \frac{\omega}{2}$$

$$\rho_p=r \operatorname{cosec} \omega \qquad \xi_p=-r \operatorname{cosec} \theta \qquad \eta_p=0 \qquad \delta_p=r \tan \frac{\theta}{2}$$

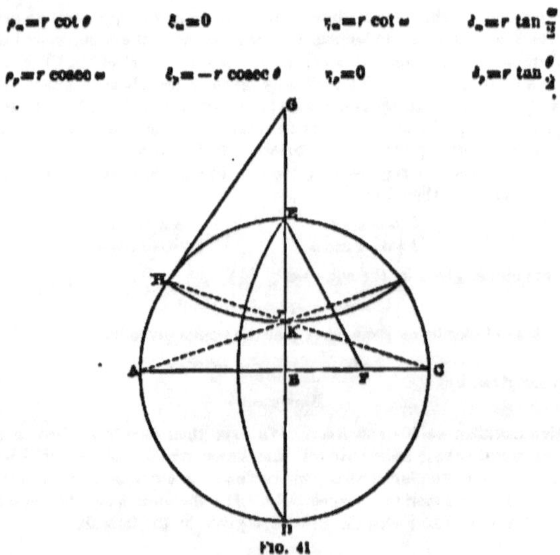

FIG. 41

We see from these that ρ^{n} is the same function of ω that ξ_{ν} is of θ; the same relation holds between η_{m} and ρ_{ν} and between δ_{m} and δ_{ν}. On this account it is only necessary to have one column in the table for each of these pairs of quantities. The following problem is solved in Part I, but is of importance, so the solution is repeated here.

TO FIND THE DISTANCE BETWEEN TWO POINTS ON THE SPHERE AND ON THE MAP.

Let δ denote the distance on the sphere between the points A and B, δ' the distance between A′, B′, their projections. Assume a point M such that

$$MA=x \qquad MB=y$$

and similarly

$$M'A'=x' \qquad M'B'=y'$$

We have thus a spherical triangle MAB, and a plane triangle M′A′B′ with the angles M and M′ equal (since the stereographic projection preserves the angles). Now, in the spherical triangle ABM we have

$$\cos\delta=\cos x\cos y+\sin x\sin y\cos M$$

and in the plane triangle

$$\delta'=x'^2+y'^2-2x'y'\cos M'$$

We know, however, that

$$x'=r\tan\frac{x}{2} \qquad y'=r\tan\frac{y}{2}$$

and therefore, after elimination of M, we obtain readily

$$\delta'=\frac{r\sin\frac{1}{2}\delta}{\cos\frac{1}{2}x\cos\frac{1}{2}y}$$

From this it follows that if x and y are constant, *e. g.*, if they are assumed to remain upon the same parallel, then is δ' proportional to $2\sin\frac{1}{2}\delta$, or to the chord of the arc AB upon the sphere, whatever be the value of M. If $M=0$,

$$\delta'=x'-y' \qquad \delta=x-y$$

and consequently the chord of δ

$$=\delta'\frac{\text{chord}\,(x-y)}{x'-y'}$$

From this the value of δ on the sphere can be found for every corresponding value of δ' on the chart. This expression cannot be employed when $x'=y'$ or when x' differs very little from y'. For this case, however, we need merely to make $M=180^{\circ}$, then

$$\text{chord}\,\delta=\delta'\frac{\text{chord}\,(x+y)}{x'+y'}$$

from which the value of δ can always be exactly obtained.

TO FIND THE LATITUDE AND LONGITUDE OF A PLACE FROM ITS POSITION ON THE CHART.

The general equation of the meridians in the stereographic projection is, as we have seen,

$$x^2+y^2-2xr\tan\alpha-2yr\frac{\cot\omega}{\cos\alpha}-r^2=0$$

that of parallels is

$$x^2+y^2+2rx\frac{\cos\alpha}{\sin\alpha+\sin\theta}+r^2\frac{(\sin\theta-\sin\alpha)}{\sin\theta+\sin\alpha}=0$$

For brevity write $x^2+y^2=\rho^2$; then from the first of these equations we have

$$\cot\omega=\frac{r^2-\rho^2}{r^2+\rho^2}\cos\alpha+\frac{x}{y}\sin\alpha$$

and from the second

$$\sin\theta=\frac{r^2-\rho^2}{r^2+\rho^2}\sin\alpha-\frac{2x}{1+\rho^2}\cos\alpha$$

These equations give us the means of finding θ and ω when x and y are known.

For the stereographic equatorial projection

$$\varepsilon=90^\circ \qquad \cot\omega=\frac{x}{y} \qquad \sin\theta=\frac{r^2-\rho^2}{r^2+\rho^2}$$

For the stereographic meridian projection $\varepsilon=0$, and consequently

$$\cot\omega=\frac{r^2-\rho^2}{2ry} \qquad \sin\theta=-\frac{2x}{1+\rho^2}$$

GNOMONIC PROJECTION.

In this projection the eye is at center of the sphere and the plane of projection is a tangent plane to the sphere. All great circles will be projected in straight lines, with the exception of the equator, which will obviously be projected in a circle at infinity. On this account the gnomonic projection can only be employed for a portion of the sphere less than a hemisphere. The general formulas, as found in Part I, for the co-ordinates of a point on the projection are

$$x=\frac{r(\sin\varepsilon\cos\theta\cos\omega-\cos\varepsilon\sin\theta)}{\cos\varepsilon\cos\theta\cos\omega+\sin\varepsilon\sin\theta} \qquad y=\frac{r\cos\theta\sin\omega}{\cos\varepsilon\cos\theta\cos\omega+\sin\varepsilon\sin\theta}$$

For the gnomonic equatorial projection we have $\varepsilon=90$, and so

$$x=r\cot\theta\cos\omega \qquad y=r\cot\theta\sin\omega$$

Elimination of θ from this gives as the equation of the meridians

$$y=x\tan\omega$$

which shows that the meridians are projected in straight lines, making the same angles with each other as the meridians themselves do on the sphere.

The equation of the parallels is

$$x^2+y^2=r^2\cot^2\theta$$

These lines are thus projected into concentric circles whose radii are proportional to the cotangents of their latitudes. The construction (see Fig. 5) is extremely simple. Divide the limiting circle of the chart into any convenient number of parts, and join the center to the points which express the latitudes counted from the diameter AA′ perpendicular to the first meridian; these radii prolonged meet the tangent TT′ parallel to this diameter, and cut off on it distances equal to the radii of the parallels.

GNOMONIC MERIDIAN PROJECTION.

For this case $\varepsilon=0$, and

$$x=-r\frac{\tan\theta}{\cos\omega} \qquad y=r\tan\omega$$

The meridians have for equation

$$y=r\tan\omega$$

which represents straight lines parallel to the axis of x.

For the parallels we have

$$x^2\cot^2\theta-y^2-a^2=0$$

This represents a series of hyperbolas having their major axes lying on the axis of x, and their minor axes perpendicular to the axis of x, which is taken as the first meridian. The major axes are given by

$$a=2r\tan\theta$$

the minor axes by

$$b=2r$$

For the construction of these hyperbolas it is most convenient to determine a series of points whose abscissas are given by

$$x=r\frac{\tan\theta}{\cos\omega}$$

and then calculate the intersections of the parallels with the meridians supposed already drawn, by giving to θ a certain value and to ω a series of values, 5°, 10°, 15°, &c., or whatever may be most convenient. For a further investigation of this projection the reader is referred to Part I.

ORTHOGRAPHIC PROJECTION.

This projection is not used for geographical representations, but has been employed in the construction of celestial charts, and is commonly employed for architectural and mechanical drawings. The point of sight in this projection is supposed at an infinite distance from the center of the sphere; this involves the writing of $c=\infty$ in the general equations for perspective projections. We have then a cylinder replacing the cone which has been used in making all of the previous projections. It is clear that all circles of the sphere will be projected as either circles, ellipses, or straight lines, according to the inclination of the plane of the circle to the axis of the projecting cylinder. On placing $c=\infty$ in our general equations, we find for the rectangular co-ordinates of any point in this projection

$$x=r(\sin\alpha\cos\theta\cos\omega-\cos\alpha\sin\theta) \qquad y=r\cos\theta\sin\omega$$

From these we have for the equation of the meridians, by eliminating θ,

$$x^2\sin^2\omega-xy\sin\alpha\sin 2\omega+y^2(1-\sin^2\alpha\sin^2\omega)-r^2\cos^2\alpha\sin^2\omega=0$$

This equation represents ellipses having their centers at the origin of co-ordinates; the ellipses have the same major axis given by

$$2a=2r$$

and have their minor axes given by

$$2b=2r\cos\alpha\sin\omega$$

For the parallels we have the equation

$$x^2+y^2+rx\cos\alpha\sin\theta-r^2\sin(\alpha-\theta)\sin(\alpha+\theta)=0$$

This denotes ellipses whose centers are on the axis of x at distances from the origin given by

$$\xi=r\cos\alpha\sin\alpha$$

and whose axes are given by

$$2a'=2r\cos\theta \qquad 2b'=2r\cos\theta\sin\alpha$$

ORTHOGRAPHIC EQUATORIAL PROJECTION.

For this case, we have, as usual, $\alpha=90°$, and consequently

$$x=r\cos\theta\cos\omega \qquad y=r\cos\theta\sin\omega$$

In this case the meridians are straight lines given by

$$y=x\tan\omega$$

and the parallels are concentric circles given by the equation

$$x^2+y^2=r^2\cos^2\theta$$

If the celestial sphere is to be projected according to this method, it will be desirable to obtain the projection of the ecliptic. This is simply a great circle whose plane makes an angle of 23° 28′ with the plane of the equator; the line of intersection has a longitude of either 0° or 180°. The required projection is simply an ellipse whose major axis is equal $2r$, and is coincident with the projection of the first meridian; the minor axis is $=r\cos 23° 28'$, and this is coincident with the projection of the meridian of 90°.

ORTHOGRAPHIC MERIDIAN PROJECTION.

As usual, we have for this case the condition $\alpha=0$, and in consequence

$$x=r\sin\theta \qquad y=r\cos\theta\sin\omega$$

For the meridians, we have the ellipses

$$\frac{x^2}{r^2}+\frac{y^2}{r^2\sin^2\omega}=1$$

whose centers are at the origin and whose axes are

$$2a=2r \qquad 2b=2r\sin\omega$$

The parallels are obviously given by

$$x=r\sin\theta$$

and are represented by right lines parallel to the axis of y or, the same thing, parallel to the equator. The ellipses will in all the preceding cases be best constructed by points; the method of doing so when the formulas are so simple is too obvious to require any explanation. The plane of projection commonly employed for celestial charts is that of the axes of the equator and ecliptic, or simply the solstitial colure. The projections of the equator and ecliptic, as also of all parallels to either, will then be right lines. The center of the projection will represent the equinoctial points, and the solstices will be projected in the extremities of the ecliptic. Declination circles of right ascension $\frac{\pi}{2}-\alpha$, and meridians of celestial longitude ω, are projected in ellipses whose major axis is $=2r$, and whose minor axes respectively equal $r\cos\alpha$ and $r\sin\omega$.

LAGRANGE'S PROJECTION.

This is an orthomorphic projection, *i. e.*, one which does not alter the angles in projecting them; it also possesses the property of representing both meridians and parallels as arcs of circles; in these respects it resembles the stereographic projection, which is indeed only a particular of Lagrange's projection. The construction of the curves being so simple, it will only be necessary to give the different formulas for finding their centers and radii. Take for axes the meridian and parallel through the center of the map. The latitude of the parallel is θ_0, its colatitude φ_0. Lay off from the center O (Fig. 42) on the axis of η the distances $PO=P'O=\lambda$; this entire distance PP' or 2λ is, of course, quite arbitrary, but, when chosen, fixes the scale of the map. It will be observed that we before used the axis of ξ as in the direction of PP', but the present plan of using η can cause no confusion. The meridians make at P and P' angles $=2\iota\omega$, ι being an arbitrary constant, and called the coefficient of the chart.

The meridians have for equation

$$\xi^2+\eta^2+2\xi\lambda\cot 2\iota\omega-\lambda^2=0$$

The center of each circle is on the axis of ξ at the point

$$\xi_0=\lambda\cot 2\iota\omega$$

on the right of η if the longitude ω is west, on the left if ω is east longitude.

For the intercepts of this circle on the axis of ξ we have $\eta=0$, and so

$$\xi'=\lambda\tan\iota\omega=OM \qquad \xi''=\lambda\cot\iota\omega=OM'$$

The radii of these circles are given by

$$\rho_m=\frac{\lambda}{\cos 2\iota\omega}$$

To draw the meridians, it is only necessary to describe on PP′ an arc containing the angle $180°-2\iota\omega$, the remainder of the circle being the arc of $180°+2\iota\omega$. The equation of the parallels is

$$\xi^2+\eta^2+2\lambda\eta\frac{k^2+1}{k^2-1}+\lambda^2=0$$

when

$$k=\tan^n\frac{\varphi}{2}\cot^n\frac{\varphi}{2}$$

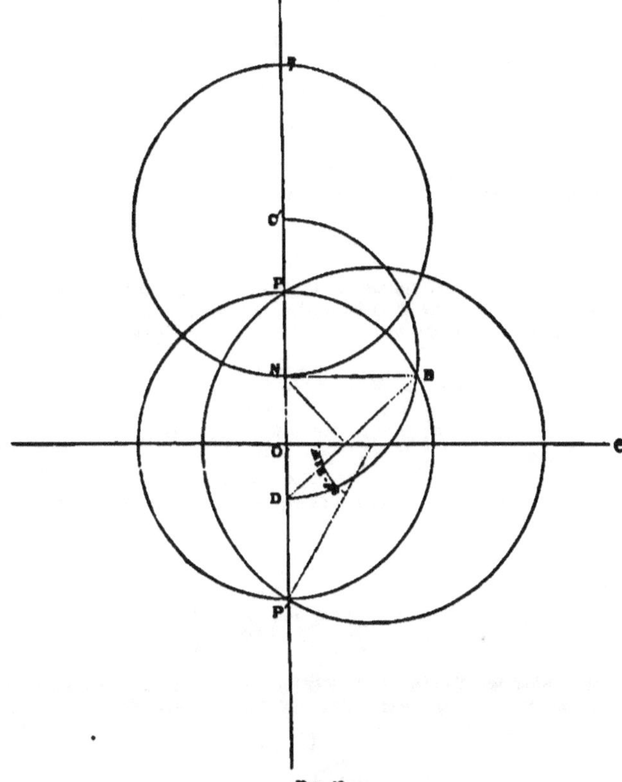

Fig. 42.

The centers of these circles are upon the axis of η, and are given by

$$\eta_0=\frac{\lambda(1+k^2)}{1-k^2}$$

The intercepts upon the axis of η are given by $\xi=0$, and are

$$\eta'=\frac{\lambda(1-k)}{1+k}=ON \qquad \eta''=\frac{\lambda(1+k)}{1-k}=ON'$$

or substituting for k its value

$$\eta'=\lambda\left[\frac{1-\cot\frac{\varphi_0}{2}\tan\frac{\varphi}{2}}{1+\cot\frac{\varphi_0}{2}\tan\frac{\varphi}{2}}\right]^{2t} \qquad \eta''=\lambda\left[\frac{1+\cot\frac{\varphi_0}{2}\tan\frac{\varphi}{2}}{1-\cot\frac{\varphi_0}{2}\tan\frac{\varphi}{2}}\right]^{2t} \qquad \rho_p=\frac{2\lambda k}{1-k^2}=\frac{\lambda^2-\eta'^2}{2\eta'}$$

The point N being given by the above formula, describe a circle upon PP″ as diameter; at the point N draw NB perpendicular to PP″ until it meets the circle in B; lay off OD=ON, and draw another circle through D and B, with its center on PP′; this center is of course formed by drawing a perpendicular to DB at its middle point, and producing it until it meets PP″; the intersection of PP′ with this circumference will be the center of the sought parallel

$$NB^2=PO^2-ON^2=ND.DC'$$

or

$$\lambda^2-\eta'^2=2\eta'\rho_p$$

It is to be remembered that, when the ellipticity of the earth is to be taken into account, we must use ζ and ζ_0 instead of φ and φ_0. The coefficient of the chart is determined by the formula

$$t=\frac{\sqrt{1+\sin^2\varphi}}{2}$$

We have found in Part I that the ratio of the corresponding elementary distances upon the chart and upon the spheroid is, for all orthomorphic projections, independent of the directions of these elements. Denoting this ratio by m, we have for the value of this quantity

$$m=\frac{-4t\lambda\sqrt{1-E^2\cos^2\varphi}}{a\left[\dfrac{\tan^{2t}\frac{\zeta}{2}}{\tan^{2t}\frac{\zeta_0}{2}}+2\cos t\omega+\dfrac{\tan^{2t}\frac{\zeta_0}{2}}{\tan^{2t}\frac{\zeta}{2}}\right]\sin\varphi}$$

where ζ denotes the polar distance of the point corrected to allow for the ellipticity of the spheroid, or for a spherical earth,

$$m=\frac{-4t\lambda}{r\left[\dfrac{\tan^{2t}\frac{\varphi}{2}}{\tan^{2t}\frac{\varphi_0}{2}}+2\cos t\omega+\dfrac{\tan^{2t}\frac{\varphi_0}{2}}{\tan^{2t}\frac{\varphi}{2}}\right]}$$

The point for which m differs least from unity is situated upon the meridian PP″, from which the longitude is measured and has its co-latitude φ_1 defined by the relation

$$\frac{2t-\cos\varphi_1}{2t+\cos\varphi_1}=\left[\frac{\tan\frac{\varphi_1}{2}}{\tan\frac{\varphi_0}{2}}\right]^{2t}$$

will be clearly advantageous in constructing a projection of this kind to assume this point as nearly as may be at the center of the map; the countries then very near the center will very approximately preserve their true form. The construction of the projection will then proceed as follows: Choose the point—some important geographical position which it is desired to place near the center of the map; its colatitude is φ_1. Assume that the longitude is measured from the meridian of this point which is represented by a right line. The coefficient of the projection is given by

$$2t = \sqrt{1 + \sin \varphi_1}$$

It is then only necessary to place this point anywhere upon the line PP″—say at K—and lay off from K the distances given by

$$\frac{PK}{P''K} = \frac{2t - \cos \varphi_1}{2t + \cos \varphi_1}$$

The points P and P″ so obtained are the poles. The distance PP″ is of course arbitrary, and depends merely upon the proposed scale of the map. For the latitude of the center of the chart we have

$$\left(\tan \frac{\varphi_0}{2}\right)^n = \frac{P''K}{PK} \left(\tan \frac{\varphi_1}{2}\right)^n$$

Knowing now the poles, the center of the chart, and its coefficient, the remainder of the construction proceeds in the manner already indicated.

PROJECTIONS BY DEVELOPMENT.

The following section has been taken almost entirely from Part I and without any material change. The considerations are all of such an elementary nature, and the projections treated are so important, that it has not seemed necessary in this practical part of the book to do more than repeat what has been given in Part I. In order that a surface may be represented upon a plane without any change of angles or areas, it must be such an one as can, by slitting it open along some line, be rolled out and made to coincide with the plane at every point. Such surfaces as the cylinders or cones obviously fulfill these conditions. The surfaces which possess this property are appropriately called developable surfaces. The sphere, however, or ellipsoid, does not satisfy this condition for exact representation, so that it is necessary to replace either of these surfaces, as nearly as may be, by developable surfaces upon which lines are drawn corresponding to the meridians and parallels. The construction of these lines upon the new surface must, of course, be of such a nature as to make them correspond in all ways as closely as possible with the original lines upon the sphere. The attempt to make projections of this kind has naturally given rise to two methods of solution: these are, first, by aid of an auxiliary cone; second, by aid of an auxiliary cylinder. Consider, first, Conical Projections. Conceive a cone passed tangent to the sphere along the parallel of latitude which is at the middle of the region to be projected. Also, imagine the planes of the different parallels and meridians to be produced until they cut the cone. We will then have upon the surface of the cone small quadrilaterals corresponding to those of the sphere; the magnitudes are different, but the angles are obviously the same. Now develop the cone upon a plane; the meridians will clearly become right lines from the vertex of the cone to the different points of the developed parallel of tangency (or any other), and the parallels will be concentric circles, the vertex of the cone being the common center. The parallel of tangency is obviously the only one unaltered by the development. The quadrilaterals upon the sphere are reproduced upon the square still as rectangular, but the magnitudes are different, as equal distances of latitude upon the sphere are represented by distances which diminish towards the pole and increase towards the equator. The differences of longitude are all greater upon the surface of the cone than upon the sphere, except for the parallel of tangency. The error in latitude may be completely (and that in longitude partially) eliminated by laying off along the middle meridian of the development the rectified lengths of the distances between the parallels, and through the points thus obtained, with the vertex of the cone as a center, describing arcs of circles. By this means we obtain for the differences in latitude their true values, and for the differences in longitude values which are more nearly correct than those given by the first method.

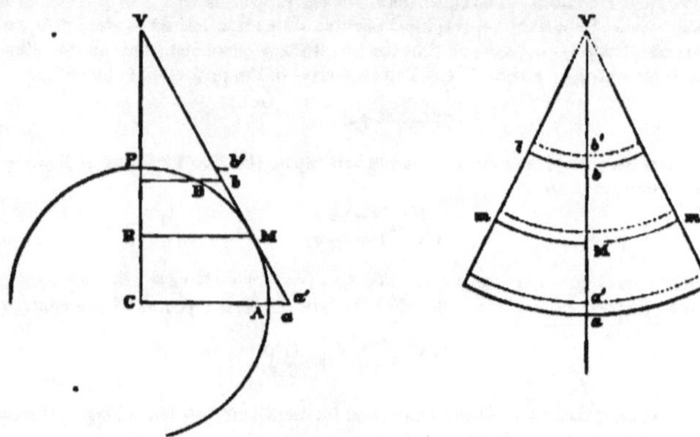

Fig. 13.

Fig. 13 shows both methods, the dotted lines corresponding to the second method. We have clearly from the first figure

$$\frac{180^\circ}{\pi r \cos \phi_0} = \frac{\kappa}{mm'}$$

where ϕ_0 is the latitude of the middle parallel RM and κ is the difference of longitude of the extreme meridians which are to be projected; also, let V denote the angle of the extreme elements of the cone which appear in the development. The radius VM of the middle parallel is given by

$$VM = r \cot \phi_0$$

and from figure (2) follows:

$$\frac{180^\circ}{\pi r \cot \phi_0} = \frac{V}{mm'}$$

Combination of these two values for mm' gives

$$V = \kappa \sin \phi_0$$

It is obvious now how to construct the projection: The angle V being determined, we have for the radius of the middle parallel $VM = r \cot \phi_0$. Lay off from M the distances Ma' and Mb' as obtained by actual rectification. If the distance ab contains n degrees,

$$ab = \frac{\pi r n}{180^\circ}$$

and Mb', Ma' each

$$= \frac{\pi r n}{90^\circ}$$

Having then the center and one point on the circumference, we can draw the circles which represent the parallels of latitude. If we call ϵ the angle between the projections of two meridians corresponding to ω upon the sphere, we have clearly

$$\frac{\epsilon}{\omega} = \frac{V}{\kappa} = \sin \phi_0$$

The radius of the parallel at latitude θ will be

$$=r[\cot\theta_0-(\theta-\theta_0)]$$

and the corresponding arc of longitude ω will be

$$=r\omega\sin\theta_0[\cot\theta_0-(\theta-\theta_0)]$$

The error for each degree of the parallel will then be

$$=r(\theta-\theta_0)\sin\theta_0$$

Euler investigated at some length the theory of conic projection, and determined a cone fulfilling the following conditions:

1. That the errors at the top and bottom extremities of the chart should be equal.
2. That they shall be equal to the greatest error which occurs near the mean parallel.

The cone in this case is obviously a secant and not a tangent cone to the sphere. Let θ_s denote the least latitude of the region to be projected, and θ_t the greatest value of the latitude; let AB, Fig. 14, denote the portion of the middle meridian comprised between these extreme latitudes. Designate by δ the length of 1° of the meridian, and let P and Q be the intersections of the cen-

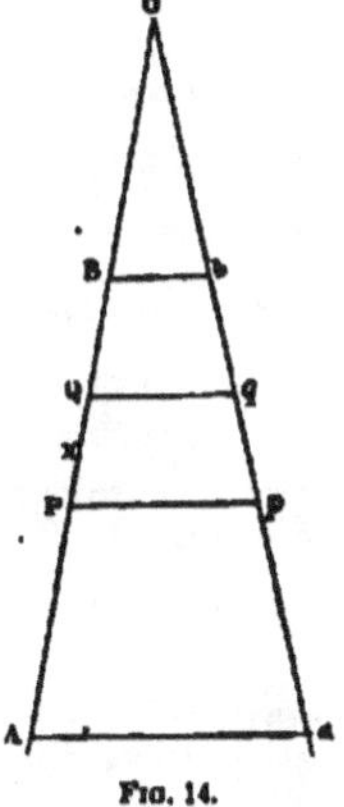

FIG. 14.

tral meridian with the parallels, along which the degrees shall preserve upon the map their exact ratio with the actual degrees of latitude; also call θ_p and θ_q the latitudes of these two parallels, upon each of which a degree of longitude has respectively the values $\delta\cos\theta_p$ and $\delta\cos\theta_q$. Lay off these two values of 1° along the lines P_p and Q_q perpendicular to AB, and join pq; this line will represent the meridian, removed one degree from AB. The point of intersection O will obviously be the common point of meeting of all the meridians and the center of all the parallels. The distance from O to any parallel is readily found; since OPp is a right angle, we have

$$\frac{Pp-Qq}{PQ}=\frac{Pp}{PO}$$

or

$$\frac{\delta(\cos\theta_p-\cos\theta_q)}{\theta_p-\theta_q}=\frac{\delta\cos\theta_p}{PO}$$

from which

$$PO=\frac{\cos\theta_p(\theta_p-\theta_q)}{\cos\theta_p-\cos\theta_q}$$

Having determined the center O, it is only necessary to draw an arc of radius OP, and upon it lay off lengths $=\delta\cos\theta_p$; these will give the points through which the meridians pass. Then laying off, along the middle meridian, distances equal to the number of degrees of latitude of the different parallels to be constructed, draw through the points thus found circles having their center at O, and the projections of the parallels will be constructed.

We will now determine the errors resulting from this construction upon the extreme parallels through A and B. Representing by ω the angle POp, we find

$$\omega=\frac{Pp}{PO}=\frac{\delta(\cos\theta_p-\cos\theta_q)}{\theta_p-\theta_q}$$

which becomes

$$\omega=\frac{\cos\theta_p-\cos\theta_q}{(\theta_p-\theta_q)\upsilon}$$

If we take $\delta=1^\circ$ and express the denominator in parts of radius, which is done by making $\upsilon=0.01745320$, the value of 1° in a circle of radius unity. Let z represent the distance in degrees from the center O to the pole. The distance from P to the pole will be $=90^\circ-\theta_p$; from P to O will be $=90^\circ-\theta_p+z$; the value of this in parts of radius will be

$$=\upsilon(90^\circ-\theta_p+z)$$

It is easy to see now that we must have

$$z=\frac{(\theta_q-\theta_p)\cos\theta_p}{\cos\theta_p-\cos\theta_q}-90^\circ+\theta_p$$

The distance of the extreme parallel A from O will be in parts of radius

$$AO=\upsilon(90^\circ-\theta_a+z)$$

Multiplying this by the value of ω, we have for the value of the degree upon this parallel

$$A_a=\frac{\delta(90^\circ-\theta_a+z)(\cos\theta_p-\cos\theta_q)}{\theta_q-\theta_p}$$

instead of $\delta\cos\theta_a$. The difference of these two values gives the error along the parallel through A. For B the error is the difference between $\delta\cos\theta_b$ and

$$\frac{\delta(90^\circ-\theta_b+z)(\cos\theta_p-\cos\theta_q)}{\theta_q-\theta_p}$$

Euler's proposition was to determine the parallels P and Q in such a manner as to make the extreme errors at A and B equal. Equating these two errors and reducing, we have

$$(\theta_a-\theta_b)(\cos\theta_p-\cos\theta_q)+(\theta_q-\theta_p)(\cos\theta_a-\cos\theta_b)=0$$

For the length of one degree upon the parallels of A and B we have

$$\upsilon(90^\circ-\theta_a+z)\omega \qquad \upsilon(90^\circ-\theta_b+z)\omega$$

We have from these

$$\upsilon(90^\circ-\theta_a+z)\omega-\cos\theta_a=\upsilon(90^\circ-\theta_b+z)-\cos\theta_b$$

from which follows

$$\omega=\frac{\cos\theta_a-\cos\theta_b}{\upsilon(\theta_b-\theta_a)}$$

Further, equate both of these errors to the greatest error which occurs between A and B, supposing in the first instance that it occurs at the point X half way from A to B. The latitude of X is

$$=\frac{\theta_a+\theta_b}{2}$$

The error there is

$$=-\left[\upsilon\left(90^\circ-\frac{\theta_a+\theta_b}{2}-z\right)\omega-\cos\frac{\theta_a+\theta_b}{2}\right]$$

its sign being opposite to the signs of the errors at A and B. The condition is now expressed by the two equations

$$\upsilon(90^\circ-\theta_a+z)\omega-\cos\theta_a=\cos\frac{\theta_a+\theta_b}{2}-\upsilon\left(90^\circ-\frac{\theta_a+\theta_b}{2}-z\right)\omega$$

$$\upsilon(90^\circ-\theta_b+z)\,\omega-\cos\theta_b=\cos\frac{\theta_a+\theta_b}{2}-\upsilon\left(90^\circ-\frac{\theta_a+\theta_b}{2}-z\right)\omega$$

Giving ω its value $\frac{\cos\theta_a-\cos\theta_b}{\upsilon(\theta_b-\theta_a)}$, we find readily

$$\frac{(180^\circ-\frac{3}{2}\theta_a-\frac{1}{2}\theta_b+2z)(\cos\theta_a-\cos\theta_b)}{\theta_b-\theta_a}=\cos\theta_a+\cos\frac{\theta_a+\theta_b}{2}$$

which reduces to

$$(180^\circ-\tfrac{3}{2}\theta_a-\tfrac{1}{2}\theta_b+2z)=\frac{\theta_b-\theta_a}{\cos\theta_a-\cos\theta_b}\left[\cos\theta_a+\cos\frac{\theta_a+\theta_b}{2}\right]$$

from which z is readily found. Applying this to the construction of a map of Russia, it is only necessary to write

$$\theta_a=40^\circ \qquad \theta_b=70^\circ \qquad \frac{\theta_a+\theta_b}{2}=55^\circ$$

The formula for ω gives now at once

$$\omega=\frac{\cos 40^\circ-\cos 70^\circ}{30\,\upsilon}=48'\,44''$$

The equation

$$(180^\circ-\tfrac{3}{2}\theta_a-\tfrac{1}{2}\theta_b+2z)\upsilon\omega=\cos\theta_a+\cos\frac{\theta_a+\theta_b}{2}$$

gives now

$$(85^\circ-2z)\upsilon\omega=1.33962$$

Now, $\upsilon\omega=0.0141$; therefore

$$2z=\frac{1.33962}{0.0141}-85^\circ=10^\circ \qquad z=5^\circ$$

So far we have assumed that the maximum error lay at the middle of AB; but we will now find the correct point, and assume that for this place the latitude is θ; the error will now be

$$\upsilon(90^\circ-\theta+z)\omega-\cos\theta$$

Differentiating this with respect to θ and equating to zero, we find for the position of maximum error

$$\sin\theta=\omega=0.8098270$$

or

$$\theta=54^\circ\,4'$$

Equating the error at θ to those of A and B,

$$\upsilon(180^\circ-\theta_a-\theta+2z)\omega=\cos\theta_a+\cos\theta$$

from which

$$z=5^\circ\,0'\,30''$$

The values of z and θ differ very little from their assumed values of 5° and 55° respectively. The errors at A and B are thus equal to

$$\upsilon\omega(90^\circ-\theta_a+z)-\cos\theta_a=0.00946$$

A degree on the parallel of 40° is then expressed by 0.77550 instead of 0.76604, its true value upon the sphere. This degree is then about $\frac{1}{81}$ greater than the true degree on the parallel of 40°, and the degree on the parallel of 70° is about $\frac{1}{36}$ too great, its true value being 0.34202.

MURDOCH'S PROJECTION.

In Fig. 15, let θ_a and θ_b denote the latitude of two extreme parallels Aa and Bb, which limit a spherical zone whose projection is to be determined. The latitude of M half way between A and B is $\frac{\theta_a+\theta_b}{2}$.

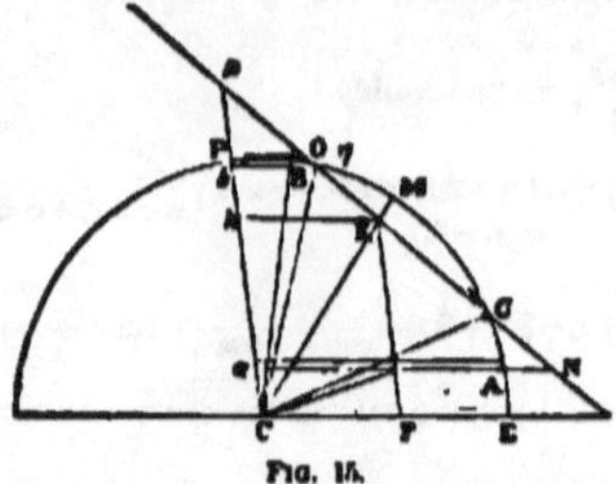

FIG. 15.

Murdoch's projection consists in making the entire area of the chart equal to the entire area of the zone to be projected. In order to effect this it will be necessary—supposing PN and po the radii of the extreme parallels of the chart (obtained by rectification)—that the surface generated by the revolution of ON(=AB) about PC shall be $=2\pi r(ab)$, where $r=$ radius of the sphere expressed in degrees. Let δ denote the equal angles qCM, GCM; we must then have

$$2\pi\, Kk\, AB = 2\pi r (ab)$$

From the similar triangles KCk and MFC we obtain

$$\frac{Kk}{FC} = \frac{KC}{MC}$$

consequently

$$Kk = r \cos\frac{\theta_a+\theta_b}{2} \cos\delta$$

and substituting this in the above equation, we have

$$\frac{\theta_b-\theta_a}{2} \cos\frac{\theta_a+\theta_b}{2} \cos\delta = \sin\theta_b - \sin\theta_a = 2\sin\frac{\theta_b-\theta_a}{2} \cos\frac{\theta_a+\theta_b}{2}$$

This gives for $\cos\delta$ the value

$$\cos\delta = \frac{\sin\frac{\theta_b-\theta_a}{2}}{\theta_b-\theta_a}$$

It is easy to see that, for the radius Kp=R of the middle parallel, we have

$$Kp = r\cos\frac{\theta_a+\theta_b}{2} \frac{\cos\delta}{\sin\frac{\theta_a+\theta_b}{2}}$$

or

$$R = r\cot\frac{\theta_a+\theta_b}{2}\cos\delta$$

The quantities which we have already denoted by n and V are here connected by the relation

$$V = n\sin\frac{\theta_a+\theta_b}{2}$$

Murdoch, in order to draw the intermediate parallels, divided the right line ON into equal parts, giving, for the radius of any parallel θ,

$$R+\frac{\theta_a+\theta_b}{2}-\theta$$

a method which, although perfectly arbitrary, had the effect of diminishing the errors in the chart. Mayer, who resumed the problem proposed by Murdoch, gave the radii ρ_ζ and ρ_η as

$$\rho_\eta=\rho K-K_\eta \qquad \rho_\zeta=\rho K+K_\zeta$$

and, as $K_\eta=K_\zeta=r\sin\delta$,

$$\rho_\eta=R-r\sin\delta=r\frac{\cos\left(\frac{\theta_a+\theta_b}{2}+\delta\right)}{\sin\frac{\theta_a+\theta_b}{2}} \qquad \rho_\zeta=R+r\sin\delta=r\frac{\cos\left(\frac{\theta_a+\theta_b}{2}-\delta\right)}{\sin\frac{\theta_a+\theta_b}{2}}$$

A second method of projection was given by Murdoch, in which the eye is placed at the center of the sphere, as in gnomonic projection, and a perspective is made which is subject to the condition of preserving the entire surface of the zone which is to be represented. Lambert was the first to indicate a method of conic development which should preserve all the angles except the one at the vertex of the cone, when the 360° having upon the sphere the pole for center will obviously be represented in different manners according to the different conditions to be fulfilled. A full account of this method is given in the chapter on orthomorphic projections.

BONNE'S PROJECTION.

This method of projection is that which has been almost universally employed for the detailed topographical maps based on the detailed trigonometrical surveys of the several states of Europe. It was originated by Bonne, was thoroughly investigated by Henry and Puissant in connection with the map of France, and tables for France were computed by Plessis. In constructing a map on this projection a central meridian and a central parallel are first assumed. A cone tangent along the central parallel is then assumed and the central meridian developed along that element of the cone which is tangent to it, and the cone is then developed on a tangent plane. The parallel falls into an arc of a circle with its center at the vertex, and the meridian becomes a graduated right line. Concentric circles are then conceived to be traced through points of this meridian at elementary distances along its length. The zones of the sphere lying between the parallels through these points are next conceived to be developed each between its corresponding arcs. Thus, all the parallel zones of the sphere are rolled out on a plane in their true relations to each other and to the central meridian, each having in projection the same width, length, and relation to the neighboring zones as on the spheroidal surface. As there are no openings between consecutive developed elements, the total area is unaltered by the development. Each meridian of the projection is so traced as to cut each parallel in the same point in which it intersected it on the sphere. If the case in hand be that involving the greatest extension of the method, or that of the projection of the entire spheroidal surface, a prime or central meridian must first be chosen, one-half of which gives the central straight line of the development, and the other half cuts the zones apart and becomes the outer boundary of the total developed figure. Next, the latitude of the governing parallel must be assumed, thus fixing the center of all the concentric circles of development. Having then drawn a straight line and graduated it from 90° north latitude to 90° south latitude, and having fixed the vertex or center of development on it, concentric arcs are drawn from this center through the different graduations. There results from this process an oblong, kidney-shaped figure which represents the entire earth's surface, and the boundary of which is the double developed lower half of the meridian first assumed. This projection preserves in all cases the areas developed without any change. The meridians intersect the central parallel at right angles, and along this as along the central meridian the map is strictly correct. For moderate areas the intersections approach tolerably to being rectangular. All distances along parallels are correct, but distances along the meridians are increased in projection in the same ratio as the cosine of the angle between the radius of the parallel and the tangent to the meridian at the point of intersection is diminished. Thus, in a full earth projection the bounding meridian is elongated to about twice its original length. While each quadrilateral of the map preserves its area unchanged, its two diagonals become unequal; one increasing and the other decreasing in receding towards the corners of the map, the greatest inequality being towards the east and west polar corners.

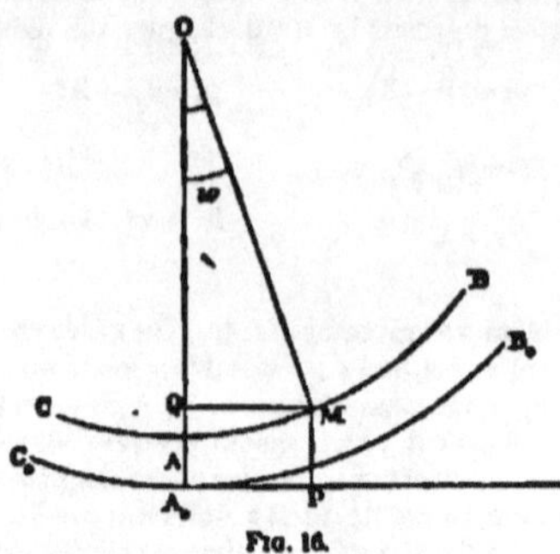

FIG. 16.

Denote the radius of the central parallel by ρ_0; then (Fig. 16)

$$OA_0=\rho_0=r\cot\theta_0$$

Denote by δ the length of the arc AA_0 and the arc passing through a given point M; θ_0, of course, denotes the latitude of the central parallel and θ that of the parallel BC. The latitude of M is $=\theta_0+\frac{\delta}{r}$ and thus

$$MA=\rho\omega=\omega r\cos\left(\theta_0+\frac{\delta}{r}\right) \qquad \rho=\rho_0-\delta=r\cot\theta_0-\delta$$

$$x=MQ=\rho\sin\omega \qquad y=MP=r\cot\theta_0-\rho\cos\omega$$

It is not difficult in this projection to take account of the spheroidal form of the earth. It is only necessary to multiply $\cot\theta_0$ by the principal normal n_0 and replace the spherical arc δ by the elliptic arc s given by

$$s=a(1-e^2)[A(\theta-\theta_0)-B\sin(\theta-\theta_0)\cos(\theta+\theta_0)+\tfrac{1}{2}C\sin^2(\theta-\theta_0)\cos^2(\theta+\theta_0)]$$

Then

$$\rho_0=\frac{a\cot\theta_0}{(1-e^2\sin^2\theta_0)^{\frac{1}{2}}}=\text{, say, }=n_0\cot\theta_0 \qquad \rho=n_0\cot\theta_0-s \qquad \omega=\frac{\omega n_0\cos\left(\theta_0+\frac{s}{a}\right)}{\rho}$$

These give the radii of the projections of the parallels, which are then readily constructed. Lay off from the central meridian upon the parallels now constructed lengths equal to one degree upon each different parallel, and through these points pass a curve, which will be the projection of the meridians. The lengths are given by the formula

$$d=\frac{2\pi}{360}n\cos\theta=\frac{\pi a\cos\theta}{180(1-e^2\sin^2\theta)^{\frac{1}{2}}}$$

The concave parts of these curves are all turned toward the central meridian.

The angle χ in Fig. 17 is the angle which the tangent to the meridian at M makes with the radius OM of the parallel through that point. This angle is also the difference between the angle

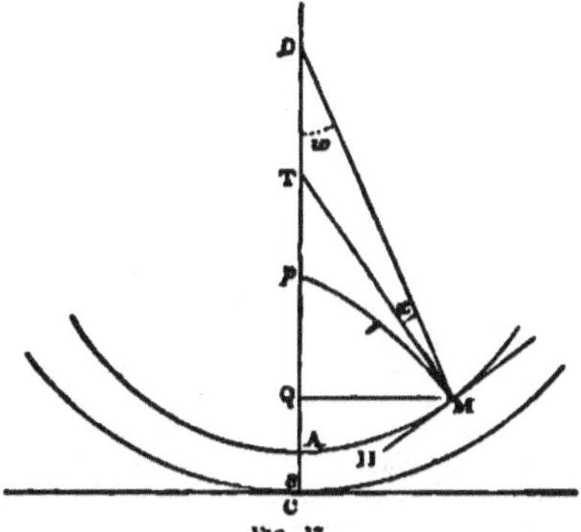

FIG. 17.

that the meridian makes with the parallel at this point and 90°. We have obviously

$$\tan\chi=\frac{\rho d\omega}{d\rho}$$

but $\rho=\rho_0+s$; therefore

$$d\rho=ds \qquad \tan\chi=\frac{\rho d\omega}{ds}$$

Now

$$\rho\omega=n\omega\cos\theta=\frac{a\omega\cos\theta}{(1-e^2\sin^2\theta)^{\frac{1}{2}}}$$

Differentiating this gives

$$\rho d\omega+\omega d\rho=\frac{a\omega\sin\theta(1-e^2)d\theta}{(1-e^2\sin^2\theta)^{\frac{3}{2}}}$$

and we have

$$\rho d\omega+\omega d\rho=\rho d\omega+\omega ds$$

But we know that

$$ds=\frac{a(1-e^2)d\theta}{(1-e^2\sin^2\theta)}$$

Consequently we may write

$$\frac{\rho d\omega}{ds}+\omega=\omega\sin\theta$$

and

$$\tan\chi=\omega\sin\theta-\omega$$

For $\theta=\theta_0$

$$\omega=\frac{a\omega\cos\theta_0}{\rho_1(1-e^2\sin^2\theta_0)^{\frac{1}{2}}} \qquad \rho_1=\frac{a\cot\theta_0}{(1-e^2\sin^2\theta_0)^{\frac{1}{2}}}$$

Combining these

$$\omega=\omega\sin\theta_0$$

and for this case $\tan\chi=0$ or $\chi=0$; which only shows what we already know, viz, that the meridians and central parallel cut at right angles.

If for the central parallel we assume the equator, the vertex of the tangent cone is removed to an infinite distance, the parallels all fall into straight lines, and we have the so-called Flamsteed's projection. The kidney-shaped Bonne projection becomes an elongated oval, with the half meridian for one axis and the whole equator for the other. The co-ordinates for any point in this projection are readily found to be

$$y=\frac{\pi}{180}a\theta \qquad\qquad x=\frac{\omega\pi}{360}a\cos\theta=K\cos\frac{y}{a}$$

The form of the equation giving x has induced M. d'Avernc to give this projection the name *sinusoidal*. This projection, which should really be called Sanson's projection, is evidently only a particular case of Bonne's method; it is based upon a division of the earth's surface into zones or rings by parallels of latitude taken at successive elementary distances laid off along the central meridian of the area to be projected. Having developed this center meridian on a straight line of the plane of projection, a series of perpendiculars is conceived to be erected at the elementary distances along this line. Between these perpendiculars the elementary zones are conceived to be developed in the correct relations to each other and the center meridian. Each zone being of uniform width, occupies a constant length along its entire developed length, and consequently the area of the plane projection is exactly equal to that of the spheroidal surface thus developed. The meridians of the developed spheroid are traced through the same points of the parallels in which they before intersected them. They all cut the parallels obliquely, and are concave towards the center meridian. Thus, while each quadrilateral between parallels and meridians contains the same area and points after development as before, the form of the configuration is considerably distorted in receding from the central meridian, and the obliquity of the intersections between parallels and meridians grows to be highly unnatural.

WERNER'S EQUIVALENT PROJECTION.

If the vertex of the cone approaches the sphere instead of receding from it, as in the preceding case, we have finally, when the tangent cone becomes a tangent plane, the projection known as Werner's Equivalent Projection. The parallels are now arcs of circles described about the pole as a center, and with radii equal to their actual distances from the pole, *i. e.*, equal to the rectified arc of the colatitudes.

The meridians are drawn by laying off on the parallels the actual distances between the meridians as they intersect the parallels on the sphere. This projection is not of enough importance to spend any time in obtaining any of the formulas connected with it.

POLYCONIC PROJECTIONS.

In all the cases of conic projection that we have treated so far, we have supposed that a narrow zone of earth was to be projected, and that for the zone was substituted a developable surface upon which the parallels and meridians were constructed according to any manner that may be desirable. We have seen that this kind of projection is only available when but a small portion of the earth is represented, and that to make a projection of a country of great extent in latitude some modification would be necessary.

The system which is used in America and in England replaces each narrow zone of the earth's surface by the corresponding conic zone in such a way as to preserve the orthogonality of the meridians and parallels. This is the projection of which we have already spoken at length in the Introduction, under the title of Polyconic Projection. As a very full account of this system has been already given, and comparisons made with the other ordinary methods of projection, we will not say anything on the subject here, but will proceed to develop the theory of the system.

The name *rectangular* polyconic projection is applied to the method in which each parallel of the spheroid is developed symmetrically from an assumed central meridian by means of the cone tangent along its circumference. Supposing each element thus developed relative to the common central meridian, it is evident that a projection results in which all parallels and meridians intersect at right angles. The parallels will be projected in circles and the meridians in curves which cut these circles at right angles. The radii of the parallels are equal to the cotangents of their latitudes (to radius supposed unity), and the centers are upon the line which has been chosen as the central meridian. Along this meridian the parallels preserve the same distances as they do upon the sphere.

In Fig. 18 let M be any point of the central meridian of which the latitude is $\theta = 90° - u$, P the pole; the arc $PM = ru$. The center of the parallel through the point M is given by $CM = r \tan u$.

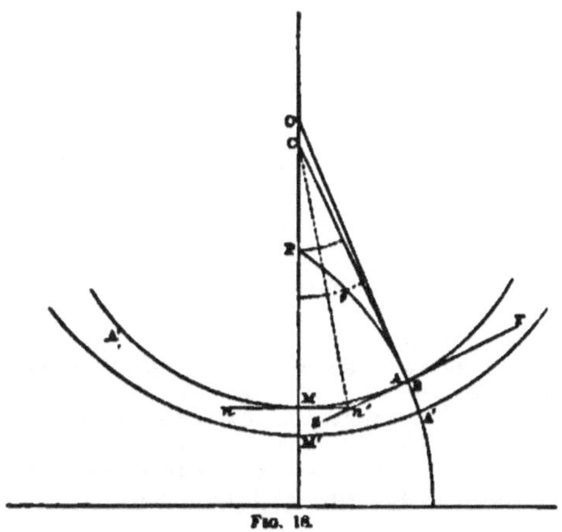

Fig. 18.

If OM′ be a point infinitely near to M, *i. e.*, $MM' = rdu$, and c' the center of the corresponding circle, we have

$$C'M' = r \tan (u + du)$$

or

$$\rho = r \tan u \qquad \rho + d\rho = r \tan (u + du)$$

Expanding the second of these we have

$$d\rho = r \sec^2 u du$$

but

$$d\rho = CC' + MM' = CC' + rdu$$

therefore

$$CC' = r \tan^2 u du$$

We have from the triangle CC′B

$$\frac{\sin \varphi}{\sin B} = \frac{C'B}{CC'}$$

or

$$-\frac{d\varphi}{\sin \varphi} = \tan s \, ds$$

and integrating

$$\log \cos u = \log \tan \frac{\varphi}{2} + \text{const.}$$

or, passing to exponentials,

$$\tan \frac{\varphi}{2} = C \cos u$$

Since

$$\tan u = \frac{\rho}{r}$$

therefore

$$\cos u = \frac{\rho}{\sqrt{r^2 + \rho^2}}$$

Substituting this in the equation for the meridians, we have

$$\frac{r\sqrt{C^2-\tan^2\frac{\varphi}{2}}}{\tan\frac{\varphi}{2}}=\rho$$

or

$$\rho=r\sqrt{C^2\cot^2\frac{\varphi}{2}-1}=\frac{r}{\sin\frac{\varphi}{2}}\sqrt{-\left[C^2+(1-C^2)\sin^2\frac{\varphi}{2}\right]}$$

The distance from any point A to the central meridian is $=\rho\sin\varphi$, or $=r\tan u\sin\varphi$; but

$$r\tan u\sin\varphi=2C\frac{r\sin u}{1+C^2\cos^2 u}$$

For $u=90°$, or at the equator, this becomes $=2Cr$. The constant C must then represent one-half the longitude of the given meridian, the equator being developed in its true length and divided into equal parts in the same manner as the central meridian. The following construction for this projection is due to Mr. O'Farrell, of the topographical department of the War Office, England. All data being as already given, draw at M the tangent nn_1 perpendicular to PM. In order to determine the point A, whose longitude is given as ω, lay off from M the lengths $Mn=Mn_1$ equal to the true length of the required arc on the parallel θ, *i. e.*, equal to the arc $\frac{\omega}{2}$ described with a radius $=r\sin u$. With n and n_1 as centers, and n_1C and nC as radii, draw arcs cutting the given parallel in the points A and A_1. Now

$$Mn=\frac{\omega}{2}\sin u=C\sin u$$

and, since $CM=r\tan u$, we have

$$\tan MCn=C\cos u=\tan\frac{\varphi}{2}$$

or, finally,

$$ACM=\varphi$$

and the distance from A to the central meridian is

$$=r\tan u\sin\varphi$$

The radius of curvature of the meridian whose longitude is ω is readily obtained. We have $AA'=ds$ and $CC'=r\tan^2 u\,du$. We have then

$$ds=r(\sec^2 u-\tan^2 u\cos\varphi)\,du=r\left[2(1+C^2)\cos^2\frac{\varphi}{2}-1\right]du$$

Also

$$\sec^2\frac{\varphi}{2}\,d\frac{\varphi}{2}=-C\sin u\,du$$

Therefore, if ρ denote the radius of curvature of the meridian, we will have

$$\rho=r\frac{1+C^2+C^2\sin^2 u}{2C\sin u}$$

Now consider the distortion in this case, and for this purpose imagine a small square described on the sphere having its sides parallel and perpendicular to the meridian. Let u and ω $(=2C)$ define its position, and let σ be the length of the side. If we differentiate the equation $\tan\frac{\varphi}{2}=C\cos u$, on the supposition that u is constant, we have

$$\sec^2\varphi\,d\varphi=\cos u\,dC$$

also the length of the representation of $2dC$ is $\tan u\, d\varphi$, or

$$\sin^2 u \cos^2 \frac{\varphi}{2} d\, 2C$$

Hence that side of the square which is parallel to the equator will be represented by a line equal to $\varepsilon \cos^2 \frac{\varphi}{2}$. Similarly the meridian side will be represented by

$$\varepsilon \cos^2 \frac{\varphi}{2} (1+C^2+C^2 \sin^2 u)$$

The square is therefore represented by a rectangle whose sides have the ratio

$$1+C^2+C^2 \sin^2 u : 1$$

and its area is increased in the ratio

$$\frac{1+C^2+C^2 \sin^2 u}{(1+C^2 \cos^2 u)^2} : 1$$

If we make this ratio = unity, there results the equation

$$C^4 \cos^4 u+3 C^2 \cos^2 u-2 C^2=0$$

which is satisfied either by $C=0$ (*i. e.*, $\omega=0$) or by

$$C^2 \cos^2 u+3 \cos^2 u-2=0$$

We see from this that there is no exaggeration of area along the meridian or along the curve given by the last equation. This curve crosses the central meridian at right angles in the latitude of about 54° 44′; it thence slowly inclines southward, and at 90° of longitude from the central meridian reaches 50° 20′ of latitude; at 180°, or the opposite meridian, it has reached 43° 40′. The areas of all tracts of countries lying on the north side of this curve will be diminished in the representation, and for all tracts of countries south of this curve the areas will be increased in the representation.

If we represent the whole surface of the globe continuously, the area of the representation is

$$r^2\left[(4+\pi^2) \tan^{-1} \frac{\pi}{2}+2\pi\right]$$

which is greater than the true surface of the globe in the ratio 8 : 5.

The perimeter of the representation is equal to the perimeter of the globe multiplied by $\sqrt{1+\pi^2}-1$ or 2.72. It is desirable in certain cases to retain the lengths of the degrees on all the parallels at the sacrifice of their perpendicularity to the meridians. We thus obtain what is known as the ordinary polyconic projection, which applied to the representation of the entire surface of the globe gives a figure with two rectangular axis and from equal quadrants, as in the rectangular polyconic projection. The central meridian alone is perpendicular to the parallels, and is developed in its true length; upon each parallel described with the cotangent of its latitude as a radius, we lay off the true lengths of the degrees of longitude and draw through the corresponding points so obtained curves which will be the projections of the meridians.

The ordinary polyconic method has been adopted by the United States Coast Survey because its operations being limited to a narrow belt along the seaboard, and not being intended to furnish a map of the country in regular uniform sheets, it is preferred to make an independent projection for each plane table and hydrographic sheet by means of its own central meridian.

The method of projection in common use in the Coast Survey Office for small areas, such as those of plane-table and hydrographic sheets, is called the equidistant polyconic projection. This is to be regarded rather as a convenient graphic approximation, admissible within certain limits, than as a distinct projection, though it is capable of being extended to the largest areas and with results quite peculiar to itself. In constructing such a projection, a central meridian and a central parallel are chosen, and they are constructed as in the rectangular polyconic method. The top or bottom parallel and a sufficient number of intermediate parallels are constructed by means of the tables prepared for the purpose, and the points of intersection of the different meridians with these parallels are then found and the meridians drawn. Then starting from the central parallel the distance to the next parallel is taken from the central meridian and laid off on each other meridian. A parallel is traced through the points thus found. Each parallel is constructed by laying off equal distances on the meridians in like manner, and the tabular auxiliary parallels are, all except the central one, erased. In fact, as only the points of intersection are required, the auxiliary parallels should not be actually drawn. From this process of construction results a projection in which equal meridian distances are intercepted everywhere between the same parallels. As large and extensive tables are required for the actual construction of this projection, and as such are already in use in the United States Coast Survey Office, the author has not thought it desirable to append such to this treatise. There are two such tables employed in the Coast Survey and the United States Navy. The first set is contained in the Report of the Superintendent of the United States Coast Survey for 1853; the second was published by the Bureau of Navigation in 1869. Both of these sets were prepared in the Coast Survey Office and both are in use in that institution. A new set is now in preparation in which the most approved values of the ellipticity of the earth (Colonel Clarke's) is employed.

CYLINDRIC PROJECTIONS.

Cylindric projections may be derived in several different ways, according as the cylinder to be developed is tangent to the sphere or is a secant cylinder. In the case of tangency the line of contact may be either the equator or any one of the meridians; if the cylinder is secant to the sphere, it may pass through either the upper or lower parallel of the zone to be projected, or (the plan usually adopted) it may be made to pass through some intermediate parallel. Consider first the case when the cylinder is tangent to the sphere.

The square projection.

Here the cylinder is tangent along the equator and the meridians and parallels are represented as equidistant generators and right sections of the cylinder; after development, both of these systems of lines will be represented as straight lines forming a network of equal squares. Distances are grossly exaggerated, particularly in an east-and-west direction, though for an elementary surface the true proportions are preserved. This projection is occasionally used to represent small areas near the equator, and for this purpose it is obviously accurate enough. The construction is so simple that no description is necessary.

Projection with converging meridians.

This is a modification of the square projection designed to conform nearly to the condition that the arcs of longitude shall appear proportional to the cosines of their respective latitudes. The straight line representing the central meridian being properly graduated, that is, the true length, by scale, of a degree of longitude (or of a minute or multiple thereof, as the case may be) having been laid off according to the scale adopted, two straight lines are drawn at right angles to the meridians to represent parallels, one near the bottom and the other near the top of the chart. These parallels are next graduated, the arcs representing degrees (multiples or subdivisions) of longitude on each having by scale the true length belonging to the latitude. The corresponding points of equal nominal angular distance from the middle meridian thus marked upon the parallels, when connected by straight lines, will produce the system of converging meridians. The disadvantages of this projection are that but two of the parallels exhibit the length of arcs of longitude in their true proportion and that the central meridian is alone at right angles to the parallels. This projection is nevertheless suitable for the representation of tolerably large areas, the above defects not being of a serious nature within ordinary limits. It also recommends itself on account of the ease with which points can be projected or taken off the chart by means of latitude and longitude.

The rectangular projection.

A less defective delineation than the square projection consists in presenting the length of degrees of longitude along the *middle* parallel of the chart in their true relation to the corresponding degrees on the sphere; they will therefore appear smaller than the length of the degrees of latitude in the proportion $1 : \cos \varphi$. In an east-and-west direction the chart is unduly expanded above and unduly contracted below the middle parallel.

The rectangular equal-surface projection.

This differs from the first in that the distances of the parallels, instead of being equal, are now drawn parallel to the equator, at distances proportional to the sine of the latitude. This gives it the distinctive property of the areas of rectangles or zones on the projection being proportional to the areas of the corresponding figures on the sphere. The distortion, however, becomes quite excessive in the higher latitudes.

Cassini's projection.

This projection makes no use of the parallels of latitude, but substitutes for them a second system of co-ordinates, viz, one at right angles to the principal or central meridian; it is consequently convenient in connection with rectangular spherical co-ordinates having their origin in the middle of the chart; the projection of Cassini's chart of France consisted of squares, and had neither meridians (excepting one) nor parallels. It would seem, however, that in this simple form it is not the projection generally distinguished by this name. It has been described as follows: Suppose a cylindrical surface with its generating line at *right angles* to the central meridian and enveloping the sphere along this meridian. This cylindric surface is supposed intersected by planes *parallel* to that of the central meridian and these intersections produce on the chart, after development of the cylinder, the straight representatives of meridians, but are in reality small circles on the sphere. Their distance from each other is defined by passing them through equal divisions of the prime vertical drawn through the center of the chart. The central meridian having been equally divided, the equidistant straight lines passing through these divisions form the prime vertical system. This projection is not now employed, as it offers no facilities for plotting positions by latitude and longitude; moreover, the distortion rapidly increases with distance from the center meridian of the chart.

In Fig. 19 let M denote the center of the sphere of radius $MA=r$; P an arbitrary point of the surface of which the latitude is $EP=AD=\theta$ and the longitude $DP=AE=\omega$; AB denotes a

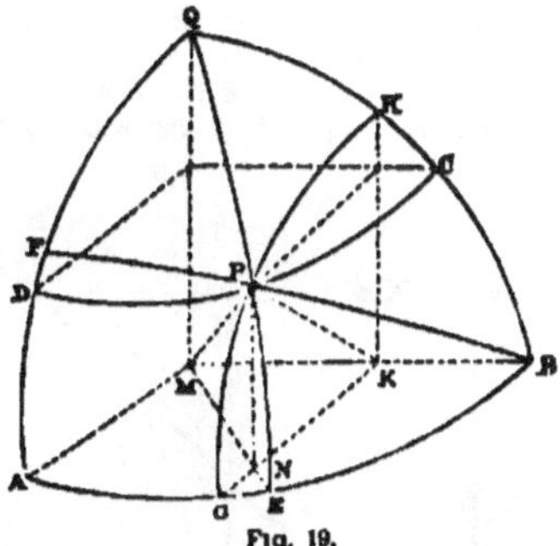

FIG. 19.

quadrant of the equator, and AQ a quadrant of the first meridian. The determination of the position of P is effected by means of the great circle passing through B and P and the circle GH, whose plane is parallel to that of the first meridian AQ. Write

$$FP=AG=\sigma \qquad GP=AF=\pi$$

then we have the following relations between all of these quantities:

$$\sin\theta_1=\cos\theta\sin\omega \qquad \sin\theta=\cos\theta_1\sin\omega_1$$

$$\cot\omega_1=\cot\theta\cos\omega \qquad \cot\omega=\cot\theta_1\cos\omega_1$$

Now, in Cassini's projection O, O′, Fig. 20, denote the center of co-ordinates

$$\eta=\mathrm{OA}=\mathrm{O'A'} \qquad \xi=\mathrm{AM}=\mathrm{A'M'}$$

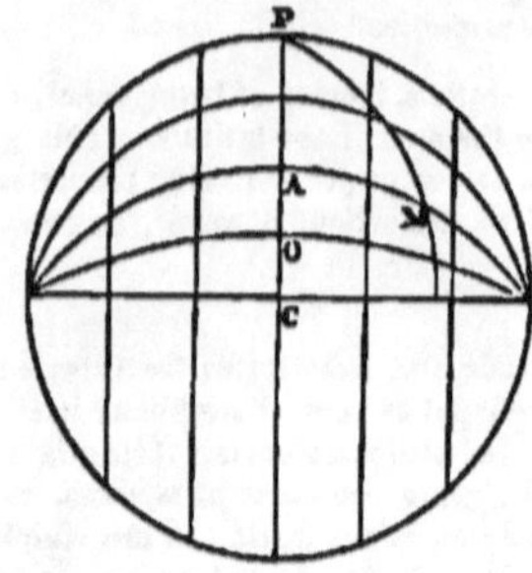

Fig. 20.

also, let θ_0 denote the latitude of O; then $\theta_0+\eta$ is the quantity denoted by ω_1 in the preceding formulas, and ξ is identical with θ_1; so we have

$$\sin\xi=\cos\theta\sin\omega \qquad \cot(\theta_0+\eta)=\cos\omega\cot\theta$$

By eliminating successively θ and ω from these formulas, we obtain the equations of the meridians and parallels respectively:

$$[\cos^2\omega+\cot^2(\theta_0+\eta)][\sin^2\omega-\sin^2\xi]=\tfrac{1}{4}\sin^2 2\omega \qquad \sin^2\xi+\cot^2(\theta_0+\eta)\sin^2\theta=\cos^2\theta$$

When the projection only represents a narrow region included between two meridians and two parallels very near together, the ratios $\frac{\xi}{r}$ and $\frac{\eta}{r}$ are very small, and so is the difference $\theta-\theta_0$. In this case it is found that the meridians are (sufficiently accurately) projected in parabolas, and the parallels in circles. A fuller mathematical investigation of this projection is given in Part I.

Write, for convenience, $\theta_0+\eta=\lambda$, and let the angle χ in Fig. 21 denote the angle which the tangent PM to the projection of a meridian makes with the axis of η; also denote by χ' the angle which the tangent PL to the projection of a parallel makes with the same axis. Then we have

$$\tan\chi=\frac{d\xi}{d\lambda}$$

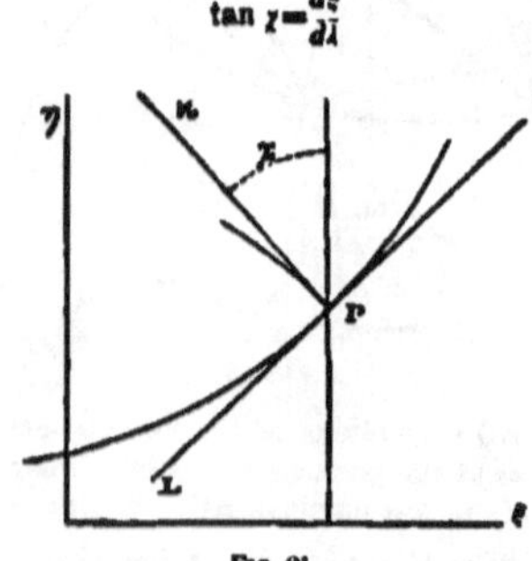

Fig. 21.

The equation of the meridians is easily thrown into the form

$$\tan \xi = \cos \lambda \tan \omega$$

and that of the parallels

$$\sin \theta = \sin \lambda \cos \xi$$

For these we may substitute

$$\cot \lambda = \cot \theta \cos \omega \qquad\qquad \sin \xi = \cos \theta \sin \omega$$

We have now

$$\frac{d\xi}{d\lambda} = -\frac{\tan \omega \sin \lambda}{1 + \cos^2 \lambda \tan^2 \omega} = \tan \chi$$

and also

$$\tan \chi = -\tan \omega \cos \xi \sin \theta$$

Again, from the equation of the parallels there results

$$\frac{d\xi}{d\lambda} = \frac{1}{\tan \lambda \tan \xi} = \tan \chi'$$

or, since $\tan \xi = \cos \lambda \tan \omega$,

$$\tan \chi' = \frac{1}{\sin \lambda \tan \omega} = \frac{\cos \xi \cot \omega}{\sin \theta}$$

Multiplying together these values of $\tan \chi$ and $\tan \chi'$ we have

$$\tan \chi \tan \chi' = \cos^2 \xi$$

The condition for orthogonality between the projections of the meridians and parallels is

$$\tan \chi \tan \chi' = 1$$

so that in general in Cassini's projection the meridians and parallels are not represented by orthogonal curves.

For $\xi = 0$, we find

$$\tan \chi \tan \chi' = 1$$

and from this it is clear that the projections of all parallels are perpendicular to the central meridian. If $\lambda = 90°$, we have $\chi = \omega$, or the projections of meridians make the same angles with each other as the meridians themselves. Obtaining from the equations for meridians and parallels the values of $\frac{d\eta}{d\xi}$ and $\frac{d^2\eta}{d\xi^2}$ and substituting in the formula

$$\rho = \frac{\left[1 + \left(\frac{d\eta}{d\xi}\right)^2\right]^{\frac{3}{2}}}{\frac{d^2\eta}{d\xi^2}}$$

for radius of curvature, we readily find for the radius of curvature of the projection of the meridians

$$\rho_m = \frac{r \sec \chi}{\sin \chi \left[\cot \lambda \cos \chi + 2 \tan \xi \sin \chi\right]}$$

and for that of the projection of the parallels

$$\rho_p = \frac{r \sin^3 \lambda \sin^3 \xi}{\cos^3 \chi' \cot \xi \left(\sin^3 \xi + \cot \lambda\right)}$$

MERCATOR'S PROJECTION.

This projection is the one most commonly employed at sea, its great convenience being that upon it all rhumb lines of the sphere are represented as straight lines. The rhumb line or loxodromic is a line upon the surface of the sphere which cuts all of the meridians at the same angle. Upon the Mercator or *reduced chart* all the meridians and parallels are given as right lines; the meridians are all parallel to each other and at equal distances apart, and the parallels are at unequal distances from each other, the distance between any two parallels increasing with the latitude. The distance between any two consecutive meridians is equal to the distance between the meridians measured on the equator, and the distance from the equator to any parallel of latitude is expressed in terms of minutes of arc of the equator. Call s the distance from the equator to the parallel of latitude θ; we have

$$s = 7915'.704674 \log \tan\left(45^\circ + \frac{\theta}{2}\right) - 3437'.7\left(e^2 \sin\theta + \frac{e^4 \sin^3\theta}{3}\right)$$

Table VIII gives the values of s for every second. It is, of course, understood that this expression has to be reduced to linear measure by multiplying it by the length on the adopted scale of a minute of arc of the equator.

The construction of the projection is now quite simple; a horizontal line is drawn to represent the equator or as much of it as the projection calls for; this is divided into equal parts, degrees or minutes, according to the chosen scale; perpendiculars are erected at the points of division to represent the meridians, and upon them are laid off the distances s taken from Table VIII, which determine the position of the parallels.

To determine the distance between any two points A and B.

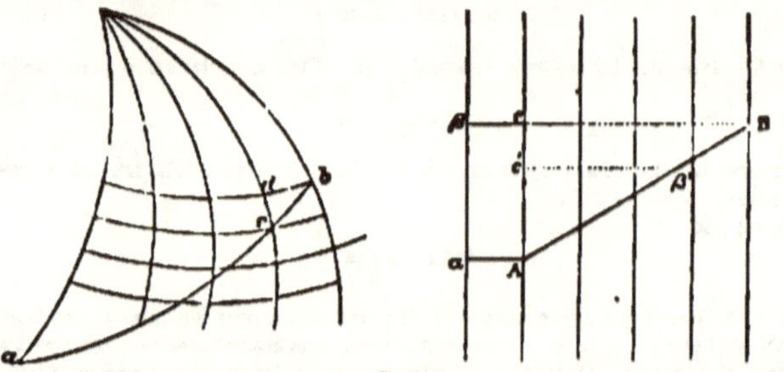

FIG. 43.

In Fig. 43, draw the corresponding loxodromic upon the sphere and divide it into equal parts; through each point of division draw a parallel and meridian; these curves then form a series of equal triangles b, c, d, since ab is equally inclined to all these meridians. The sum of the meridian sides of the triangles is equal $ab \times$ sine of the angle which the loxodromic makes with the meridians. This sum is also equal to the difference of latitude of the points a and b, multiplied by the radius of the sphere (returning again to the sphere for simplicity and a sufficient degree of exactness). Now, upon the scale of equal parts we take a length equal to the difference of latitude of the two points under consideration, *i. e.*, as many minutes of the equator as there are minutes in the α, β, determined by drawing parallels through A and B. Lay off this distance upon the meridian through A, obtaining, suppose, AC'; through C' draw the parallel B'C'; the distance AB' evaluated in minutes upon the scale of equal parts will be the distance required; multiplying by $1855^m.1$ the length of one minute of the equator will give the distance in meters.

In following a loxodromic curve upon the sphere the navigator obviously does not follow the shortest path between any two points, for, as we know, the geodesics upon a sphere are the great circles. In order, then, that a vessel shall take the shortest distance between any two points, she must follow the arc of a great circle joining those points, or at least must follow successive portions of different loxodromic curves which coincide as nearly as may be with this arc of a great circle. It will be necessary now to determine the rhumb line or continuous series of rhumb lines which the navigator must follow in passing from one end of this arc of a great circle to the other. Let A and B respectively represent the points of departure and arrival, φ and φ' the colatitudes of these points, and ω the difference of longitude.

We wish now to find the angle made by the loxodromic curve with this arc of a great circle AB at the point A; then sailing along an infinitesimal distance on this loxodromic tangent we find, knowing the latitude and longitude of the points arrived at, the angle which a new loxodromic tangent makes with the arc of a great circle passing through this point and the final point B, and so on until arriving at B. Suppose that we have a tetrahedron with vertex at the center of the

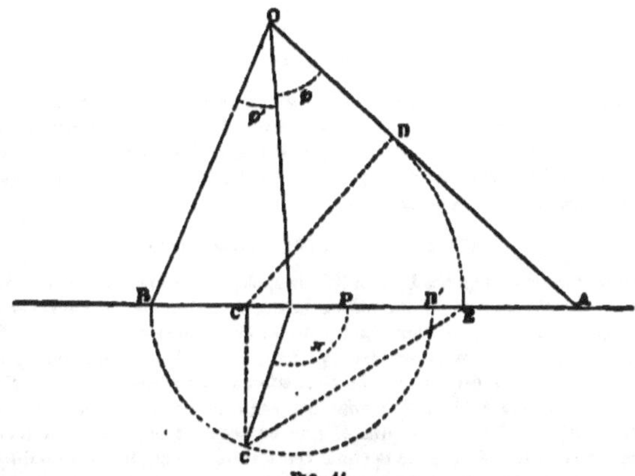

FIG. 44.

sphere and the faces which meet at this point cutting from the sphere arcs of great circles which are respectively the polar distances of A and B and the geodesic distance between A and B. Two of the plane angles at the vertex of this tetrahedron are respectively $=\varphi$ and φ', and we also know the diedral angle π, which is the difference of longitude of A and B. It is required to find the diedral angle PAB opposite the face φ'. Draw two right lines BA and PO at right angles to each other; from O draw OB, making the angle POB$=\varphi'$ and also draw OA, making POA$=\varphi$; from the point P describe a circle with PB as radius; draw PC, making the angle APC$=\omega$, the difference of longitude of the points of arrival and departure; draw CC' perpendicular to AB and then C'D perpendicular to OA; with C' center and C'D as radius, describe an arc of a circle DE cutting AB in E, and join EC; the angle BEC will be the angle sought.*

The geographical co-ordinates of the points of intersection of an arc of a great circle with the meridians chosen arbitrarily can be readily effected with a sufficient degree of exactness by constructing an auxiliary projection upon which the meridians and the arc required are easily traced; the points of intersection will thus be found and their latitudes and longitudes readily determined. Applying these so-found latitudes and longitudes to the reduced chart, we will be able to trace the representation of the arc of a great circle by merely drawing a certain series of short straight lines.

EQUIVALENT PROJECTIONS.

The condition to be fulfilled in this class of projections is the equivalence of an elementary quadrilateral upon the spheroid with the corresponding quadrilateral upon the map. Let ρ represent the radius of parallel of latitude θ, s the meridional distance of a point from the pole, and as usual let

* This construction is given by Germain, Traité des Projections, p. 288.

ω denote longitude; then for the area of an indefinitely small quadrilateral included between two meridians and two parallels infinitely near together, we have

$$r\rho\, ds\, d\omega$$

Since ρ and s are functions of θ, write

$$r\rho\, ds = \Theta d\theta$$

when Θ is a function of θ; the element of area is now $= \Theta\, d\theta\, d\omega$. For the earth we have readily, ϵ denoting the ellipticity,

$$\Theta = \frac{r^2 (1-\epsilon^2) \cos\theta}{(1-\epsilon^2 \sin^2\theta)^2}$$

Denoting by ξ and η the co-ordinates of a point on the map, we have for the condition of this class of projections

$$\frac{d\xi}{d\omega}\frac{d\eta}{d\theta} - \frac{d\xi}{d\theta}\frac{d\eta}{d\omega} = \Theta$$

This equation is studied at length in Part I and numerous applications made to different projections, but here we are only going to take up two or three of the most important of these equivalent projections and treat them in full; the others, which have been examined in Part I, are rather more curious than useful. The central equivalent projection is treated so simply by Collignon that I have here reproduced most of his work.

CENTRAL EQUIVALENT PROJECTION.

The projection that we designate by this title is spoken of by Germain as the "zenithal equivalent," but in adopting the above title the author has preferred to choose a term as nearly as possible like that adopted by Collignon when he described the projection; this was, "Système central d'égale superficie."[1] This system is founded upon the principle of elementary geometry that the area of a zone is equal to the product of the circumference of a great circle by the height of the zone. The same law of area holding for a spherical segment or zone of one base, we have (representing by h the altitude of the zone) area of zone or segment $= \pi 2rh$. But $2rh =$(chord of half the arc)^2; therefore the area of the zone is equal the area of the circle whose radius is equal to the rectilinear distance from the pole of the zone to the circumference which serves as a base. If from the pole of the zone we draw two arcs of great circles including a certain definite angle, and from the center of the equivalent circle two radii including the same angle, the portion of the zone bounded by its base and these two arcs will be equal to the sector of the circle cut out by the two corresponding radii. This gives us, then, an obvious manner of representing any portion of a given spherical surface without alteration of area. Any point can be assumed upon the sphere as center; so, for simplicity, the pole of the equator is chosen. The parallels are seen to be transformed into concentric circles, and the meridians into straight lines passing through the common center.

Taking now the projection of the principal meridian as the axis of ξ, and writing $\varphi = 90° - \theta$, we have for the equation of the meridians

$$\eta = \xi \tan\varphi$$

and for the parallels

$$\xi^2 + \eta^2 = 4r^2 \sin^2\frac{\varphi}{2}$$

from which

$$\xi = 2r \sin\frac{\varphi}{2} \cos\omega \qquad \eta = 2r \sin\frac{\varphi}{2} \sin\omega$$

and consequently

$$\frac{d\xi}{d\omega} = -2r \sin\frac{\varphi}{2} \sin\omega \qquad \frac{d\xi}{d\theta} = -r \cos\frac{\varphi}{2} \cos\omega$$

$$\frac{d\eta}{d\omega} = 2r \sin\frac{\varphi}{2} \cos\omega \qquad \frac{d\eta}{d\theta} = -r \cos\frac{\varphi}{2} \sin\omega$$

[1] Journal de l'École polytechnique, cahier 41: Représentation de la surface du globe terrestre; E. Collignon.

Substituting these in our general differential equation,

$$\frac{d\xi}{d\omega}\frac{d\eta}{d\theta}-\frac{d\xi}{d\theta}\frac{d\eta}{d\omega}=\theta$$

we find

$$\frac{d\xi}{d\omega}\frac{d\eta}{d\theta}-\frac{d\xi}{d\theta}\frac{d\eta}{d\omega}=2\,r^2\sin\frac{\varphi}{2}\cos\frac{\varphi}{2}\sin^2\omega+2\,r^2\sin\frac{\varphi}{2}\cos\frac{\varphi}{2}\cos^2\omega=r^2\sin\varphi=r^2\cos\theta$$

which verifies our supposition of equal areas. It is also easy to see that

$$\frac{d\xi}{d\omega}\frac{d\xi}{d\theta}+\frac{d\eta}{d\omega}\frac{d\eta}{d\theta}=0$$

or the meridians and parallels cut at right angles on the chart as on the sphere.

ALTERATION OF ANGLES.

The alteration of angles is zero at the center of the chart. At any point whatever, M, of the chart, Fig. 30, draw a line MM′ such that the corresponding direction upon the sphere shall make an angle θ with the meridian; we wish to find the angle V upon the chart made by this line with the projection of the meridian, *i. e.*, with the line drawn from M to the center O. Let (θ, ω) represent the geographical co-ordinates of M, and $(\theta+d\theta, \omega+d\omega)$ the geographical co-ordinates of M′, infinitely near to M; then

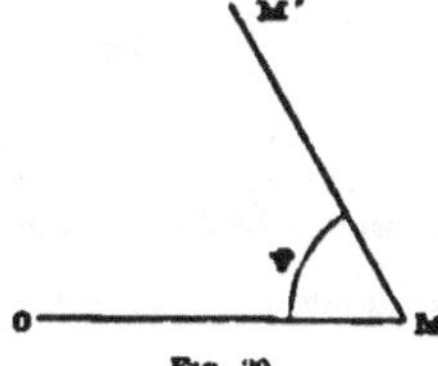

FIG. 30.

$$\tan\theta=\cos\theta\frac{d\omega}{d\theta}=\sin\varphi\frac{d\omega}{d\theta}\qquad\qquad\tan V=2\tan\frac{\varphi}{2}\frac{d\omega}{d\theta}$$

From which

$$\tan V=\frac{\tan\theta}{\cos^2\frac{\varphi}{2}}$$

The maximum of alteration $V-\theta$, or ϕ, corresponds to the direction for which

$$V+\theta=\frac{\pi}{2}$$

For

$$\tan\phi=\frac{\tan V-\tan\theta}{1+\tan V\tan\theta}=\frac{\tan\theta\left(1-\cos^2\frac{\varphi}{2}\right)}{\cos^2\frac{\varphi}{2}+\tan^2\theta}$$

and in seeking for the maximum of this, since $1-\cos^2\frac{\varphi}{2}$ is constant, we need only consider the factor

$$\frac{\tan\theta}{\cos^2\frac{\varphi}{2}+\tan^2\theta}$$

Equating to zero the derivative of this with respect to θ, there results simply

$$\tan\theta = \pm\cos\frac{\varphi}{2}$$

Consequently

$$\tan\Psi = \pm\frac{1}{\cos\frac{\varphi}{2}}$$

the upper signs being taken together, and also the lower ones. From these follows

$$\tan\theta\tan\Psi = 1$$

which, as in a former case, excluding negative arcs and arcs greater than $\frac{\pi}{2}$, gives

$$\theta + \Psi = \frac{\pi}{2}$$

We can deduce from this that the maximum deviation for the direction OM is given by

$$\tan(\Psi - \theta) = \tfrac{1}{2}(\tan\Psi - \tan\theta) = \tfrac{1}{2}\tan\frac{\varphi}{2}\sin\frac{\varphi}{2}$$

The angle θ, upon the sphere, of maximum deviation is $=45°$ for $\varphi = 0$, *i. e.*, at the center of the chart; θ then decreases while Ψ, and consequently θ, increases. When $\varphi = \frac{\pi}{2}$

$$\tan\theta = \frac{1}{\sqrt{2}} \qquad \tan\Psi = \sqrt{2}$$

The angular alteration is thus seen to increase continuously from the center to that point of the sphere which is diametrically opposite the assumed center. It is evidently useless to prolong the chart so far as that, and, indeed, the custom is in this projection to represent the map in two parts, one for each hemisphere.

ALTERATION OF LENGTHS.

In the direction OM the projection substitutes for the arc on the sphere the chord of the same arc. As usual, let φ represent the angular distance OM, then the length of this line upon the sphere is $=r\varphi$, and its length upon the chart, *i. e.*, the length of the chord of the arc OM, is $=2r\sin\frac{\varphi}{2}$. Differentiation of each of these gives us the lengths of the element of the meridian upon the sphere and upon the chart; these are $rd\varphi$ and $r\cos\frac{\varphi}{2}\,d\varphi$. Thus the meridional elements are reduced upon the chart in the ratio $\cos\frac{\varphi}{2}:1$. The converse is true concerning the elements of the parallels; they are augmented in the ratio $1:\cos\frac{\varphi}{2}$; this is obvious on account of the necessity for conserving the areas. Suppose now that upon the sphere we take any element ds, making the angle θ with the meridian OM; its projection upon MO will $=ds\cos\theta$, and perpendicular to MO will be $=ds\sin\theta$; similarly, if $d\sigma$ correspond upon the chart to ds upon the sphere, $d\sigma\cos\Psi$ will be the projection of $d\sigma$ upon the radius OM, and $d\sigma\sin\Psi$ will be the projection of the same element in the direction perpendicular to OM. Now, since the projection does not alter the right angle at which $ds\cos\theta$ and $ds\sin\theta$ cut each other, we will have

$$ds\cos\theta\cos\frac{\varphi}{2} = d\sigma\cos\Psi \qquad ds\sin\theta\frac{1}{\cos\frac{\varphi}{2}} = d\sigma\sin\Psi$$

from which by squaring and adding

$$d\sigma^2 = ds^2\left(\cos^2\theta\cos^2\frac{\varphi}{2} + \sin^2\theta\cdot\frac{1}{\cos^2\frac{\varphi}{2}}\right)$$

Now, the expression in parenthesis reduces to unity when, upon the sphere,

$$\tan\theta = \cos\frac{\varphi}{2}$$

or when upon the chart

$$\tan \Upsilon = \frac{1}{\cos\frac{\varphi}{2}}$$

that is for the direction of maximum deviation. This direction then possesses the remarkable property of conserving the lengths. Now, through any given point upon the sphere, and upon the chart, as M, we can draw two curves which shall cut all the meridians MO of the sphere and the radii MO of the chart under the angles θ and Υ in such a way that the distances on these two curves between any two corresponding points shall be the same. The curves so constructed are called by Collignon isoperimetric curves. The curve upon the sphere passes through O', the antipodal point to O, and winding round the sphere becomes indefinitely near to O, a logarithmic spiral which cuts the meridians at an angle of 45°. Upon the chart the isoperimetric curve, for small values of φ, that is, for points near the center, is very nearly the logarithmic spiral which cuts the radii under the angle of 45°; for increasing values of φ, θ also increases, and is $=90°$ for $\varphi=180°$; the curve then touches the circle into which the point O' has been transformed, and is continued beyond this point in a branch symmetrical to the first.

To obtain the polar equation of the isoperimetric curve upon the chart, take $\rho=$OM and α the angle between ρ and some fixed axis. Now

$$\rho\frac{d\alpha}{d\rho} = \tan\Upsilon = \frac{1}{\cos\frac{\varphi}{2}}$$

but

$$\rho = 2r\sin\frac{\varphi}{2}$$

therefore

$$d\alpha = \frac{d\rho}{\rho\sqrt{1-\frac{\rho^2}{4r^2}}}$$

the differential equation of the sought curve. For the integration observe that we have

$$d\rho = r\cos\frac{\varphi}{2}d\varphi$$

which substituted in the first written equation gives

$$d\alpha = \frac{\frac{d\varphi}{2}}{\sin\frac{\varphi}{2}}$$

and by integration

$$\alpha = \log\tan\frac{\varphi}{4} + c$$

This equation joined with

$$\rho = 2r\sin\frac{\varphi}{2}$$

gives the means of constructing the curve.

For the element of arc of the isoperimetric curve we have obviously

$$ds = \sqrt{d\rho^2+\rho^2 d\alpha^2} = d\varphi\sqrt{r^2\cos^2\frac{\varphi}{2} + 4r^2\sin^2\frac{\varphi}{2}\cdot\frac{1}{4\sin^2\frac{\varphi}{2}}}$$

or

$$ds = r\,d\varphi\sqrt{1+\cos^2\frac{\varphi}{2}}$$

If we write $\frac{\varphi}{2}=\theta$, this equation becomes very simply

$$ds=\sqrt{2}r\sqrt{1-\tfrac{1}{2}\sin^2\theta}\,d\theta$$

or

$$s=\sqrt{2}r\int\Delta(k\theta)d\theta \qquad k=\sqrt{\tfrac{1}{2}}$$

an elliptic integral of the second kind, which gives the rectification of the arc of the ellipse, whose eccentricity is $=\sqrt{\tfrac{1}{2}}$. The element of area of the isoperimetric curves is, in polar co-ordinates,

$$\tfrac{1}{2}\rho^2 d\alpha=r^2\sin\frac{\varphi}{2}d\varphi$$

and the integral of this is

$$=\text{const.}-2r^2\cos\frac{\varphi}{2}$$

TRANSFORMATION OF A GREAT CIRCLE.

The angle between the planes of two great circles on the sphere is measured by the arc of a great circle joining their poles. This property affords the means of determining the differential equation of the curve upon chart, which represents the great circle on the sphere.

Take O, Fig. 31, for the central point, and P for the pole of a great circle which passes through a point M. The same letters accented denote the corresponding points upon the chart. It is proposed at M′ to draw a tangent to the curve which passes through this point and represents the great circle through M. Join O′M′, and call $O'M'=\rho$ and $M'O'P'=\alpha$, the line O′P′ being taken as the initial line. Let S, upon the sphere, denote the pole of the great circle OM, which passes through the center O and cuts the given circle at M; this point S will be found in the plane of a great circle OS perpendicular to that of OM at the point O; the angle V is measured by the arc SP. We have now in the spherical triangle OSP

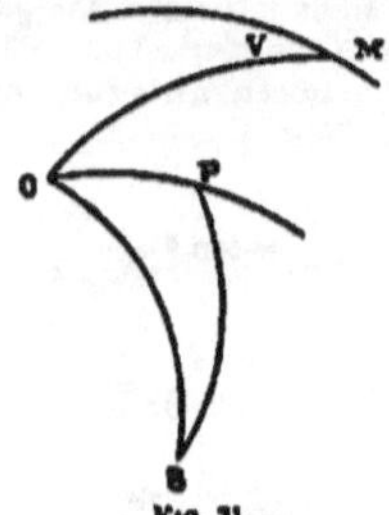

FIG. 31.

$$\cos SP=\cos OS\cos OP+\sin OS\sin OP\cos POS$$

or, since OS is a quadrant,

$$\cos SP=\sin OP\cos POS=\sin OP\sin\alpha$$

OP is a constant arc that we may call λ; then we have

$$\cos V=\sin\lambda\sin\alpha$$

The angle V on the sphere of course corresponds with V upon the chart, and the connecting relation is

$$\tan V=\frac{\tan V'}{\cos^2\frac{\varphi}{2}}$$

φ being the angular distance OM. But

$$\tan V' = \frac{\rho d\alpha}{d\rho} \qquad \rho = 2 \sin \frac{\varphi}{2}$$

taking the radius of the sphere as unity. Eliminating V, V′, and φ between these four equations we will arrive at the differential equation sought. The first of these equations affords the relation

$$\sin V = \sqrt{1 - \sin^2 \lambda \sin^2 \alpha}$$

and consequently

$$\tan V = \frac{\sqrt{1 - \sin^2 \lambda \sin^2 \alpha}}{\sin \lambda \sin \alpha}$$

and then

$$\frac{\rho d\alpha}{d\rho} = \frac{\sqrt{1 - \sin^2 \lambda \sin^2 \alpha}}{\sin \lambda \sin \alpha} \frac{1}{\sqrt{1 - \frac{\rho^2}{4}}}$$

The constant of integration will be determined by observing that the great circle, of which P is the pole, passes through the pole of the great circle OP; so that for $\theta = \frac{\pi}{2}$ or $\frac{3\pi}{2}$ we should have $\rho = \sqrt{2}$.

The equation of the projected great circle can be better arrived at in another manner, to the the explanation of which we shall now proceed.

Conceive first that a stereographic projection has been made—that is, the parallels and meridians have been constructed—with the point of sight at the center, or the antipodal point to the center, of the proposed central equivalent projection. Let E_1, Fig. 32, denote any point of the stereographic projection; O_1, the center, or point of sight, represented on the central equivalent

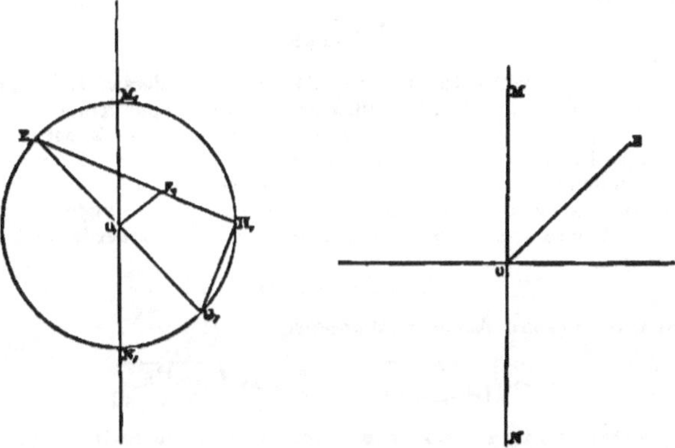

FIG. 32.

projection by $O_1M_1N_1$, the meridian through O represented on the other chart by MN; O_1M_1 is equal to the radius of the sphere. Required to find the position of the point on the central equivalent projection represented by E_1 on the stereographic projection. Lay off at O the angle MOE $= M_1O_1E_1$; the point sought is on the line OE, and the distance OE is in consequence all that has to be determined. Draw the diameter F_1G_1 perpendicular to O_1E_1; join F_1E_1 and produce it to H_1; join G_1H_1, then G_1H_1 is the distance required.

Another method for constructing the central equivalent from a stereographic projection is as follows:

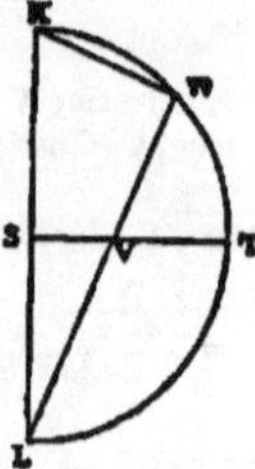

FIG. 33.

In Fig. 33 the length KW is the distance on the central equivalent, corresponding to SV on the stereographic projection. The similar triangles KWL and VSL give

$$SV : KW :: LV : KL$$

Dividing through by ST, the radius, and observing that $LV = \sqrt{SV^2 + SL^2}$, we find

$$\frac{SV}{ST} : \frac{KW}{ST} = \sqrt{1+\left(\frac{SV}{ST}\right)^2} : 2$$

which gives the ratio $\frac{KW}{ST}$ as a function of $\frac{SV}{ST}$; calling the former of these ratio C and the latter S, we have for the formula of transformation from the stereographic or S, projection to the central equivalent or C projection

$$C = \sqrt{\frac{2S}{1+S^2}}$$

If we write $S = \tan \varphi$, we have $C = 2 \sin \varphi$. Table XIV gives the values of S and C, and affords a rapid and very easy and exact method of constructing the central equivalent projection.

We are now prepared to solve a much more general problem than the one proposed above, viz, to find the equation of the central equivalent projection of any circle of the sphere, whether great or small. Denoting as usual by ξ, η, the rectangular co-ordinates on the required projection, let ξ', η' denote rectangular co-ordinates on the auxiliary stereographic projection. The circle of the sphere will be a circle upon the stereographic chart, and if its center is at (α', β'), its radius ρ' will be given by

$$(\xi'-\alpha')^2+(\eta'-\beta')^2=\rho'^2$$

But, according to the proposed plan of transformation,

$$\xi' = \xi\sqrt{\frac{r^2}{4r^2-(\xi^2+\eta^2)}} \qquad \eta' = \eta\sqrt{\frac{r^2}{4r^2-(\xi^2+\eta^2)}}$$

consequently, we have, for the equation of the curve on the central equivalent projection which represents a circle on the sphere,

$$\left(\xi\sqrt{\frac{r^2}{4r_1-(\xi_1+\eta^2)}}-\alpha'\right)^2+\left(\eta\sqrt{\frac{r^2}{4r^2-(\xi^2+\eta^2)}}-\beta'\right)^2=\rho'^2$$

or, transforming to polars by means of the formulas $\xi = \rho \cos \theta$, $\eta = \rho \sin \theta$,

$$\left(\frac{r\rho \cos \theta}{\sqrt{4r^2-\rho^2}}-\alpha'\right)^2+\left(\frac{r\rho \sin \theta}{\sqrt{4r^2-\rho^2}}-\beta'\right)^2=\rho'^2$$

A still further simplification is possible by writing

$$k=\sqrt{a'^2+\beta'^2} \qquad \varphi=\tan^{-1}\frac{\beta'}{a'}$$

The equation becomes, now,

$$\left(\frac{r\rho}{\sqrt{4r^2-\rho^2}}\right)^2-\frac{2r\rho k}{\sqrt{4r^2-\rho^2}}\cos(\theta-\varphi)+(k^2-\rho'^2)=0$$

This is merely the polar equation of the circle, in which the stereographic radius vector ρ, has been replaced by its value $\frac{r\rho}{\sqrt{4r^2-\rho^2}}$ as a function of the radius vector in the central equivalent system. The equation in Cartesian co-ordinates shows that the curve is of the fourth degree. The equation in polar co-ordinates enables us readily to determine the condition that the curve shall represent a great circle of the sphere. Make $\theta=\varphi$; then

$$\frac{r\rho}{\sqrt{4r^2-\rho^2}}=k\pm\rho'$$

from which we obtain

$$\rho=\pm\sqrt{\frac{4(k\pm\rho')^2r^2}{r^2+(k\pm\rho')^2}}$$

This affords four real values for ρ. The signs + and − in the numerator and denominator of this quantity are to be taken in this manner:

$$\frac{+}{+} \qquad \frac{-}{-}$$

Now, in order that the polar equation shall represent a great circle of the sphere, it is necessary and sufficient that the sums of the squares of the two values of ρ, obtained by taking ρ' first with the + and second with the − sign under the radical, shall be equal to $4r^2$, or that we shall have

$$\frac{(k+\rho')^2}{r^2+(k+\rho)^2}+\frac{(k-\rho')^2}{r^2+(k-\rho')^2}=1$$

That this is a correct formula is easily seen from the following simple geometrical considerations: Let C denote the center, and AMBNP the orthographic projection of the sphere; P, Fig. 34, is the point of sight of the orthographic projection, and the plane MN parallel to the tangent plane at P is the plane of this projection; let AB denote the trace of a plane cutting a great circle

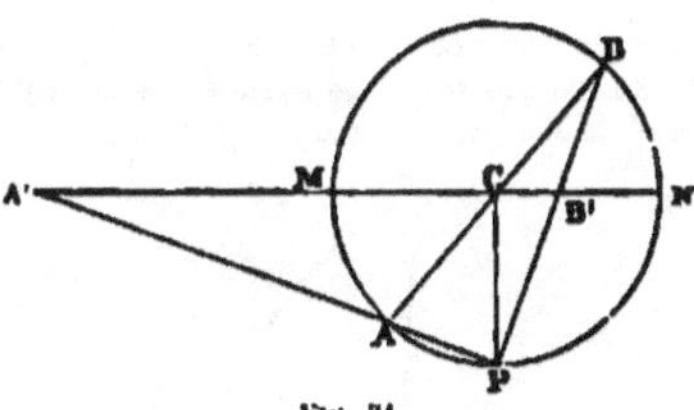

FIG. 34.

from the sphere; and, finally, let A'B' denote the projection of this great circle; then we have

$$CA'=K+\rho' \qquad CB'=k-\rho'$$

and, also, since $PA'^2=CP^2+CA'^2$,

$$\frac{(k+\rho')^2}{r^2+(k+\rho')^2}=\cos^2 PA'C \qquad \frac{(k-\rho')^2}{r^2+(k-\rho')^2}=\cos^2 PB'C$$

but APB=A'PB' is a right angle, and consequently PB'C and PA'C are complementary angles, and the sum of the squares of their cosines is equal to unity. *Q. E. D.*

LOXODROMIC CURVES.

The pole being taken as center, it is very easy to obtain the loxodromic curve. Denote by θ the angle made on the sphere by such a curve with a meridian; then θ' denoting the corresponding angle on the chart, we have

$$\tan\theta'=\tan\theta\cdot\frac{1}{\cos^2\frac{\varphi}{2}}$$

Now, $\tan\theta$ is constant, and, for $r=1$, $2\sin\frac{\varphi}{2}=\rho$, and also $\tan\theta'=\frac{\rho d\alpha}{d\rho}$. The differential equation of the curve is then

$$\frac{\rho d\alpha}{d\rho}=\frac{\tan\theta}{1-\frac{\rho^2}{4}}$$

from which follows

$$d\alpha=\tan\theta\frac{d\rho}{\rho\left(1-\frac{\rho^2}{4}\right)}$$

and integrating

$$\alpha=\tan\varphi\log\frac{\rho}{\sqrt{4-\rho^2}}+C$$

PROJECTION UPON THE PLANE OF A MERIDIAN.

We will now take up the case of the projection upon the plane of any meridian of the parallels and meridians of the terrestrial sphere. The center will be upon the equator, and the given meridional plane will cut the equator in two points distant each 90° from the center.

A few definitions will be adopted, both for brevity and clearness of language.

The *central station* is the point of the sphere chosen as center of the map; this we shall designate by O upon the sphere and by O' upon the projection.

The *central distance* of a point M of the sphere is the ratio of the length of the arc MO of a great circle to the radius of the sphere; this we shall denote by δ. It is the quantity that, in the case of the pole being taken as central station, we have heretofore denoted by φ.

The *radius vector* ρ of the point M' upon the chart is the distance O'M' of this point from the central station. As usual, r denoting the radius of the sphere, we have

$$\rho=2r\sin\tfrac{1}{2}\delta$$

The *azimuthal angle* of the point M upon the sphere is the angle ω formed by the arc OM with the meridian through O. Upon the chart it is the equal angle formed by the right line O'M' with the meridian through O', which is also, as we know, a right line.

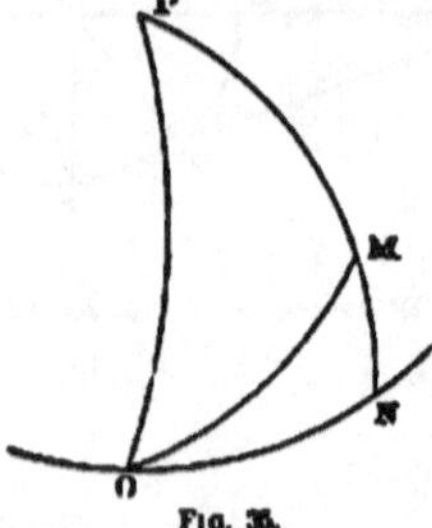

FIG. 35.

Now, having given the position of O, we wish to determine the values of ρ and ω in terms of the geographical co-ordinates (θ, ω) of any point whatever, as M. We have already resolved the problem for the case where O is assumed as the pole of the sphere, and a very simple transformation of co-ordinates enables us to resolve it for this more general case where O is taken upon the equator. Take OP, Fig. 35, for principal meridian; ω is the longitude of M with respect to this meridian; the portion ON of the equator included between O and the point of intersection of the equator with the meridian through M is measured by ω, and the arc MN is measured by θ; the angle MON is the complement of α, and finally OM is $=\lambda$. Now, since N is a right angle, we have in the triangle OMN

$$\cos\lambda=\cos\omega\cos\theta \qquad \tan\alpha=\sin\omega\cot\theta$$

which determine λ and α; ρ is determined by

$$\rho=2\,r\sin\tfrac{1}{2}\lambda$$

It is obvious that λ, and consequently ρ, remains the same for all values of ω and θ which give the same value for $\cos\omega\cos\theta$, for example, for the two points of which the latitude of the one equals the longitude of the other. Take for the axes of ξ and η the right lines representing respectively the equator and the first meridian, and we have, in consequence,

$$\xi=\rho\sin\alpha \qquad \eta=\rho\cos\alpha$$

or

$$\rho=\sqrt{\xi^2+\eta^2} \qquad \tan\alpha=\frac{\xi}{\eta}$$

But

$$\rho=2\,r\sin\tfrac{1}{2}\lambda \qquad \cos\lambda=\cos\omega\cos\theta \qquad \tan\alpha=\sin\omega\cot\theta$$

The second of these relations gives

$$\sin\tfrac{1}{2}\lambda=\sqrt{\frac{1-\cos\lambda}{2}}=\sqrt{\frac{1-\cos\omega\cos\theta}{2}}$$

so that

$$\xi^2+\eta^2=2\,r^2(1-\cos\omega\cos\theta) \qquad \xi=\eta\sin\omega\cot\theta$$

These are the formulas of transformation from polar to rectilinear co-ordinates. The elimination of θ between these equations gives us the equation of the meridian whose longitude is ω, and the elimination of ω in like manner gives the equation of the parallel of latitude θ.

EQUATION OF THE MERIDIANS.

The result of the elimination of θ is the equation

$$\xi^2+\eta^2=2\,r^2\left(1-\frac{\xi\cos\omega}{\sqrt{\xi^2+\eta^2\sin^2\omega}}\right)$$

By clearing of fractions and radicals this becomes

$$\xi^6+(2+\sin^2\omega)\xi^4\eta^2+(1+2\sin^2\omega)\xi^2\eta^4+\sin^2\omega\,\eta^6-4r^2(\xi^4+\eta^4\sin^2\omega)$$
$$-4\,r^2(1+\sin^2\omega)\xi^2\eta^2+4\,r^4(\xi^2+\eta^2)\sin^2\omega=0$$

This equation of the sixth degree is easily factored into

$$(\xi^2+\eta^2)\left\{\xi^4+[(1+\sin^2\omega)\eta^2-4r^2]\xi^2+(\eta^4-4r^2\eta^2+4r^4)\sin^2\omega\right\}=0$$

The factor to be suppressed here is obviously the binomial $\xi^2+\eta^2$, as equating that to zero would only result in giving an imaginary locus (or infinitely small circle) and in consequence would be of no practical use. We have then remaining a biquadratic equation in ξ and η.

If we write

$$\xi^2=\xi' \qquad \eta^2=\eta',$$

the equation becomes one of the second degree in ξ' and η', viz:

$$\xi'^2+(1+\sin^2\omega)\xi'\eta'+\eta'^2\sin^2\omega-4r^2\xi'-4r^2\sin^2\omega\,\eta'+4r^4\sin^2\omega=0$$

This last is the equation of an hyperbola whose center is at the intersection of the lines

$$2\xi'+(1+\sin^2\omega)\eta'-4r^2=0 \qquad (1+\sin^2\omega)\xi'+2\sin^2\omega\eta'-4r^2\sin^2\omega=0$$

or at the point

$$\xi'=-4r^2\tan^2\omega \qquad \eta'=4r^2\sec^2\omega$$

Representing by m_1 and m_2 the angular coefficients which determine the asymptotes, these quantities are obtained as the roots of the equation

$$m^2\sin^2\omega+(1+\sin^2\omega)m+1=0$$

from which

$$m_1=-\frac{1}{\sin^2\omega} \qquad m_2=-1$$

Confining ourselves to the region where ξ' and η' are both positive, we can readily construct this hyperbola, on any chosen scale, for each value of ω; then construct the required curve whose co-ordinates, measured on the same scale, are the square roots of ξ' and η', the co-ordinates of each point on the hyperbola. For $\eta=0$ we have

$$\xi=\pm 2r\sin\frac{\omega}{2} \qquad \xi=\pm 2r\cos\frac{\omega}{2}$$

For $\xi=0$ we have

$$\eta=\pm r\sqrt{2}$$

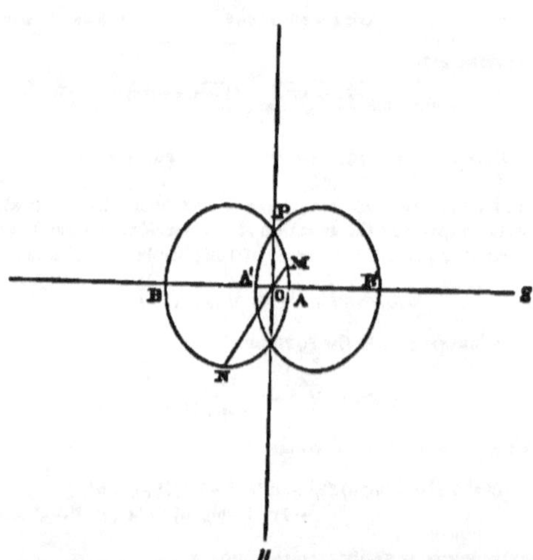

Since $\sin^2\omega=\sin^2(-\omega)$ and the equation of the curve contains only $\sin^2\omega$, the equation represents at the same time projection of the meridians of longitude ω and $-\omega$, respectively; these two curves will be symmetrically situated the one to the other with respect to the axis of η. If upon the axis of ξ we take, Fig. 30,

$$OA=2r\sin\frac{\omega}{2} \qquad OB=2r\cos\frac{\omega}{2}$$

and upon the axis of η take

$$OP=OP''=r\sqrt{2}$$

the curve will pass through the four points, A, P, B, P'', and the entire locus will be composed of this curve and the curve A'PBP'', symmetrical to the first with respect to the axis of η.

EQUATION OF A PARALLEL.

To obtain this equation we eliminate ω by the relation

$$\sin\omega = \frac{\xi\tan\theta}{\eta}$$

and obtain

$$\xi^2+\eta^2=2r^2\left(1-\cos\theta\sqrt{\frac{\eta^2-\xi^2\tan^2\theta}{\eta^2}}\right)$$

By clearing this of fractions and radicals we arrive at an equation of the sixth degree in η and of the fourth in ξ, which will contain only the even powers of the variables; as in the case of the equation of a meridian, this will contain the factor $\xi^2+\eta^2$, and dividing out by this factor we obtain as the resulting equation of a parallel

$$\eta^4+(\xi^2-4r^2)\eta^2+4r^4\sin^2\theta=0$$

Substitute again

$$\xi^2=\xi' \qquad\qquad \eta^2=\eta'$$

and we are conducted to the equation

$$\eta'^2+\xi'\eta'-4r^2\eta'+4r^4\sin^2\theta=0$$

of the second degree in ξ', η' and, as in the former case, representing an hyperbola. The center of the hyperbola is on the axis of ξ and is given by $\xi'=4r^2$; one asymptote is parallel to, and therefore coincident with, the axis of ξ. The same construction being made as before, we obtain for the projection of the parallel of latitude θ the curve ARBS, Fig. 37, and of latitude $-\theta$ the

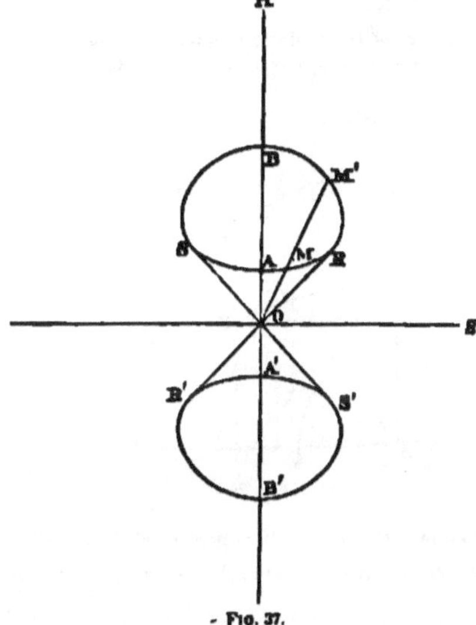

FIG. 37.

curve A'B'R'S'. These two curves are symmetrically situated with respect to the axis of ξ, and the sum of the squares of the intercepts made by any line OM' with one branch of the curve is constant and equal to the square of the diameter of the sphere, *i. e.*,

$$OM'^2+OM^2=4r^2$$

The truth of this is easily seen if we transform the equation of the parallels into polar co-ordinates; that is, write

$$\xi = \rho \cos \chi \qquad \eta = \rho \sin \chi$$

The equation then becomes

$$\rho^4 \sin^4 \chi + (\rho^2 \cos^2 \chi - 4r^2)\, \rho^2 \sin^2 \chi + 4r^4 \sin^2 \theta$$

Making the obvious reductions, this is

$$\rho^4 - 4r^2 \rho^2 + \frac{4r^4 \sin^2 \theta}{\sin^2 \chi} = 0$$

Calling the roots of this ρ_1 and ρ_2, we have $\rho_1^2 = OM'^2$ and $\rho_2^2 = OM''^2$, and from the known principles of the theory of equations,

$$\rho_1^2 + \rho_2^2 = 4r^2.$$

MOLLWEIDE'S PROJECTION.

This projection was invented by Prof. C. B. Mollweide, of Halle, in 1805, and in 1857 a number of applications of it were made by Babinet, whose name thus became attached to it, the projection being known commonly as Babinet's homalographic projection. The problem proposed for solution here is to represent the entire surface of the earth in an ellipse the ratio of whose major and minor axes, represented by the equator and first meridian, respectively, shall be 2 : 1; the parallels to be projected in parallel right lines and the meridians in ellipses, all of which pass through two fixed points—the poles and each zone of the sphere to be represented upon the chart in its true size.

Let b and $2b$ denote the axes of the limiting ellipse, then the including area will be $= 2b^2\pi$; but this is to equal the entire area of the sphere, or $4\pi r^2$; this condition, then, gives us for the axes of this ellipse

$$b = \sqrt{2}\, r \qquad 2b = 2\sqrt{2}\, r$$

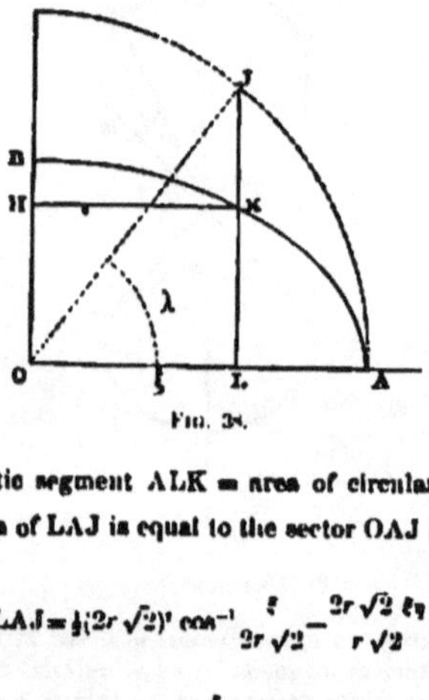

FIG. 38.

The area (Fig. 38) of the elliptic segment ALK = area of circular segment LAJ multiplied by $\frac{OB}{OA}$, that is, by $\frac{1}{2}$; now the area of LAJ is equal to the sector OAJ minus the triangle OLJ, or

$$LAJ = \tfrac{1}{2}(2r\sqrt{2})^2 \cos^{-1} \frac{\xi}{2r\sqrt{2}} - \frac{2r\sqrt{2}\,\xi\eta}{r\sqrt{2}}$$

or

$$LAJ = 4r^2 \cos^{-1} \frac{\xi}{2r\sqrt{2}} - 2\xi\eta$$

and then for the elliptic segment we have only to divide this by 2; add to this result the area of the rectangle OLKH, or $\xi\eta$, and we obtain finally

$$OAKH = 2r^2 \cos^{-1} \frac{\xi}{2r\sqrt{2}} + \tfrac{1}{2}\xi\eta$$

Assume for the angle AOJ the symbol λ, then follows

$$\cos^{-1} \frac{\xi}{2r\sqrt{2}} = \lambda \qquad \xi = 2r\sqrt{2} \cos \lambda \qquad \eta = r\sqrt{2} \sin \lambda$$

and consequently

$$OAKH = 2r^2\lambda + r^2 \sin 2\lambda$$

This surface is, however, to be equal to the area of the semi-zone between the equator and parallel of θ, or equal to $\pi r^2 \sin \theta$. Equating these and we have for the fundamental equation of the Mollweide projection

$$\pi \sin \theta = \sin 2\lambda + 2\lambda$$

The values of λ or $\sin \lambda$ have to be obtained from this equation for each given value of θ. Lay off, then, on the semi-minor axis of the ellipse the lengths $r\sqrt{2} \sin \chi$ measured from the center, and the points so obtained will be the points of intersection of each parallel with the principal meridian or minor axis of the limiting ellipse; through these points draw parallels to the equator, and they will represent the parallels. For the construction of the meridians by points it is only necessary to divide the equator and parallels in parts which correspond exactly to the points of division of these lines on the sphere. For example, if it is desired to draw the meridians of every ten degrees we have only to divide the entire equator and also the meridians of the chart into 36 equal parts, and through the corresponding points thus obtained draw the ellipses representing the meridians.

For the computation of λ from the above equation the following method of approximation answers very well. Assume a value λ' such that

$$\sin 2\lambda' + 2\lambda' = \pi \sin \theta'$$

where θ' differs but little from θ; let δ represent the correction to λ', i. e., $\lambda' + \delta = \lambda$; then

$$\sin 2(\lambda' + \delta) + 2(\lambda' + \delta) = \pi \sin \theta$$

Subtracting the first of these equations from the second gives

$$\sin 2(\lambda' + \delta) - \sin 2\lambda' + 2\delta = \pi (\sin \theta - \sin \theta')$$

or

$$2 \cos (2\lambda' + \delta) + 2\delta = \pi (\sin \theta - \sin \theta')$$

As δ will be a very small quantity, we can write $\sin \delta = \delta$, and

$$\cos (2\lambda' + \delta) = \cos 2\lambda'$$

Writing, then, for $\sin \theta'$ its value, we obtain for δ the approximate value

$$\delta = \frac{\pi \sin \theta - (\sin 2\lambda' + 2\lambda')}{2(1 + \cos 2\lambda')}$$

This method of approximation can of course be carried as far as we choose, or until we reach any required degree of exactness. Table X gives the value of $\sin \lambda$ for values of θ differing by 30'; this was computed by Jules Bourdin, and is more accurate and extended than the one computed by Mollweide himself for the values of λ.

In conclusion, we will examine briefly a projection proposed by M. Collignon, in which he represents the central equivalent projection in the form of a square. Suppose that, as in Mollweide's projection, the parallels are parallel right lines and that the meridians are also right lines parting from a common point, the pole. Call h the ordinate of the point taken as pole; then the equation of the meridians will be in the form

$$\xi = (h - \eta) f(\omega)$$

The origin is supposed placed at the foot of the perpendicular from the pole upon the equator; the function $f(\omega)$ is independent of θ. As in the Mollweide projection, η is a function of θ only or

$$\frac{d\eta}{d\omega}=0$$

and consequently

$$\frac{d\xi}{d\omega}=(h-\eta)f'(\omega)$$

The condition for the conservation of surfaces now becomes

$$(h-\eta)f'(\omega)\frac{d\eta}{d\theta}=r^2\cos\theta$$

This would give $f'(\omega)$ as a function of θ, which is contradictory to the previous assumption made concerning $f(\omega)$; the interpretation of this is, since $f'(\omega)$ does not contain ω, that $f'(\omega)=m$, a constant, and so $f(\omega)$ is a linear function of the longitude ω, or

$$f(\omega)=m\omega+n$$

The equation of condition is thus

$$m(h-\eta)d\eta=r^2\cos\theta d\theta$$

From which, by integration, follows

$$mh\eta-\frac{\eta^2}{2}=C+r^2\sin\theta$$

Since for $\theta=\frac{\pi}{2}$ we have $\eta=h$, we find

$$C=\tfrac{1}{2}mh^2-r^2$$

And again, since $\theta=0$ gives $\eta=0$,

$$C=0$$

or

$$\tfrac{1}{2}mh^2=r^2$$

Finally, since we wish the extreme meridians which limit the chart to form a square, it will be necessary for $\theta=0$ and $\omega=\pm\frac{\pi}{2}$ that we have $\xi=\pm h$, the corresponding signs to be taken together; but

$$\xi=(h-\eta)(m\omega+n)$$

in this making $\xi=\pm a$. Remembering that, for $\theta=0$, $\eta=0$, there follows

$$m\frac{\pi}{2}+n=1 \qquad -m\frac{\pi}{2}+n=1$$

Solution of these equations gives

$$m=\frac{2}{\pi} \qquad n=0$$

and so, by virtue of the relation $h^2=\frac{2r^2}{m}$,

$$h=r\sqrt{\pi}$$

and, finally, the equation connecting θ and η is

$$\eta^2-2r\sqrt{\pi}\eta+\pi r^2\sin\theta=0$$

The projection need, of course, only be constructed for the positive values of θ, and then repeated symmetrically below the equator for the negative values of θ.

§ XII.

TABLES.

The following tables have been nearly all extracted from the original memoirs of the writers on different parts of the subject of projections. In many cases, however, an author's own tables have been improved upon by others, in which case of course only the latter are given. In cases where it has been necessary to take into account the ellipticity of the earth it will be observed that the same value has not been used throughout; the explanation of that is obvious, as the tables have been constructed at so many different periods of time; the reason for not correcting to one value of the ellipticity is to be found in the fact that but very slight—in fact, almost inappreciable—changes would be thus introduced; and inasmuch as final values have not yet been agreed upon for the figure and size of the earth, it would seem to be a mere waste of time to make any corrections to tables that for practical purposes are already correct enough. Some of the tables given are only useful for the purpose of comparing different methods of projection, but these will, in every case, explain themselves. The names at the beginning of each table and the references in the text make it quite unnecessary to go into a detailed account of their use; the formulas are furthermore arranged so that it is a very simple matter to make the necessary substitutions.

The accompanying plates, which are referred to in their proper places in the text, serve as illustrations of the principal projections in actual use. Many which are treated of in the text it was not thought necessary to preserve in the collection of plates, inasmuch as they do not at the present day subserve any useful purpose.

TABLE I.

Values of the degrees of longitude and the lengths of the sides of cones tangent to the sphere along a parallel of latitude.

The degree on the equator is taken for unity. Radius of the sphere = 57.295,779.

Lat.	$D_p = \cos\theta = \sin(90° - \theta)$	$G = r\cot\theta$	Lat.	$D_p = \cos\theta = \sin(90° - \theta)$	$G = r\cot\theta$	Lat.	$D_p = \cos\theta = \sin(90° - \theta)$	$G = r\cot\theta$
θ	D_p	G	θ	D_p	G	θ	D_p	G
0	1.000000	∞	31	0.85717	[illegible]	61	0.48481	[illegible]
1	0.99985	3282.473	32	84805	[illegible]	62	46947	[illegible]
2	[illegible]	[illegible]	33	[illegible]	[illegible]	63	[illegible]	[illegible]
3	[illegible]	[illegible]	34	[illegible]	[illegible]	64	[illegible]	[illegible]
4	[illegible]	[illegible]	35	81915	[illegible]	65	[illegible]	[illegible]
5	[illegible]	[illegible]	36	[illegible]	[illegible]	66	[illegible]	[illegible]
6	[illegible]	[illegible]	37	[illegible]	[illegible]	67	[illegible]	[illegible]
7	[illegible]	[illegible]	38	[illegible]	[illegible]	68	[illegible]	[illegible]
8	[illegible]	[illegible]	39	77715	[illegible]	69	[illegible]	[illegible]
9	[illegible]	[illegible]	40	[illegible]	[illegible]	70	[illegible]	[illegible]
10	[illegible]	[illegible]	41	75471	[illegible]	71	[illegible]	[illegible]
11	[illegible]	[illegible]	42	74314	[illegible]	72	[illegible]	[illegible]
12	97815	[illegible]	43	[illegible]	61.442	73	[illegible]	17.517
13	[illegible]	248.178	44	71934	[illegible]	74	27564	[illegible]
14	[illegible]	[illegible]	45	70711	57.296	75	[illegible]	[illegible]
15	[illegible]	[illegible]	46	[illegible]	[illegible]	76	[illegible]	[illegible]
16	[illegible]	[illegible]	47	[illegible]	[illegible]	77	[illegible]	[illegible]
17	[illegible]	[illegible]	48	[illegible]	[illegible]	78	[illegible]	[illegible]
18	[illegible]	[illegible]	49	[illegible]	[illegible]	79	[illegible]	[illegible]
19	[illegible]	[illegible]	50	64279	48.077	80	17365	[illegible]
20	[illegible]	157.419	51	[illegible]	[illegible]	81	[illegible]	[illegible]
21	[illegible]	[illegible]	52	61566	44.764	82	13917	[illegible]
22	[illegible]	[illegible]	53	[illegible]	43.175	83	[illegible]	[illegible]
23	[illegible]	[illegible]	54	58779	[illegible]	84	[illegible]	[illegible]
24	91355	[illegible]	55	[illegible]	40.119	85	[illegible]	5.013
25	[illegible]	[illegible]	56	55919	[illegible]	86	[illegible]	4.007
26	[illegible]	117.474	57	[illegible]	[illegible]	87	[illegible]	[illegible]
27	[illegible]	[illegible]	58	[illegible]	[illegible]	88	[illegible]	[illegible]
28	[illegible]	107.730	59	[illegible]	34.427	89	[illegible]	[illegible]
29	[illegible]	[illegible]	60	[illegible]	[illegible]	90	0.00000	[illegible]
30	[illegible]	[illegible]						

Table II.

Polar distance corrected for an ellipticity = $\frac{1}{299.15}$

Distances from the pole upon the sphere, co-latitude.	Corrections.	First differences.	Distances from the pole upon the sphere, co-latitude.	Corrections.	First differences.	Distances from the pole upon the sphere, co-latitude.	Corrections.	First differences.
° °	° ′ ″	″	° °	° ′ ″	″	° °	° ′ ″	″
1 or 89	0 0 26.8	......	16 or 74	0 6 4.7	20.6	31 or 59	0 10 7.7	11.6
2 or 88	0 46.8	24.8	17 or 73	6 26.0	20.3	32 or 58	10 18.7	11.0
3 or 87	1 11.8	23.9	18 or 72	6 44.6	19.7	33 or 57	10 29.2	10.1
4 or 86	1 25.8	23.9	19 or 71	7 2.4	19.2	34 or 56	10 38.3	9.4
5 or 85	1 50.5	23.7	20 or 70	7 22.5	18.7	35 or 55	10 46.8	8.5
6 or 84	2 12.1	23.6	21 or 69	7 40.6	18.1	36 or 54	10 54.7	7.9
7 or 83	2 46.5	23.4	22 or 68	7 56.2	17.6	37 or 53	11 1.7	7.0
8 or 82	3 8.7	23.2	23 or 67	8 15.1	17.0	38 or 52	11 7.8	6.2
9 or 81	3 30.7	23.0	24 or 66	8 31.5	16.4	39 or 51	11 13.2	5.4
10 or 80	3 56.4	22.7	25 or 65	8 47.3	15.8	40 or 50	11 17.6	4.4
11 or 79	4 17.8	22.4	26 or 64	9 2.4	15.1	41 or 49	11 21.6	3.6
12 or 78	4 39.2	22.1	27 or 63	9 16.8	14.4	42 or 48	11 24.6	3.0
13 or 77	5 1.7	21.8	28 or 62	9 30.6	13.7	43 or 47	11 26.7	2.1
14 or 76	5 23.1	21.4	29 or 61	9 43.7	13.2	44 or 46	11 27.8	1.1
15 or 75	5 44.1	21.0	30 or 60	9 56.1	12.4	45 or 45	11 28.2	0.4

Table III.

Transformation from geographical to zenithal co-ordinates.

Values of the central distance α expressed in degrees, $\cos \alpha = \cos \omega \cos \xi$.

	φ=0°	5°	10°	15°	20°	25°	30°	35°	40°	45°	50°	55°	60°	65°	70°	75°	80°	85°	90°
° ′	° ′	° ′	° ′	° ′	° ′	° ′	° ′	° ′	° ′	° ′	° ′	° ′	° ′	° ′	° ′	° ′	° ′	° ′	° ′
ω=0	0 0	5 0	10 0	15 0	20 0	25 0	30 0	35 0	40 0	45 0	50 0	55 0	60 0	65 0	70 0	75 0	80 0	85 0	90 0
5	5 0	7 4	11 10	15 46	20 35	25 25	30 21	35 19	40 16	45 13	50 11	55 8	60 5	65 9	70 4	75 2	80 2	85 1	90 0
10	10 0	11 10	14 8	17 56	22 16	26 46	31 20	36 13	41 2	45 44	50 41	55 26	60 30	65 24	70 18	75 14	80 8	85 4	90 0
15	15 0	15 [illegible]	17 56	21 6	24 56	28 45	33 14	37 42	42 18	46 56	51 37	56 21	61 7	65 58	70 42	75 31	80 21	85 10	90 0
20	20 0	20 35	22 16	24 45	27 50	31 27	35 22	39 46	43 57	48 22	52 50	57 33	61 16	66 36	71 15	75 45	80 37	85 16	90 0
25	25 0	25 25	26 46	28 54	31 37	34 47	38 28	42 6	46 3	[illegible]	[illegible]	[illegible]	[illegible]	[illegible]	[illegible]	[illegible]	[illegible]	[illegible]	90 0
30	30 0	30 21	31 28	33 16	35 32	38 30	41 35	44 40	48 25	[illegible]	[illegible]	[illegible]	[illegible]	[illegible]	[illegible]	[illegible]	[illegible]	[illegible]	90 0
35	35 0	35 19	36 18	37 42	39 40	42 6	44 40	47 53	51 2	[illegible]	[illegible]	[illegible]	[illegible]	[illegible]	[illegible]	[illegible]	[illegible]	[illegible]	90 0
40	40 0	[illegible]	[illegible]	[illegible]	[illegible]	[illegible]	[illegible]	[illegible]	[illegible]	[illegible]	[illegible]	[illegible]	[illegible]	[illegible]	[illegible]	[illegible]	[illegible]	[illegible]	90 0
45	45 0	[illegible]	[illegible]	[illegible]	[illegible]	[illegible]	[illegible]	[illegible]	[illegible]	[illegible]	[illegible]	[illegible]	[illegible]	[illegible]	[illegible]	[illegible]	[illegible]	[illegible]	90 0
50	50 0	[illegible]	[illegible]	[illegible]	[illegible]	[illegible]	[illegible]	[illegible]	[illegible]	[illegible]	[illegible]	[illegible]	[illegible]	[illegible]	[illegible]	[illegible]	[illegible]	[illegible]	90 0
55	55 0	[illegible]	[illegible]	[illegible]	[illegible]	[illegible]	[illegible]	[illegible]	[illegible]	[illegible]	[illegible]	[illegible]	[illegible]	[illegible]	[illegible]	[illegible]	[illegible]	[illegible]	90 0
60	60 0	[illegible]	[illegible]	[illegible]	[illegible]	[illegible]	[illegible]	[illegible]	[illegible]	[illegible]	[illegible]	[illegible]	[illegible]	[illegible]	[illegible]	[illegible]	[illegible]	[illegible]	90 0
65	65 0	[illegible]	[illegible]	[illegible]	[illegible]	[illegible]	[illegible]	[illegible]	[illegible]	[illegible]	[illegible]	[illegible]	[illegible]	[illegible]	[illegible]	[illegible]	[illegible]	[illegible]	90 0
70	70 0	[illegible]	[illegible]	[illegible]	[illegible]	[illegible]	[illegible]	[illegible]	[illegible]	[illegible]	[illegible]	[illegible]	[illegible]	[illegible]	[illegible]	[illegible]	[illegible]	[illegible]	90 0
75	75 0	[illegible]	[illegible]	[illegible]	[illegible]	[illegible]	[illegible]	[illegible]	[illegible]	[illegible]	[illegible]	[illegible]	[illegible]	[illegible]	[illegible]	[illegible]	[illegible]	[illegible]	90 0
80	80 0	[illegible]	[illegible]	[illegible]	[illegible]	[illegible]	[illegible]	[illegible]	[illegible]	[illegible]	[illegible]	[illegible]	[illegible]	[illegible]	[illegible]	[illegible]	[illegible]	[illegible]	90 0
85	85 0	[illegible]	[illegible]	[illegible]	[illegible]	[illegible]	[illegible]	[illegible]	[illegible]	[illegible]	[illegible]	[illegible]	[illegible]	[illegible]	[illegible]	[illegible]	[illegible]	[illegible]	90 0
90	90 0	90 0	90 0	90 0	90 0	90 0	90 0	90 0	90 0	90 0	90 0	90 0	90 0	90 0	90 0	90 0	90 0	90 0	90 0

Transformation from geographical to zenithal co-ordinates—Continued.

Values of the azimuthal angle β expressed in degrees, $\tan\beta = \sin w \cot\theta$.

	θ = 0°	5°	10°	15°	20°	25°	30°	35°	40°	45°	50°	55°	60°	65°	70°	75°	80°	85°	90°
	° ′	° ′	° ′	° ′	° ′	° ′	° ′	° ′	° ′	° ′	° ′	° ′	° ′	° ′	° ′	° ′	° ′	° ′	° ′
w = 0		0 0	0 0	0 0	0 0	0 0	0 0	0 0	0 0	0 0	0 0	0 0	0 0	0 0	0 0	0 0	0 0	0 0	0 0
5	90 0	45 [illegible]	[illegible] 10	18 1	13 [illegible]	10 [illegible]	[illegible]	7 6	6 [illegible]	4 [illegible]	4 11	3 [illegible]	[illegible]	2 13	1 [illegible]	1 [illegible]	0 [illegible]	0 [illegible]	0 0
10	90 0	[illegible] 10	44 34	[illegible] 57	25 [illegible]	20 25	10 44	13 [illegible]	11 42	9 51	8 17	6 [illegible]	5 44	4 [illegible]	4 2	[illegible]	1 45	0 [illegible]	0 0
15	90 0	71 [illegible]	[illegible] 44	44 0	[illegible]	[illegible] 7	34 [illegible]	[illegible] 17	17 14	14 31	12 15	10 16	8 [illegible]	6 [illegible]	5 [illegible]	3 [illegible]	2 [illegible]	1 14	0 0
20	90 0	75 [illegible]	[illegible] 44	[illegible]	43 13	[illegible] 15	[illegible]	[illegible] 2	22 11	19 [illegible]	16 1	13 [illegible]	11 10	9 4	7 6	5 14	3 [illegible]	1 [illegible]	0 0
25	90 0	79 4	67 21	[illegible]	40 16	42 11	[illegible] 17	[illegible] 7	[illegible] 44	[illegible]	19 31	16 [illegible]	13 [illegible]	11 0	[illegible]	6 [illegible]	4 10	2 [illegible]	0 0
30	90 0	[illegible] 0	70 54	61 [illegible]	53 57	47 0	40 51	35 [illegible]	30 47	[illegible] 34	[illegible]	19 18	16 0	13 7	10 19	7 [illegible]	5 7	2 [illegible]	0 0
35	90 0	81 [illegible]	73 [illegible]	[illegible]	57 [illegible]	[illegible]	44 [illegible]	[illegible]	34 [illegible]	[illegible]	[illegible] 42	[illegible]	13 10	14 [illegible]	11 [illegible]	[illegible] 44	5 [illegible]	2 [illegible]	0 0
40	90 0	[illegible] 15	74 [illegible]	67 [illegible]	[illegible]	54 7	48 4	[illegible]	37 37	32 44	[illegible] 21	24 14	20 [illegible]	[illegible]	13 10	9 [illegible]	6 [illegible]	3 13	0 0
45	90 0	[illegible] 57	78 0	[illegible] 15	[illegible]	[illegible]	[illegible]	45 17	[illegible] 7	[illegible]	[illegible] 41	[illegible] 21	[illegible] 13	[illegible] 18	14 [illegible]	[illegible] 44	7 7	3 [illegible]	0 0
50	90 0	[illegible]	77 2	70 43	64 53	[illegible]	[illegible] 0	47 [illegible]	42 34	37 27	[illegible] 44	[illegible] 12	[illegible] 51	[illegible]	[illegible]	11 [illegible]	7 42	3 [illegible]	0 0
55	90 0	[illegible] 54	77 54	71 [illegible]	[illegible] 3	[illegible] 21	54 [illegible]	[illegible]	44 [illegible]	[illegible]	[illegible]	[illegible]	[illegible] 19	[illegible] 54	[illegible] 16	12 [illegible]	[illegible] 13	4 0	0 0
60	90 0	84 14	78 [illegible]	72 [illegible]	67 13	[illegible] 42	56 10	51 3	45 34	[illegible] 54	[illegible] 9	31 14	[illegible] 34	31 [illegible]	17 [illegible]	12 1	9 41	4 [illegible]	0 0
65	90 0	84 [illegible]	78 [illegible]	73 [illegible]	[illegible] 7	[illegible] 47	57 [illegible]	[illegible]	47 [illegible]	42 11	37 14	[illegible]	27 27	[illegible]	[illegible] 18	13 [illegible]	9 [illegible]	4 [illegible]	0 0
70	90 0	84 41	70 [illegible]	74 5	[illegible]	[illegible]	[illegible]	[illegible] 19	[illegible] 14	43 13	[illegible] 15	[illegible] 31	[illegible]	[illegible]	[illegible]	14 [illegible]	[illegible]	4 [illegible]	0 0
75	90 0	84 [illegible]	79 [illegible]	74 [illegible]	[illegible] 21	[illegible] 14	[illegible] 0	[illegible] 4	[illegible] 13	44 [illegible]	[illegible] 1	34 4	[illegible]	[illegible] 18	19 [illegible]	14 31	[illegible]	4 [illegible]	0 0
80	90 0	84 [illegible]	79 31	74 47	[illegible] 43	[illegible]	[illegible] 37	54 [illegible]	[illegible] 34	44 34	[illegible] 34	[illegible]	[illegible] 37	34 [illegible]	19 [illegible]	14 47	[illegible]	4 [illegible]	0 0
85	90 0	84 [illegible]	79 [illegible]	74 [illegible]	[illegible]	[illegible]	[illegible] 54	54 [illegible]	[illegible] 34	44 [illegible]	[illegible]	[illegible]	[illegible]	[illegible]	19 [illegible]	14 [illegible]	[illegible]	4 [illegible]	0 0
90	90 0	85 0	80 0	75 0	70 0	65 0	60 0	55 0	50 0	45 0	40 0	35 0	30 0	25 0	20 0	15 0	10 0	5 0	0 0

TABLE IV.

Construction of the stereographic equatorial projection.

Vide page —.

θ	ρ	θ	ρ
0°	1.[illegible]	45°	0.41421
5	0.[illegible]	50	0.[illegible]
10	0.[illegible]	55	0.[illegible]
15	0.[illegible]	60	0.[illegible]
20	0.[illegible]	65	0.[illegible]
23° 27′ 30″	0.[illegible]	66° 32′ 30″	0.[illegible]
25°	0.[illegible]	70°	0.17[illegible]
30	0.[illegible]	75	0.[illegible]
35	0.[illegible]	80	0.[illegible]
40	0.[illegible]	85	0.[illegible]

TABLE V.

Stereographic meridian projection.

θ or ω	R or ρ	R' or q	δ or δ'	θ or ω
0°			0.00000	0°
5	11.47371	11.43005	0.04366	5
10	5.75877	5.67128	0.08749	10
15	3.86370	3.73205	0.13165	15
20	2.92380	2.74748	0.17633	20
20° 27' 30"	2.81304	2.59642	0.20702	20° 27' 30"
25°	2.36620	2.14451	0.22169	25°
30	2.00000	1.73205	0.26795	30
35	1.74345	1.42815	0.31530	35
40	1.55572	1.19173	0.36397	40
45	1.41421	1.00000	0.41421	45
50	1.30541	0.83910	0.46631	50
55	1.22077	0.70021	0.52057	55
60	1.15470	0.57735	0.57735	60
65	1.10338	0.46631	0.63707	65
65° 32' 30"	1.09960	0.45506	0.65616	65° 32' 30"
70°	1.06418	0.36397	0.70021	70°
75	1.03528	0.26795	0.76733	75
80	1.01543	0.17633	0.83910	80
85	1.00382	0.08749	0.91633	85
90	1.00000	0.00000	1.00000	90

TABLE VI.

La Grange's projection.

For this case $c = \frac{1}{2}$, and the whole surface of the globe is represented by a complete circumference.

ζ	ψ	ζ	ψ
90°....................	0.00000	40°....................	0.26795
80°....................	0.04362	30°....................	0.31763
70°....................	0.08749	20°....................	0.42224
60°....................	0.13240	10°....................	0.54346
50°....................	0.18844	0°....................	1.00000

TABLE VII.

La Grange's projection.

For this case $k = 1$.

ζ	ψ'	ψ''	ζ	ψ'	ψ''
10	0.926	1.081	100	0.050	16.760
20	0.836	1.196	110	0.048	20.460
30	0.761	1.340	120	0.163	6.130
40	0.647	1.545	130	0.285	3.513
50	0.562	1.812	140	0.426	2.410
60	0.467	2.166	150	0.585	1.855
70	0.366	2.772	160	0.763	1.471
80	0.262	3.882	170	0.930	1.142
90	0.165	6.071	180	1.000	1.000

TABLE VIII.

Mercator's projection—Table for the calculation of increasing latitudes.

Ellipticity = $\frac{1}{[illegible]}$. $N = 7915'.704674 \log \tan\left(45° + \frac{\theta}{2}\right) - 3437.7\left(e^2 \sin\theta + \frac{e^4 \sin^3\theta}{3}\right)$.

Lats.	Increasing latitudes N.	Lats.	Increasing latitudes N.	Lats.	Increasing latitudes N.	Lats.	Increasing latitudes N.	Lats.	Increasing latitudes N.
° ′	′	° ′	′	° ′	′	° ′	′	° ′	′
0 0	0.0	28 20	1763.0	38 0	2454.2	47 40	3244.9	57 20	4200.3
30	29.8	30	1774.4	10	2466.9	50	3259.7	30	4218.9
1 0	59.6	40	1785.7	20	2479.6	48 0	3274.6	40	4237.5
30	89.4	50	1797.0	30	2492.3	10	3289.6	50	4256.2
2 0	119.2	29 0	1808.4	40	2505.0	20	3304.5	58 0	4275.0
30	149.1	10	1819.9	50	2517.8	30	3319.5	10	4293.8
3 0	178.9	20	1831.2	39 0	2530.6	40	3334.6	20	4312.8
30	208.7	30	1842.9	10	2543.4	50	3349.7	30	4331.9
4 0	238.6	40	1854.1	20	2556.2	49 0	3364.9	40	4351.1
30	268.5	50	1865.5	30	2569.2	10	3380.1	50	4370.3
5 0	298.4	30 0	1877.0	40	2582.1	20	3395.4	59 0	4389.6
30	328.3	10	1888.5	50	2595.1	30	3410.7	10	4409.1
6 0	358.3	20	1900.0	40 0	2608.1	40	3426.1	20	4428.6
30	388.3	30	1911.5	10	2621.1	50	3441.5	30	4448.2
7 0	418.3	40	1923.1	20	2634.1	50 0	3457.0	40	4467.9
30	448.3	50	1934.7	30	2647.2	10	3472.6	50	4487.7
8 0	478.4	31 0	1946.3	40	2660.3	20	3488.2	60 0	4507.7
30	508.5	10	1957.6	50	2673.5	30	3503.9	10	4527.7
9 0	538.7	20	1969.9	41 0	2686.7	40	3519.5	20	4547.9
30	568.9	30	1981.3	10	2699.9	50	3535.3	30	4568.2
10 0	599.1	40	1992.9	20	2713.1	51 0	3551.1	40	4588.5
30	629.4	50	2004.6	30	2726.4	10	3567.0	50	4609.0
11 0	659.7	32 0	2016.3	40	2739.7	20	3582.9	61 0	4629.3
30	690.1	10	2028.1	50	2753.1	30	3598.9	10	4650.0
12 0	720.6	20	2039.8	42 0	2766.5	40	3615.0	20	4670.7
30	751.1	30	2051.9	10	2779.9	50	3631.1	30	4691.6
13 0	781.7	40	2063.4	20	2793.4	52 0	3647.3	40	4712.6
30	812.3	50	2075.3	30	2806.9	10	3663.5	50	4733.7
14 0	843.0	33 0	2087.1	40	2820.4	20	3679.8	62 0	4754.9
30	873.7	10	2099.0	50	2834.0	30	3696.1	10	4776.3
15 0	904.6	20	2110.6	43 0	2847.9	40	3712.6	20	4797.7
30	935.5	30	2122.8	10	2861.3	50	3729.0	30	4819.2
16 0	966.5	40	2134.8	20	2875.0	53 0	3745.6	40	4840.9
30	997.5	50	2146.9	30	2888.8	10	3762.2	50	4862.7
17 0	1028.6	34 0	2158.8	40	2902.6	20	3778.9	63 0	4884.7
30	1059.9	10	2170.9	50	2916.3	30	3795.6	10	4906.7
18 0	1091.2	20	2182.9	44 0	2930.0	40	3812.4	20	4928.9
30	1122.6	30	2194.0	10	2943.9	50	3829.3	30	4951.3
19 0	1154.1	40	2207.9	20	2957.8	54 0	3846.2	40	4973.7
30	1185.7	50	2219.1	30	2971.7	10	3863.2	50	4996.3
20 0	1217.4	35 0	2231.2	40	2985.7	20	3880.3	64 0	5019.0
30	1249.1	10	2243.4	50	2999.8	30	3897.5	10	5041.9
21 0	1281.0	20	2255.9	45 0	3013.8	40	3914.7	20	5064.8
30	1313.0	30	2267.8	10	3028.0	50	3932.0	30	5087.9
22 0	1345.2	40	2280.0	20	3042.1	55 0	3949.3	40	5111.2
30	1377.4	50	2292.3	30	3056.3	10	3966.8	50	5134.6
23 0	1409.7	36 0	2304.6	40	3070.6	20	3984.3	65 0	5158.2
30	1442.2	10	2316.9	50	3084.9	30	4001.8	10	5181.9
24 0	1474.9	20	2329.3	46 0	3099.2	40	4019.5	20	5205.7
30	1507.8	30	2341.7	10	3113.5	50	4037.2	30	5229.8
25 0	1540.4	40	2354.1	20	3128.0	56 0	4055.0	40	5253.9
30	1573.4	50	2366.5	30	3142.4	10	4072.9	50	5278.2
26 0	1606.5	37 0	2378.9	40	3156.9	20	4090.9	66 0	5302.7
30	1639.8	10	2391.4	50	3171.5	30	4109.0	10	5327.4
27 0	1673.2	20	2403.9	47 0	3186.1	40	4127.0	20	5352.2
30	1706.8	30	2416.5	10	3200.7	50	4145.3	30	5377.1
28 0	1740.5	40	2429.0	20	3215.4	57 0	4163.5	40	5402.3
10	1751.7	50	2441.9	30	3230.1	10	4181.9	50	5427.6

Mercator's projection—Table for the calculation of increasing latitudes—Continued.

Lats.	Increasing latitudes N.	Lats.	Increasing latitudes N.	Lats.	Increasing latitudes N.	Lats.	Increasing latitudes N.	Lats.	Increasing latitudes N.
° ′	′	° ′	′	° ′	′	° ′	′	° ′	′
67 0	5452.1	71 40	6248.1	76 20	7271.6	81 0	8716.0	85 40	11234.4
10	5478.7	50	6281.0	30	7314.3	10	8781.1	50	11369.4
20	5504.5	72 0	6313.2	40	7357.3	20	8846.5	86 0	11509.6
30	5530.9	10	6345.7	50	7400.9	30	8913.6	10	11656.2
40	5556.6	20	6378.5	77 0	7445.0	40	8982.2	20	11809.2
50	5583.1	30	6411.5	10	7489.7	50	9051.8	30	11969.2
68 0	5609.7	40	6444.9	20	7535.0	82 0	9122.9	40	12137.0
10	5636.5	50	6478.6	30	7580.0	10	9195.3	50	12313.8
20	5663.4	73 0	6512.7	40	7627.4	20	9269.0	87 0	12500.4
30	5690.6	10	6547.0	50	7674.5	30	9345.4	10	12698.0
40	5718.0	20	6581.7	78 0	7722.3	40	9422.9	20	12904.5
50	5745.5	30	6616.7	10	7770.7	50	9502.1	30	13128.4
69 0	5772.8	40	6652.1	20	7819.8	83 0	9585.7	40	13369.7
10	5801.3	50	6687.6	30	7869.0	10	9669.3	50	13630.3
20	5829.5	74 0	6723.8	40	7920.1	20	9751.3	88 0	13913.7
30	5857.9	10	6760.3	50	7971.4	30	9840.0	10	14192.0
40	5886.6	20	6797.1	79 0	8023.4	40	9929.1	20	14559.0
50	5915.4	30	6834.5	10	8076.2	50	10019.9	30	14963.6
70 0	5944.5	40	6871.6	20	8129.9	84 0	10114.2	40	15327.7
10	5973.9	50	6909.8	30	8184.3	10	10211.2	50	15746.8
20	6003.4	75 0	6948.4	40	8239.5	20	10311.1	89 0	16267.8
30	6033.2	10	6987.2	50	8295.7	30	10413.0	10	16890.0
40	6063.3	20	7026.4	80 0	8352.6	40	10520.8	20	17670.7
50	6093.6	30	7066.2	10	8410.9	50	10628.1	30	18630.7
71 0	6124.7	40	7106.3	20	8469.9	85 0	10741.8	40	20053.0
10	6155.6	50	7146.9	30	8529.6	10	10858.6	50	22436.5
20	6186.1	76 0	7188.9	40	8591.0	20	10978.4	90 0	Infinity.
30	6217.5	10	7229.5	50	8653.3	30	11104.5		

TABLE IX.

Werner's projection.

$\zeta = 57^\circ.29578\,\frac{\cos\theta}{\theta} = 57^\circ\,(57.29578)\,\frac{\cos\theta}{\theta}$

N. lat., θ	ζ	S. lat., θ	ζ
	° ′		° ′
90	90 0	0	57 18
80	89 30	10	50 45
70	88 11	20	44 8
60	85 57	30	37 13
50	82 53	40	30 23
40	79 1	50	23 40
30	74 26	60	17 10
20	69 12	70	11 1
10	63 27	80	5 10
0	57 18	90	0 0

TABLE X.

Construction of the homalographic projection.

$\pi \sin \theta = 2A + \sin 2A$

Values of θ for every 30'.		Cos A	Sin A	Differences of sin A for every 30'.	Values of θ for every 30'.		Cos A	Sin A	Differences of sin A for every 30'.
°	'				°	'			
0	0	1.0000000	0.00000000	654431	22	0	0.9043490	0.29871850	[illegible]
0	30	0.9999787	0.00654431	[illegible]	22	30	0.9022274	0.30387380	[illegible]
1	0	[illegible]	0.01370613	[illegible]	23	0	[illegible]	0.31291040	[illegible]
1	30	0.9997981	[illegible]	[illegible]	23	30	0.9478704	[illegible]	[illegible]
2	0	[illegible]	0.2741435	[illegible]	24	0	0.9456170	[illegible]	[illegible]
2	30	[illegible]	[illegible]	[illegible]	24	30	[illegible]	[illegible]	[illegible]
3	0	[illegible]	0.04111710	[illegible]	25	0	[illegible]	[illegible]	[illegible]
3	30	[illegible]	[illegible]	[illegible]	25	30	[illegible]	[illegible]	[illegible]
4	0	[illegible]	[illegible]	[illegible]	26	0	[illegible]	[illegible]	[illegible]
4	30	[illegible]	0.06168115	[illegible]	26	30	[illegible]	[illegible]	[illegible]
5	0	[illegible]	[illegible]	[illegible]	27	0	[illegible]	[illegible]	[illegible]
5	30	0.9971572	[illegible]	[illegible]	27	30	[illegible]	[illegible]	[illegible]
6	0	[illegible]	[illegible]	[illegible]	28	0	[illegible]	[illegible]	[illegible]
6	30	[illegible]	[illegible]	[illegible]	28	30	[illegible]	[illegible]	[illegible]
7	0	[illegible]	[illegible]	[illegible]	29	0	[illegible]	[illegible]	[illegible]
7	30	[illegible]	[illegible]	[illegible]	29	30	0.9176219	[illegible]	[illegible]
8	0	[illegible]	[illegible]	[illegible]	30	0	[illegible]	[illegible]	[illegible]
8	30	[illegible]	[illegible]	[illegible]	30	30	[illegible]	[illegible]	647010
9	0	[illegible]	[illegible]	[illegible]	31	0	[illegible]	[illegible]	[illegible]
9	30	[illegible]	[illegible]	[illegible]	31	30	[illegible]	[illegible]	[illegible]
10	0	[illegible]	[illegible]	681195	32	0	[illegible]	[illegible]	[illegible]
10	30	[illegible]	[illegible]	[illegible]	32	30	[illegible]	[illegible]	[illegible]
11	0	[illegible]	[illegible]	[illegible]	33	0	[illegible]	[illegible]	[illegible]
11	30	[illegible]	[illegible]	[illegible]	33	30	[illegible]	[illegible]	[illegible]
12	0	[illegible]	[illegible]	[illegible]	34	0	[illegible]	[illegible]	[illegible]
12	30	[illegible]	[illegible]	[illegible]	34	30	[illegible]	[illegible]	[illegible]
13	0	[illegible]	[illegible]	[illegible]	35	0	[illegible]	[illegible]	[illegible]
13	30	[illegible]	[illegible]	[illegible]	35	30	[illegible]	[illegible]	[illegible]
14	0	[illegible]	[illegible]	[illegible]	36	0	[illegible]	[illegible]	[illegible]
14	30	[illegible]	[illegible]	[illegible]	36	30	[illegible]	[illegible]	[illegible]
15	0	[illegible]	[illegible]	[illegible]	37	0	[illegible]	[illegible]	[illegible]
15	30	[illegible]	0.21147380	[illegible]	37	30	[illegible]	[illegible]	[illegible]
16	0	[illegible]	[illegible]	[illegible]	38	0	[illegible]	[illegible]	[illegible]
16	30	[illegible]	[illegible]	674115	38	30	[illegible]	0.51231080	[illegible]
17	0	[illegible]	[illegible]	[illegible]	39	0	[illegible]	[illegible]	[illegible]
17	30	[illegible]	[illegible]	[illegible]	39	30	[illegible]	[illegible]	[illegible]
18	0	[illegible]	[illegible]	[illegible]	40	0	[illegible]	[illegible]	[illegible]
18	30	[illegible]	[illegible]	[illegible]	40	30	[illegible]	[illegible]	[illegible]
19	0	[illegible]	[illegible]	[illegible]	41	0	[illegible]	[illegible]	[illegible]
19	30	[illegible]	[illegible]	[illegible]	41	30	[illegible]	[illegible]	[illegible]
20	0	[illegible]	[illegible]	[illegible]	42	0	[illegible]	[illegible]	[illegible]
20	30	[illegible]	[illegible]	[illegible]	42	30	[illegible]	[illegible]	[illegible]
21	0	[illegible]	[illegible]	[illegible]	43	0	[illegible]	[illegible]	[illegible]
21	30	[illegible]	[illegible]	[illegible]	43	30	0.8100142	[illegible]	[illegible]
22	0	[illegible]	[illegible]		44	0	[illegible]	[illegible]	

Construction of the homalographic projection—Continued.

Values of θ for every 30'.		Cos λ	Sin λ	Differences of sin λ for every 30'.	Values of θ for every 30'.		Cos λ	Sin λ	Differences of sin λ for every 30'.
°	′				°	′			
44	0	0.8146320	0.5797710	604380	67	0	0.5525820	0.8334960	482800
44	30	0.8102900	0.5862916	602380	67	30	0.5451794	0.8382960	479800
45	0	0.8059656	0.5920270	600800	68	0	0.5377970	0.8431124	475400
45	30	0.8014604	0.5980470	599410	68	30	0.5302971	0.8478520	471040
46	0	0.7969494	0.6040170	596960	69	0	0.5228461	0.8525640	468900
46	30	0.7924040	0.6099530	594340	69	30	0.5144715	0.8572740	464320
47	0	0.7877940	0.6158970	592320	70	0	0.5070913	0.8619100	460420
47	30	0.7831270	0.6218180	589230	70	30	0.4991341	0.8665140	456640
48	0	0.7784035	0.6277410	588130	71	0	0.4911428	0.8710780	452900
48	30	0.7736225	0.6336540	586270	71	30	0.4830814	0.8756000	448160
49	0	0.7687835	0.6395360	583880	72	0	0.4749167	0.8800600	443800
49	30	0.7638865	0.6453480	581680	72	30	0.4666962	[illegible]	439540
50	0	0.7589440	0.6511680	578910	73	0	0.4583813	0.8888940	435280
50	30	0.7539817	0.6569470	577080	73	30	0.4500146	0.8932780	430720
51	0	0.7489343	0.6627250	574850	74	0	0.4409511	0.8975480	426160
51	30	0.7437375	0.6684720	572310	74	30	0.4329850	0.9018410	421440
52	0	0.7385313	0.6741971	571800	75	0	0.4231814	[illegible]	416800
52	30	0.7332854	0.6798916	567820	75	30	0.4141136	0.9102430	412100
53	0	0.7279985	0.6854740	566440	76	0	0.4048824	0.9143320	407800
53	30	0.7226082	0.6912180	562880	76	30	0.3956136	0.9184600	401480
54	0	0.7172058	0.6968130	560420	77	0	0.3861534	0.9224400	398550
54	30	0.7117173	0.7024540	557880	77	30	0.3765400	0.9263900	391160
55	0	0.7061670	0.7080460	555870	78	0	0.3667706	0.9303150	387710
55	30	0.7005550	0.7135880	553820	78	30	0.3568222	0.9341080	380700
56	0	0.6948780	0.7191350	550770	79	0	0.3467166	0.9379790	374350
56	30	0.6891300	0.7246280	547630	79	30	0.3374137	0.9417141	368130
57	0	0.6833242	0.7301870	543860	80	0	0.3250734	0.9453980	361880
57	30	0.6774641	0.7355870	541280	80	30	0.3152258	0.9489130	355430
58	0	0.6713085	0.7407870	538400	81	0	0.3048189	0.9523740	349830
58	30	0.6656270	0.7462550	536470	81	30	0.2931735	0.9558040	342190
59	0	0.6596300	0.7517460	533040	82	0	0.2817780	0.9590400	336840
59	30	0.6535232	0.7570800	530870	82	30	0.2701670	0.9623040	327470
60	0	0.6471101	0.7623670	528480	83	0	0.2581516	0.9653470	319470
60	30	0.6408456	0.7676000	525170	83	30	0.2454837	0.9686490	311130
61	0	0.6345010	0.7728320	522180	84	0	0.2332727	0.9724160	302880
61	30	0.6280283	0.7781431	519140	84	30	0.2202700	0.9754840	294060
62	0	0.6216801	0.7833644	516070	85	0	0.2069345	0.9783780	284080
62	30	0.6150447	0.7884620	512880	85	30	0.1929140	0.9812152	273450
63	0	0.6084076	0.7932470	509820	86	0	0.1794407	0.9838970	261900
63	30	0.6016201	0.7977220	506610	86	30	0.1653412	0.9862870	249860
64	0	0.5948143	0.8037980	503470	87	0	0.1476433	0.9890470	236190
64	30	0.5884519	0.8088200	500120	87	30	0.1358888	0.9914250	220970
65	0	0.5811197	0.8138430	496830	88	0	0.1136372	0.9935470	203970
65	30	0.5740884	0.8187620	493410	88	30	0.0973843	0.9954840	189820
66	0	0.5668978	0.8237260	490040	89	0	0.0718850	0.9974770	132080
66	30	0.5599024	0.8286200	486160	89	30	0.0447815	0.9988977	109230
67	0	0.5525600	0.8334960		90	0	0.0000000	1.0000000	

TABLE XI.

Central equivalent projection.

Values of the central distance λ expressed in degrees.

Values of the latitude θ.	Values of the longitude ω.									
	0°	10°	20°	30°	40°	50°	60°	70°	80°	90°
	° ′ ″	° ′ ″	° ′ ″	° ′ ″	° ′ ″	° ′ ″	° ′ ″	° ′ ″	° ′ ″	° ′ ″
0°	00 00 00	10 00 00.0	20 00 00.0	30 00 00.0	40 00 00.0	50 00 00.0	60 00 00.0	70 00 00.0	80 00 00.0	90 00 00
10	10 00 00	14 0 21.0	22 18 7.5	31 [illegible]	41 1 33.6	50 43 [illegible]	60 30 3.7	70 19 38.4	[illegible] 0 12.4	90 00 00
20	20 00 00	22 10 7.5	27 50 27.3	35 31 [illegible]	43 57 [illegible]	52 50 [illegible]	61 [illegible]	71 13 18.0	80 [illegible] 31.4	90 00 00
30	30 00 00	31 [illegible]	[illegible] 31 [illegible]	41 31 34.7	48 26 [illegible]	56 18 27.0	64 33 [illegible]	72 46 14.2	81 21 3.1	90 00 00
40	40 00 00	41 1 [illegible]	43 57 [illegible]	48 26 [illegible]	54 4 5.0	60 [illegible] 4.7	67 [illegible] 44.4	74 40 [illegible]	82 21 [illegible]	90 00 00
50	50 00 00	50 43 35.0	[illegible] 50 [illegible]	56 10 27.0	60 30 4.7	65 35 [illegible]	71 13 19.0	77 14 00.0	83 35 [illegible]	90 00 00
60	60 00 00	60 30 3.7	61 50 [illegible]	64 33 [illegible]	67 29 44.4	71 15 18.0	75 31 21.0	80 0 2.5	85 1 8.7	90 00 00
70	70 00 00	70 19 38.4	71 13 18.0	72 46 14.2	74 46 [illegible]	77 14 00.0	80 0 2.5	82 10 [illegible]	88 35 44.2	90 00 00
80	80 00 00	80 0 12.4	80 30 21.4	81 21 3.1	82 21 [illegible]	83 35 [illegible]	85 1 8.7	86 35 44.2	88 16 18.3	90 00 00
90	90 00 00	90 00 00.0	90 00 00.0	90 00 00.0	90 00 00.0	90 00 00.0	90 00 00.0	90 00 00.0	90 00 00.0	90 00 00

TABLE XII.

Central equivalent projection.

Values of the radius vector ρ expressed in degrees.

Values of the latitude θ.	Values of the longitude ω.									
	0°	10°	20°	30°	40°	50°	60°	70°	80°	90°
0°	0.[illegible]	0.17432	0.34730	0.51764	0.[illegible]	0.84524	1.[illegible]	1.14716	1.[illegible]	1.41422
10	0.17432	0.[illegible]	0.[illegible]	0.54348	0.[illegible]	0.[illegible]71	1.[illegible]	1.15167	1.[illegible]	1.41422
20	0.34730	0.[illegible]	0.[illegible]	0.[illegible]	0.76454	0.[illegible]	1.[illegible]	1.[illegible]	1.[illegible]	1.41421
30	0.51764	0.54348	0.[illegible]	0.[illegible]	0.[illegible]47	0.[illegible]	1.[illegible]	1.[illegible]	1.[illegible]	1.41422
40	0.[illegible]	0.[illegible]	0.74854	0.[illegible]47	0.[illegible]	1.[illegible]	1.[illegible]	1.21[illegible]	1.31[illegible]	1.41422
50	0.84524	0.[illegible]71	0.[illegible]	0.94[illegible]	1.[illegible]	1.[illegible]	1.16[illegible]	1.[illegible]	1.[illegible]	1.41422
60	1.[illegible]	1.[illegible]	1.02472	1.[illegible]	1.11[illegible]	1.16[illegible]	1.[illegible]474	1.[illegible]750	1.35[illegible]	1.41422
70	1.14716	1.15167	1.18[illegible]	1.18[illegible]	1.21[illegible]	1.[illegible]	1.[illegible]730	1.[illegible]	1.37[illegible]	1.41422
80	1.[illegible]	1.[illegible]	1.[illegible]	1.[illegible]	1.31[illegible]	1.[illegible]	1.35[illegible]	1.37150	1.[illegible]	1.41422
90	1.41422	1.41422	1.41422	1.41422	1.41422	1.41422	1.41422	1.41422	1.41422	1.41422

TABLE XIII.

Central equivalent projection.

Values of the azimuthal angle α expressed in degrees.

Values of the latitude θ.	Values of the longitude ω.									
	0°	10°	20°	30°	40°	50°	60°	70°	80°	90°
	° ′ ″	° ′ ″	° ′ ″	° ′ ″	° ′ ″	° ′ ″	° ′ ″	° ′ ″	° ′ ″	° ′ ″
0°	Indeterm.	90 00 00.0	90 00 00.0	90 00 00.0	90 00 00.0	90 00 00.0	90 00 00.0	90 00 00.0	90 00 00.0	90 00 00
10	0 0 0	44 [illegible] 41.2	[illegible] 43 [illegible]	70 34 [illegible]	74 [illegible]	77 2 13.1	78 [illegible] 29.0	79 22 [illegible].7	79 [illegible]	80 00 00
20	0 0 0	25 [illegible]	43 13 [illegible]	53 [illegible] 51.4	[illegible] 47.4	64 [illegible] 18.5	67 12 14.0	[illegible] 37.7	[illegible]	70 00 00
30	0 0 0	16 41 [illegible]	[illegible] 32.4	40 [illegible]	48 4 11.0	[illegible] 42.0	[illegible] 19 [illegible].7	[illegible] 25 [illegible]	[illegible] 27 [illegible].7	60 00 00
40	0 0 0	11 41 31.4	[illegible] 10 [illegible]	[illegible] 47 [illegible]	37 27 13.4	42 [illegible]	45 54 16.0	48 14 13.1	49 34 3.1	50 00 00
50	0 0 0	8 17 26.4	10 0 44.4	27 45 37.7	[illegible]	32 43 [illegible]	30 0 18.7	[illegible] 15 [illegible]	[illegible] 34 7.2	40 00 00
60	0 0 0	5 43 [illegible]	11 10 12.8	10 6 6.3	[illegible] 21 [illegible]	25 51 31.2	[illegible] 54.1	[illegible]	[illegible] 27 17.0	30 00 00
70	0 0 0	3 [illegible]	7 5 45.6	10 10 [illegible]	13 10 4.2	16 [illegible] 45.6	17 [illegible] 42.8	10 22 34.2	10 43 11.1	20 00 00
80	0 0 0	1 46 18.6	3 27 4.2	0 3 18.1	0 37 [illegible]	7 41 [illegible]	6 40 55.0	0 24 [illegible]	0 31 3.1	10 00 00
90	0 0 0	0 00 00.0	0 00 00.0	0 00 00.0	0 00 00.0	0 00 00.0	0 00 00.0	0 00 00.0	0 00 00.0	00 00 00

TABLE XIV.

Transformation of the stereographic system into the central equivalent system.

Values of S.	Values of C.	Differences.	Values of S.	Values of C.	Differences.
0.00	0.000	100	50	894	70
05	100	99	55	964	65
10	199	98	60	1.029	61
15	297	95	65	1.090	57
20	392	93	70	1.147	53
25	485	90	75	1.200	49
30	575	86	80	1.249	46
35	661	82	85	1.295	43
40	743	78	90	1.338	40
45	821	73	95	1.374	38
50	894		1.00	1.414	

TABLE XV.

Central equivalent projection.

Distance in degrees from the center of the map ϕ	Angle of maximum deviation— On the sphere.			Angle of maximum deviation— On the map.			Maximum deviation.		
	°	′	″	°	′	″	°	′	″
0°	45	0	0.0	45	0	0.0	0	0	0.0
10	44	53	26.8	45	6	33.2	0	13	6.4
20	44	32	41.1	45	26	18.9	0	52	37.8
30	43	0	25.2	45	59	34.7	1	58	9.4
40	43	13	9.8	46	46	51.0	3	33	42.0
50	42	11	56.5	47	48	40.5	5	37	39.6
60	40	53	36.2	49	6	23.8	8	12	47.6
70	39	19	21.6	50	40	38.4	11	21	16.8
80	37	27	21.4	52	32	46.6	15	5	25.2
90	35	15	51.6	54	44	8.2	19	28	16.4

TABLE XVI.

Central equivalent projection.

Distance in degrees from the center of the map.	Distance in parts of radius from center of map. Along the arc of a great circle.	Distance in parts of radius from center of map. Along the chord or upon the map.	Coefficient of the reduction of an element in the direction of the radius vector.	Coefficient of the increase of an element in the direction perpendicular to the radius vector.
10°	0.174533	0.17432	0.99619	1.00382
20	0.349066	0.34730	0.98481	1.01542
30	0.523599	0.51764	0.96593	1.03527
40	0.698132	0.68404	0.93969	1.06417
50	0.872665	0.84524	0.90631	1.10338
60	1.047198	1.00000	0.86603	1.15470
70	1.221731	1.14715	0.81915	1.22077
80	1.396264	1.28558	0.76604	1.30541
90	1.570796	1.41422	0.70711	1.41422

TABLE XVII.

Cylindric equivalent projection.

Values of η: $\eta = \tan^{-1}\frac{\tan\theta}{\cos\omega}$

θ	Longitude ω.									
	0°	10°	20°	30°	40°	50°	60°	70°	80°	90°
90°	1.57080	1.57080	1.57080	1.57080	1.57080	1.57080	1.57080	1.57080	1.57080	1.57080
80	1.39626	1.[illegible]	1.40[illegible]	1.41025	1.43570	1.45788	1.48341	1.[illegible]	1.54018	1.57080
70	1.22173	1.22643	1.30175	1.25843	1.28[illegible]	1.34097	1.39078	1.46[illegible]	1.[illegible]	1.57080
60	1.04720	1.0[illegible]	1.07370	1.10712	1.1[illegible]	1.21348	1.29077	1.37[illegible]	1.47[illegible]	1.57080
50	0.87266	1.[illegible]	0.90311	0.94[illegible]	0.99851	1.07017	1.17[illegible]	1.[illegible]	1.47811	1.57080
40	0.69813	0.70[illegible]	0.72[illegible]	0.76[illegible]	0.8[illegible]	0.94[illegible]	1.0[illegible]	1.18173	1.[illegible]	1.57080
30	0.52360	0.53[illegible]	0.55[illegible]	0.5[illegible]	0.64[illegible]	0.73182	0.85767	1.0[illegible]	1.27[illegible]	1.57080
20	0.34907	0.35401	0.36[illegible]	0.39[illegible]	0.44255	0.51[illegible]	0.6[illegible]	0.81[illegible]	1.[illegible]	1.57080
10	0.17453	0.17717	0.18[illegible]	0.20[illegible]	0.22[illegible]	0.27[illegible]	0.33[illegible]	0.4[illegible]	0.7[illegible]	1.57080
0	0.00000	0.00000	0.00000	0.00000	0.00000	0.00000	0.00000	0.00000	0.00000	Indeterm.

TABLE XVIII.

Cylindric equivalent projection.

Values of ξ: $\xi = \sin\omega\cos\theta$

θ	Longitude ω.									
	0°	10°	20°	30°	40°	50°	60°	70°	80°	90°
90°	0	0.00000	0.00000	0.00000	0.00000	0.00000	0.00000	0.00000	0.00000	0.00000
80	0	0.03015	0.05[illegible]	0.08[illegible]	0.11162	0.13302	0.15[illegible]	0.16317	0.17101	0.17365
70	0	0.05[illegible]	0.11[illegible]	0.17101	0.21985	0.26140	0.29[illegible]	0.32139	0.33[illegible]	0.34[illegible]
60	0	0.08[illegible]	0.17101	0.25[illegible]	0.32139	0.38[illegible]	0.43301	0.46[illegible]	0.4[illegible]	0.50000
50	0	0.11[illegible]	0.21[illegible]	0.32139	0.41317	0.49[illegible]	0.5[illegible]	0.60[illegible]	0.6[illegible]	0.64279
40	0	0.13[illegible]	0.26[illegible]	0.38[illegible]	0.49[illegible]	0.58[illegible]	0.66341	0.71[illegible]	0.75441	0.76[illegible]
30	0	0.15[illegible]	0.29[illegible]	0.43301	0.55[illegible]	0.66341	0.75[illegible]	0.81[illegible]	0.85[illegible]	0.86[illegible]
20	0	0.16317	0.32139	0.46[illegible]	0.60[illegible]	0.71985	0.81[illegible]	0.8[illegible]	0.9[illegible]	0.93[illegible]
10	0	0.17101	0.33[illegible]	0.49[illegible]	0.63[illegible]	0.75441	0.85[illegible]	0.9[illegible]	0.9[illegible]	0.9[illegible]
0	0	0.17365	0.34[illegible]	0.50000	0.64279	0.76[illegible]	0.8[illegible]	0.9[illegible]	0.98[illegible]	1.00000

TABLE XIX.

Orthomorphic-conic projection.

Values of the radii ρ of the parallels for $\phi = 90° - \theta$.

ϕ	$\rho = \frac{1}{3}$	$\rho = \frac{1}{2}$	$\rho = \frac{2}{3}$	$\rho = \frac{3}{4}$
0°	0.000	0.000	0.000	0.0000
10	0.444	0.296	0.197	0.1[illegible]
20	0.561	0.420	0.316	0.27[illegible]
30	0.645	0.518	0.416	0.37[illegible]
40	0.714	0.[illegible]	0.520	0.4[illegible]
50	0.779	0.[illegible]	0.[illegible]1	0.[illegible]
60	0.[illegible]	0.760	0.[illegible]	0.[illegible]
70	0.[illegible]	0.837	0.7[illegible]	0.7[illegible]
80	0.943	0.916	0.[illegible]	0.876[illegible]
90	1.000	1.000	1.000	1.0000
100	1.0[illegible]	1.0[illegible]	1.1[illegible]	1.1[illegible]
110	1.1[illegible]	1.195	1.2[illegible]	1.2[illegible]
120	1.201	1.316	1.443	1.[illegible]
130	1.[illegible]	1.464	1.[illegible]	1.77[illegible]
140	1.401	1.657	1.[illegible]	2.1[illegible]
150	1.561	1.932	2.4[illegible]	2.[illegible]4
160	1.783	2.3[illegible]	3.173	3.67[illegible]
170	2.2[illegible]	3.3[illegible]	5.674	8.21[illegible]

TABLE XX.

Lambert's orthomorphic-cylindric projection.

Values of ξ: $\xi = \frac{1}{2M} \log \frac{1 + \sin \omega \cos \phi}{1 - \sin \omega \cos \phi}$

	Longitude ω.									
φ	0°	10°	20°	30°	40°	50°	60°	70°	80°	90°
90°	0	0.00000	0.00000	0.00000	0.00000	0.00000	0.00000	0.00000	0.00000	0.00000
80	0	[illegible]	[illegible]	0.08704	[illegible]	[illegible]	[illegible]	[illegible]	0.17271	0.17543
70	0	[illegible]	[illegible]	0.17271	[illegible]	[illegible]	[illegible]	[illegible]	[illegible]	[illegible]
60	0	0.08704	0.17271	[illegible]	[illegible]	[illegible]	[illegible]	[illegible]	[illegible]	[illegible]
50	0	[illegible]	[illegible]	[illegible]	[illegible]	[illegible]	[illegible]	[illegible]	0.74464	[illegible]
40	0	[illegible]	[illegible]	[illegible]	[illegible]	[illegible]	[illegible]	[illegible]	[illegible]	[illegible]
30	0	[illegible]	[illegible]	[illegible]	[illegible]	[illegible]	[illegible]	[illegible]	[illegible]	[illegible]
20	0	[illegible]	[illegible]	[illegible]	[illegible]	[illegible]	[illegible]	[illegible]	[illegible]	[illegible]
10	0	0.17271	[illegible]	[illegible]	[illegible]	[illegible]	[illegible]	[illegible]	[illegible]	[illegible]
0	0	0.17543	[illegible]	[illegible]	[illegible]	[illegible]	[illegible]	[illegible]	[illegible]	Infinite.

TABLE XXI.

Lambert's orthomorphic-cylindric projection.

Values of η: $\eta = \tan^{-1} \frac{\tan \phi}{\cos \omega}$

	Longitude ω.									
φ	0°	10°	20°	30°	40°	50°	60°	70°	80°	90°
90°	1.57080	1.57080	1.57080	1.57080	1.57080	1.57080	[illegible]	[illegible]	[illegible]	[illegible]
80	[illegible]	[illegible]	[illegible]	[illegible]	[illegible]	[illegible]	[illegible]	[illegible]	[illegible]	[illegible]
70	1.22173	[illegible]	[illegible]	[illegible]	[illegible]	[illegible]	[illegible]	[illegible]	[illegible]	[illegible]
60	1.04720	[illegible]	[illegible]	[illegible]	[illegible]	[illegible]	[illegible]	[illegible]	[illegible]	[illegible]
50	0.87266	[illegible]	[illegible]	[illegible]	[illegible]	[illegible]	[illegible]	[illegible]	[illegible]	1.57080
40	0.69813	[illegible]	[illegible]	[illegible]	[illegible]	[illegible]	[illegible]	[illegible]	[illegible]	[illegible]
30	[illegible]	[illegible]	[illegible]	[illegible]	[illegible]	[illegible]	[illegible]	[illegible]	[illegible]	1.57080
20	0.34907	[illegible]	[illegible]	[illegible]	[illegible]	[illegible]	[illegible]	[illegible]	[illegible]	1.57080
10	0.17453	0.17717	[illegible]	[illegible]	[illegible]	[illegible]	[illegible]	[illegible]	[illegible]	1.57080
0	0.00000	0.00000	0.00000	0.00000	0.00000	0.00000	0.00000	0.00000	0.00000	Indeterm.

TABLE XXII.

Boole's Projection.

Polar distance.	$n = 1$		$n = \frac{1}{4}$	
	Sphere.	Spheroid.	Sphere.	Spheroid.
10°	.0875	0.0880	.8430	.8447
20	.1763	.1774	.6440	[illegible]
30	.2679	.2694	.7165	.7205
40	.3640	.3656	.7767	.7777
50	.4663	.4663	.8384	.8373
60	.5774	.5792	.8717	.8734
70	.7002	.7017	.9145	.9153
80	.8391	.8406	.9571	.9574
90	1.0000	1.0000	1.0000	1.0000
100	1.1918	1.1904	1.0448	1.0445
110	1.4281	1.4230	1.0902	[illegible]
120	1.7321	1.7305	1.1473	1.1463
130	2.1445	2.1357	1.2101	[illegible]
140	2.7475	2.7340	1.2973	[illegible]
150	3.7321	3.7134	1.3800	[illegible]
160	5.6713	5.6373	1.5432	1.5400
170	11.4301	11.3561	1.8307	1.8856

TABLE XXIII.

Sir John Herschel's Projection.

	$n=1$	$n=\frac{2}{3}$	$n=\frac{1}{2}$	$n=\frac{1}{3}$		$n=1$	$n=\frac{2}{3}$	$n=\frac{1}{2}$	$n=\frac{1}{3}$
$\phi =$ °	$\rho =$	$\rho =$	$\rho =$	$\rho =$	$\phi =$ °	$\rho =$	$\rho =$	$\rho =$	$\rho =$
0	0.000	0.000	0.000	0.000	80	0.839	[illegible]	0.916	0.943
10	0.087	0.107	[illegible]	0.444	90	1.000	1.000	1.000	1.000
20	0.176	0.214	0.420	0.541	100	1.192	1.134	1.092	1.060
30	0.268	0.416	0.518	0.645	110	1.428	1.288	1.195	1.128
40	0.364	0.540	[illegible]	0.714	120	1.732	1.449	1.316	1.201
50	0.466	0.691	0.693	0.776	130	2.144	1.665	1.464	1.289
60	0.577	[illegible]	0.760	0.832	140	2.747	1.962	1.657	1.401
70	0.700	0.793	0.837	[illegible]	150	3.732	2.408	1.932	1.551
80	0.839	[illegible]	0.916	0.943	160	5.671	3.173	2.381	1.759

TABLE XXIV.

Sir Henry James's projection.

Lat.	Radius for parallel.	Degree of longitude.	Lat.	Radius for parallel.	Degree of longitude.	Lat.	Radius for parallel.	Degree of longitude.	Lat.	Radius for parallel.	Degree of longitude.
0°	∞	1.00000	23°	[illegible]	.92050	46°	[illegible]	.69466	69°	[illegible]	.35837
1	[illegible]	.99985	24	[illegible]	.91355	47	[illegible]	.68200	70	[illegible]	.34202
2	[illegible]	.99939	25	[illegible]	.90631	48	[illegible]	.66913	71	[illegible]	.32557
3	[illegible]	.99863	26	117.474	.89879	49	[illegible]	.65606	72	18.817	.30902
4	[illegible]	.99756	27	112.449	.89101	50	48.077	.64279	73	17.317	.29237
5	[illegible]	.99619	28	[illegible]	.88295	51	[illegible]	.62932	74	[illegible]	.27564
6	[illegible]	.99452	29	[illegible]	.87462	52	44.764	.61566	75	[illegible]	.25882
7	[illegible]	.99255	30	[illegible]	.86603	53	43.175	.60182	76	[illegible]	.24192
8	[illegible]	.99027	31	[illegible]	.85717	54	41.629	.58779	77	[illegible]	.22495
9	[illegible]	.98769	32	[illegible]	.84805	55	40.119	.57358	78	12.179	.20791
10	[illegible]	.98481	33	[illegible]	.83867	56	[illegible]	.55919	79	[illegible]	.19081
11	[illegible]	.98163	34	[illegible]	.82904	57	[illegible]	.54464	80	[illegible]	.17365
12	[illegible]	.97815	35	[illegible]	.81915	58	[illegible]	.52992	81	[illegible]	.15643
13	[illegible]	.97437	36	[illegible]	.80902	59	34.437	.51504	82	[illegible]	.13917
14	[illegible]	.97030	37	[illegible]	.79864	60	[illegible]	.50000	83	[illegible]	.12187
15	[illegible]	.96593	38	[illegible]	.78801	61	[illegible]	.48481	84	[illegible]	.10453
16	[illegible]	.96126	39	[illegible]	.77715	62	[illegible]	.46947	85	[illegible]	.08716
17	[illegible]	.95630	40	[illegible]	.76604	63	[illegible]	.45399	86	[illegible]	.06976
18	[illegible]	.95106	41	65.911	.75471	64	27.945	.43837	87	[illegible]	.05234
19	[illegible]	.94552	42	[illegible]	.74314	65	26.717	.42262	88	[illegible]	.03490
20	[illegible]	.93969	43	[illegible]	.73135	66	25.510	.40674	89	[illegible]	.01745
21	[illegible]	.93358	44	[illegible]	.71934	67	[illegible]	.39073	90	0.000	.00000
22	[illegible]	.92718	45	[illegible]	.70711	68	[illegible]	.37461			

Degree of equator = degree of meridian = 1. Radius of sphere = 57.2958.

TABLE XXV.

Capt. Clarke's comparison of "Balance of Errors," Sir Henry James's, and "Equal Radial" Projections.

a	Balance of Errors.		Sir H. James's.		Equal Radial.		a
	U	U sin a	U	U sin a	U	U sin a	
0°	[illegible]	.0000	[illegible]	.0000	[illegible]	.0000	0°
5	.1164	[illegible]	.1173	[illegible]	[illegible]	[illegible]	5
10	.1149	[illegible]	.1147	[illegible]	.0919	[illegible]	10
15	[illegible]	[illegible]	[illegible]	[illegible]	[illegible]	[illegible]	15
20	[illegible]	.0374	[illegible]	[illegible]	.0771	[illegible]	20
25	[illegible]	.0444	[illegible]	.0414	.0737	[illegible]	25
30	[illegible]	.0501	[illegible]	[illegible]	[illegible]	[illegible]	30
35	[illegible]	[illegible]	[illegible]	[illegible]	[illegible]	[illegible]	35
40	[illegible]	[illegible]	[illegible]	[illegible]	[illegible]	[illegible]	40
45	.0912	.0874	[illegible]	[illegible]	[illegible]	[illegible]	45
50	[illegible]	[illegible]	[illegible]	[illegible]	[illegible]	[illegible]	50
55	[illegible]	[illegible]	[illegible]	.0331	[illegible]	[illegible]	55
60	[illegible]	.0516	[illegible]	[illegible]	[illegible]	.0374	60
65	[illegible]	[illegible]	.0370	[illegible]	[illegible]	[illegible]	65
70	[illegible]	[illegible]	[illegible]	[illegible]	[illegible]	[illegible]	70
75	.0471	[illegible]	[illegible]	[illegible]	[illegible]	[illegible]	75
80	[illegible]	[illegible]	[illegible]	[illegible]	[illegible]	[illegible]	80
85	[illegible]	.0573	[illegible]	[illegible]	[illegible]	[illegible]	85
90	[illegible]	[illegible]	[illegible]	[illegible]	.1041	.1041	90
95	[illegible]	[illegible]	.1461	[illegible]	[illegible]	[illegible]	95
100	[illegible]	[illegible]	[illegible]	[illegible]	[illegible]	[illegible]	100
105	.2442	[illegible]	.2771	[illegible]	.3013	[illegible]	105
110	[illegible]	[illegible]	[illegible]	[illegible]	[illegible]	[illegible]	110
114	[illegible]	[illegible]	[illegible]	[illegible]	[illegible]	[illegible]	114

TABLE XXVI.

$$\rho = \frac{1.[illegible] \sin \alpha}{1.[illegible] + \cos \theta}$$

α	ρ	α	ρ	α	ρ
0°	0.0000	40°	0.5009	80°	1.0625
5	0.0613	45	0.5666	85	1.1395
10	0.1227	50	0.6335	90	1.2187
15	0.1844	55	0.7016	95	1.2995
20	0.2464	60	0.7710	100	1.3713
25	0.3090	65	0.8417	105	1.4484
30	0.3722	70	0.9136	110	1.5259
35	0.4361	75	0.9874	115	1.6045

TABLE XXVII.

Radial distances from the center of the map, for different great-circle distances α from center of reference.

α	Equal radial degrees.	Unchanged areas.	Stereographic.	Sir H. James.	Balance of errors.
5°	0.08727	0.08724	0.08732	0.08729	0.08730
10	0.17453	0.17431	0.17498	0.17471	0.17465
15	0.26180	0.26105	0.26331	0.26349	0.26214
20	0.34907	0.34730	0.35265	0.35047	0.34907
25	0.43634	0.43298	0.44339	0.43887	0.43811
30	0.52360	0.51764	0.53590	0.52831	0.52678
35	0.61087	0.60141	0.63060	0.61930	0.61549
40	0.69814	0.68404	0.72794	0.70915	0.70427
45	0.78540	0.76537	0.82843	0.80094	0.72850
50	0.87267	0.84524	0.93262	0.89275	[illegible]
55	0.95994	0.92350	1.04123	0.98702	0.98171
60	1.04720	1.00000	1.15470	1.08253	1.07565
65	1.13447	1.07460	1.27414	1.17640	1.17170
70	1.22174	1.14716	1.40042	1.27585	1.27040
75	1.30901	1.21752	1.53465	1.37290	1.37087
80	1.39628	1.28558	1.67820	1.47105	1.47434
85	1.48354	1.35118	1.83280	1.56915	1.58157
90	1.57080	1.41422	2.00000	1.66685	1.69315
95	1.65807	1.47456	2.18330	1.76275	1.81092
100	1.74504	1.53209	2.38351	1.85622	1.93342
105	1.83261	1.58671	2.60645	1.94550	2.06408
110	1.91988	1.63830	2.85630	2.02873	2.20056
115	2.00714	1.68678	3.13897	2.10365	2.36219
120	2.09440	1.73205	3.46410	2.16765	2.53042
125	2.18167	1.77402	3.84100	2.21062	2.72850
130	2.26894	1.81262	4.28901	2.23412	2.94778
135	2.35620	1.84776	4.82843	After this the	3.20596
140	2.44347	1.87939	5.49495	radius	3.51347
145	2.53074	1.90743	6.34319	diminishes.	3.88804
150	2.61801	1.93185	7.46410	....	4.44621
155	2.70528	1.95259	9.02143	....	5.18860
160	2.79255	1.96962	11.34256	....	6.26650

Table XXVIII.

Exaggeration, as shown by the proportions of projected area to original area, for different great-circle distances α from the center of reference.

α	Equal radial degrees.	Unchanged areas.	Stereographic	Sir H. James.	Balance of errors.
5°	1.00127	1.00000	1.00382	1.00230	1.00191
10	1.00508	1.00000	1.01537	1.00917	1.00767
15	1.01152	1.00000	1.03488	1.02079	1.01735
20	1.02060	1.00000	1.06315	1.03708	1.03137
25	1.03245	1.00000	1.10071	1.05805	1.04961
30	1.04544	1.00000	1.14875	1.08465	1.07275
35	1.06201	1.00000	1.20071	1.11674	1.10171
40	1.08610	1.00000	1.28330	1.15432	1.13685
45	1.11072	1.00000	1.37258	1.19780	1.17728
50	1.13919	1.00000	1.46017	1.24774	1.22605
55	1.17185	1.00000	1.61343	1.30413	1.28049
60	1.20920	1.00000	1.77778	1.36719	1.35543
65	1.25274	1.00000	1.97044	1.43682	1.43284
70	1.30014	1.00000	2.22007	1.51302	1.53040
75	1.35517	1.00000	2.52430	1.59579	1.65092
80	1.41790	1.00000	2.90891	1.68043	1.80097
85	1.48929	1.00000	3.39430	1.76730	1.98777
90	1.57080	1.00000	4.00000	1.85185	2.21960
95	1.66420	1.00000	4.80028	1.92941	2.48060
100	1.77235	1.00000	5.85774	1.99095	2.80651
105	1.89724	1.00000	7.26120	1.99969	3.20027
110	2.04307	1.00000	9.25021	1.88910	3.67170
115	2.21462	1.00000	11.96061	1.82302	4.24228
120	2.41840	1.00000	16.00000	1.56254	5.06125
125	2.66993	1.00000	21.96771	1.09762	7.09205
130	2.96163	1.00000	31.24770	0.39540	10.33547
135	3.33216	1.00000	46.62740	After this the projection fails.	14.80065
140	3.80135	1.00000	73.67941		22.26300
145	4.41223	1.00000	122.38176		35.60621
150	5.23599	1.00000	214.46940	....	70.08547
155	6.40910	1.00000	455.67282	....	128.01008
160	8.16480	1.00000	1098.81278	....	338.05120

Table XXIX.

Distortion, as shown by the proportions of the transverse side to the radial side, in the projection of an area originally square, for different great-circle distances α from the centre of reference.

α	Equal radial degrees.	Unchanged areas.	Stereographic.	Sir H. James.	Balance of errors.
5°	1.00127	1.00191	1.00000	1.00076	1.00004
10	1.00510	1.00765	1.00000	1.00307	[illegible]
15	1.01152	1.01733	1.00000	[illegible]	[illegible]
20	1.02060	1.03109	1.00000	1.01232	[illegible]
25	1.03245	1.04915	1.00000	[illegible]	[illegible]
30	1.04720	1.07180	1.00000	1.02814	1.03446
35	1.06501	1.09941	1.00000	[illegible]	[illegible]
40	1.08610	1.13247	1.00000	1.06443	1.06127
45	1.11072	1.17157	1.00000	1.07167	1.07775
50	1.13919	1.21744	1.00000	[illegible]	[illegible]
55	1.17186	1.27099	1.00000	1.11481	1.11617
60	1.20920	1.33333	1.00000	[illegible]	[illegible]
65	1.25174	1.40586	1.00000	[illegible]	1.16171
70	1.30014	1.49030	1.00000	1.21744	[illegible]
75	1.35517	1.58879	1.00000	[illegible]	1.21331
80	1.41780	1.70409	1.00000	[illegible]	[illegible]
85	1.48920	1.83966	1.00000	[illegible]	[illegible]
90	1.57080	2.00000	1.00000	[illegible]	[illegible]
95	1.66440	2.19095	1.00000	[illegible]	[illegible]
100	1.77225	2.42028	1.00000	1.79851	1.34743
105	1.89724	2.69840	1.00000	[illegible]	[illegible]
110	2.04307	3.03961	1.00000	2.57752	[illegible]
115	2.21462	3.46391	1.00000	[illegible]	1.40271
120	2.41840	4.00000	1.00000	[illegible]	[illegible]
125	2.66332	4.69017	1.00000	[illegible]	[illegible]
130	2.96188	5.59891	1.00000	23.30304	[illegible]
135	3.33216	6.82843	1.00000	After this the projection fails.	[illegible]
140	3.80135	8.54863	1.00000		[illegible]
145	4.41219	11.05901	1.00000		[illegible]
150	5.23599	[illegible]	1.00000	....	[illegible]
155	6.40119	21.34646	1.00000	....	[illegible]
160	8.16480	33.16345	1.00000	....	[illegible]

TABLE XXX.

Rectangular co-ordinates for construction of the "quincuncial projection."

x (for longitudes in upper line). *y* (for longitudes in lower line).

Lat.	0° 90	5° 85	10° 80	15° 75	20° 70	25° 65	30° 60	35° 55	40° 50	45° 45	50° 40	55° 35	60° 30	65° 25	70° 20	75° 15	80° 10	85° 5	Lat.
85°	[illegible]	[illegible]	[illegible]	[illegible]	[illegible]	[illegible]	[illegible]	[illegible]	[illegible]	.024	.021	[illegible]	.017	.014	.011	[illegible]	[illegible]	[illegible]	85°
80	.067	[illegible]	[illegible]	[illegible]	[illegible]	[illegible]	[illegible]	[illegible]	.051	.047	.043	[illegible]	[illegible]	[illegible]	[illegible]	.017	.012	[illegible]	80
75	[illegible]	.100	[illegible]	[illegible]	[illegible]	.091	[illegible]	[illegible]	.077	.071	[illegible]	[illegible]	[illegible]	[illegible]	[illegible]	[illegible]	.017	[illegible]	75
70	[illegible]	.134	[illegible]	[illegible]	.127	[illegible]	.117	.110	[illegible]	[illegible]	[illegible]	.077	.067	.057	[illegible]	[illegible]	[illegible]	[illegible]	70
65	[illegible]	[illegible]	[illegible]	[illegible]	[illegible]	.154	.147	[illegible]	[illegible]	[illegible]	[illegible]	[illegible]	[illegible]	.073	[illegible]	.044	[illegible]	.015	65
60	[illegible]	[illegible]	[illegible]	[illegible]	[illegible]	.186	.177	[illegible]	[illegible]	.145	.131	.117	[illegible]	[illegible]	.070	[illegible]	[illegible]	.015	60
55	.241	.240	[illegible]	[illegible]	[illegible]	.218	[illegible]	.197	[illegible]	.170	[illegible]	[illegible]	.130	[illegible]	[illegible]	[illegible]	.042	[illegible]	55
50	.278	.277	.274	[illegible]	.261	[illegible]	.240	.227	.212	[illegible]	.175	[illegible]	[illegible]	.117	[illegible]	.072	[illegible]	[illegible]	50
45	.317	.316	.312	[illegible]	.297	[illegible]	.273	[illegible]	.241	[illegible]	[illegible]	.161	.156	.134	[illegible]	[illegible]	[illegible]	[illegible]	45
40	.357	[illegible]	.351	.344	.334	.321	.307	[illegible]	.270	[illegible]	[illegible]	.204	.179	.151	[illegible]	[illegible]	[illegible]	[illegible]	40
35	.400	[illegible]	[illegible]	.384	.373	[illegible]	.341	[illegible]	.301	.279	.254	[illegible]	[illegible]	.170	[illegible]	[illegible]	.071	[illegible]	35
30	.446	.443	.437	.427	.413	[illegible]	.377	.356	[illegible]	[illegible]	.291	[illegible]	.223	[illegible]	.155	.119	[illegible]	.041	30
25	[illegible]	[illegible]	[illegible]	.471	[illegible]	.435	.414	[illegible]	[illegible]	[illegible]	[illegible]	.279	[illegible]	.211	.174	[illegible]	[illegible]	[illegible]	25
20	.548	.545	.534	.518	[illegible]	.476	[illegible]	[illegible]	[illegible]	[illegible]	[illegible]	[illegible]	.273	[illegible]	.195	.151	[illegible]	[illegible]	20
15	[illegible]	.604	[illegible]	[illegible]	.544	.517	.480	[illegible]	[illegible]	[illegible]	[illegible]	[illegible]	[illegible]	[illegible]	.219	.173	[illegible]	[illegible]	15
10	[illegible]	.672	.649	[illegible]	[illegible]	[illegible]	[illegible]	[illegible]	[illegible]	.454	[illegible]	[illegible]	[illegible]	[illegible]	[illegible]	[illegible]	[illegible]	.076	10
5	.778	[illegible]	.712	.673	[illegible]	[illegible]	[illegible]	[illegible]	[illegible]	[illegible]	[illegible]	[illegible]	[illegible]	[illegible]	[illegible]	.234	.177	[illegible]	5
0	1.000	[illegible]	.774	[illegible]	[illegible]	[illegible]	[illegible]	[illegible]	[illegible]	[illegible]	[illegible]	[illegible]	[illegible]	[illegible]	[illegible]	.277	[illegible]	[illegible]	0

TABLE XXXI.

Preceding table enlarged for the spaces surrounding infinite points.

x (for longitudes in upper line). *y* (for longitudes in lower line).

Lat.	0° 90	1° 89	2° 88	3° 87	4° 86	5° 85	6° 84	7° 83	10° 80	12½° 77½	15 75	75° 15	77½° 12½	80° 10	82° 8	84° 6	85° 5	86° 4	87° 3	88° 2	89° 1	Lat.
15°	[illegible]	[illegible]	[illegible]	[illegible]	[illegible]	[illegible]	[illegible]	[illegible]	[illegible]	[illegible]	[illegible]	.173	.147	.121	[illegible]	.074	[illegible]	[illegible]	[illegible]	[illegible]	.012	15
12½	.643	.643	.642	.641	[illegible]	[illegible]	.634	.627	.616	[illegible]	[illegible]	[illegible]	[illegible]	.131	.107	[illegible]	[illegible]	[illegible]	.043	[illegible]	.014	12½
10	[illegible]	[illegible]	[illegible]	[illegible]	.675	.672	[illegible]	[illegible]	[illegible]	[illegible]	[illegible]	[illegible]	.173	.143	.118	[illegible]	.079	[illegible]	.047	.031	[illegible]	10
8	.715	.714	[illegible]	.710	.706	.702	.697	[illegible]	.674	[illegible]	.641	[illegible]	[illegible]	[illegible]	[illegible]	.100	[illegible]	[illegible]	[illegible]	[illegible]	[illegible]	8
6	[illegible]	[illegible]	[illegible]	[illegible]	.741	.735	[illegible]	.714	[illegible]	[illegible]	[illegible]	[illegible]	[illegible]	[illegible]	.142	.112	[illegible]	.075	[illegible]	[illegible]	[illegible]	6
5	.775	.774	.770	[illegible]	[illegible]	[illegible]	.745	[illegible]	.718	[illegible]	[illegible]	[illegible]	[illegible]	.177	[illegible]	.119	[illegible]	[illegible]	[illegible]	.044	[illegible]	5
4	[illegible]	.797	[illegible]	[illegible]	[illegible]	.770	[illegible]	[illegible]	[illegible]	.704	[illegible]	[illegible]	.213	[illegible]	[illegible]	.128	.119	[illegible]	.071	[illegible]	[illegible]	4
3	[illegible]	[illegible]	.817	[illegible]	[illegible]	[illegible]	.775	[illegible]	[illegible]	.715	[illegible]	[illegible]	.234	.194	[illegible]	.137	[illegible]	.101	.079	[illegible]	[illegible]	3
2	[illegible]	[illegible]	.843	[illegible]	.819	[illegible]	.794	.773	[illegible]	[illegible]	[illegible]	[illegible]	[illegible]	[illegible]	[illegible]	[illegible]	[illegible]	.112	[illegible]	[illegible]	[illegible]	2
1	[illegible]	[illegible]	.872	.854	[illegible]	.824	[illegible]	[illegible]	[illegible]	.757	.713	[illegible]	.243	[illegible]	[illegible]	.161	.144	[illegible]	[illegible]	.079	[illegible]	1
0	1.000	[illegible]	[illegible]	.877	[illegible]	.841	[illegible]	[illegible]	.774	.767	[illegible]	.277	[illegible]	[illegible]	[illegible]	.175	[illegible]	[illegible]	[illegible]	[illegible]	.071	0

www.ingramcontent.com/pod-product-compliance
Lightning Source LLC
Chambersburg PA
CBHW021348230426
43666CB00006B/448

* 9 7 8 3 7 4 3 3 5 4 2 0 3 *